METHODS IN MOLECULAR BIOLOGY

Series Editor
John M. Walker
School of Life and Medical Sciences
University of Hertfordshire
Hatfield, Hertfordshire, UK

For further volumes:
http://www.springer.com/series/7651

For over 35 years, biological scientists have come to rely on the research protocols and methodologies in the critically acclaimed *Methods in Molecular Biology* series. The series was the first to introduce the step-by-step protocols approach that has become the standard in all biomedical protocol publishing. Each protocol is provided in readily-reproducible step-by-step fashion, opening with an introductory overview, a list of the materials and reagents needed to complete the experiment, and followed by a detailed procedure that is supported with a helpful notes section offering tips and tricks of the trade as well as troubleshooting advice. These hallmark features were introduced by series editor Dr. John Walker and constitute the key ingredient in each and every volume of the *Methods in Molecular Biology* series. Tested and trusted, comprehensive and reliable, all protocols from the series are indexed in PubMed.

Meiosis

Methods and Protocols

Edited by

Jesús A. Carballo

CSIC - USAL, Instituto de Biología Funcional y Genómica, Salamanca, Spain

Humana Press

Editor
Jesús A. Carballo
CSIC - USAL
Instituto de Biología Funcional y Genómica
Salamanca, Spain

ISSN 1064-3745 ISSN 1940-6029 (electronic)
Methods in Molecular Biology
ISBN 978-1-0716-3905-4 ISBN 978-1-0716-3906-1 (eBook)
https://doi.org/10.1007/978-1-0716-3906-1

This Humana imprint is published by the registered company Springer Science+Business Media, LLC, part of Springer Nature.
The registered company address is: 1 New York Plaza, New York, NY 10004, U.S.A.

Paper in this product is recyclable.

Preface

The generation of healthy gametes in sexually reproducing organisms necessitates a specialized form of cell division known as meiosis. This evolutionarily conserved developmental process typically entails a singular round of DNA replication, succeeded by an initial G2/prophase stage wherein homologous recombination of the DNA occurs. The events of replication, DNA recombination, along with sister chromatid/centromeric cohesion, collectively enable meiotic cells to undergo two consecutive rounds of chromosome segregation. These sequential steps, denoted as meiosis I and meiosis II, are both preceded by G2/prophase interphase-like stages commonly referred to as prophase I and prophase II. DNA duplication and meiotic recombination take place during prophase I, a stage characterized by its complexity and extended length when compared to prophase II or mitotic prophase.

Owing to profound morphological transformations in the structure, shape, and composition of meiotic chromosomes, various sub-stages have been employed for their differentiation. Consequently, prophase I is delineated into four distinct sub-stages, namely, leptotene, zygotene, pachytene, and diplotene. Simultaneously, during these phases, many organisms manifest an evolutionarily conserved proteinaceous structure known as the synaptonemal complex (SC). The SC comprises a tripartite structure formed between already paired homologous chromosomes. This complex consists of three primary elements, with the lateral elements constituting axial structures, such as the chromosome scaffold, cohesins, and several other meiosis-specific axial proteins, including those important for double-strand break (DSB) formation. The third component is the central element, encompassing transverse filaments and other proteins situated within the central region of the SC. Even in organisms lacking the SC, a distinct reorganization of the chromosome axis transpires during prophase I as an integral part of meiosis. Meiotic recombination typically occurs within the context of the synaptonemal complex (SC). It is well-established that anomalies in the assembly of the SC can impact the meiotic recombination process, and vice versa.

However, as in many natural processes, exceptions abound, and various organisms exhibit unique adaptations, enriching the landscape of meiosis with diversity. This edition of Methods in Molecular Biology aims to explore the latest advancements in meiosis research, offering a collection of protocols that provide to laboratories of varying sizes and resources.

The compilation spans diverse model organisms, from yeast to human, including those less commonly found in previous literature. This diversity ensures a comprehensive exploration of meiosis, capturing the nuances specific to different organisms and shedding light on less conserved features. The presented protocols will investigate into genetics, genomics, cell cultures, cell isolation, microscopy, live imaging, and automated computational tools, offering a general perspective on the multifaceted field of meiotic research.

We express our gratitude to the contributing authors for their valuable insights and expertise. It is our hope that this volume serves as a practical guide for researchers at various stages of their careers, facilitating their exploration of the fascinating world of meiosis.

Salamanca, Spain
December 2024

Jesús A. Carballo

Contents

Contributors

IAN R. ADAMS • *MRC Human Genetics Unit, Institute of Genetics and Cancer, The University of Edinburgh, Western General Hospital, Edinburgh, UK*

RACHAL M. ALLISON • *Genome Damage and Stability Centre, School of Life Sciences, University of Sussex, Brighton, UK*

DAVID ÁLVAREZ-MELO • *Louvain Institute of Biomolecular Science and Technology, Université Catholique de Louvain, Louvain-La-Neuve, Belgium*

ODEI BARREÑADA • *Magee-Womens Research Institute, Department of Obstetrics, Gynecology & Reproductive Sciences, University of Pittsburgh, Pittsburgh, PA, USA*

INÉS BERENGUER • *Departamento de Biología, Universidad Autónoma de Madrid (UAM), Madrid, Spain; Department of Molecular Neuropathology, Centro de Biología Molecular Severo Ochoa (CBMSO), Madrid, Spain*

MIGUEL ANGEL BRIEÑO-ENRÍQUEZ • *Magee-Womens Research Institute, Department of Obstetrics, Gynecology & Reproductive Sciences, University of Pittsburgh, Pittsburgh, PA, USA*

GEORGE G. B. BROWN • *Genome Damage and Stability Centre, School of Life Sciences, University of Sussex, Brighton, UK*

AGUSTIN CARBAJAL • *Cell Cycle and Cancer Biology Research Program, Oklahoma Medical Research Foundation, Oklahoma City, OK, USA*

JESÚS A. CARBALLO • *CSIC - USAL, Instituto de Biología Funcional y, Salamanca, Spain*

YAO CHEN • *State Key Laboratory of Molecular Biology, Shanghai Key Laboratory of Molecular Andrology, Center for Excellence in Molecular Cell Science, University of Chinese Academy of Sciences, Chinese Academy of Sciences, Shanghai, China*

HOA H. CHUONG • *Oklahoma Medical Research Foundation, Oklahoma City, OK, USA*

MARÍA CORREAS • *ITACA Institute, Universitat Politècnica de València, Valencia, Spain*

JAMES H. CRICHTON • *MRC Human Genetics Unit, Institute of Genetics and Cancer, The University of Edinburgh, Western General Hospital, Edinburgh, UK; RILD Building, University of Exeter, Royal Devon and Exeter Hospital, Exeter, Devon, UK*

WAYNE CRISMANI • *DNA Repair & Recombination Laboratory, St Vincent's Institute of Medical Research, Fitzroy, VIC, Australia; The Faculty of Medicine, Dentistry and Health Science, The University of Melbourne, Parkville, VIC, Australia*

DEAN S. DAWSON • *Oklahoma Medical Research Foundation, Oklahoma City, OK, USA; Department of Cell Biology, University of Oklahoma Health Science Center, Oklahoma City, OK, USA*

JULIA GONZÁLEZ DE ALEDO • *Departamento de Biología, Universidad Autónoma de Madrid (UAM), Madrid, Spain*

LUCIANA PREVIATO DE ALMEIDA • *Cell Cycle and Cancer Biology Research Program, Oklahoma Medical Research Foundation, Oklahoma City, OK, USA*

RODRIGO O. DE CASTRO • *Cell Cycle and Cancer Biology Research Program, Oklahoma Medical Research Foundation, Oklahoma City, OK, USA*

MARIO GARCÍA DE LACOBA • *Centro de Investigaciones Biológicas Margarita Salas, CSIC, Madrid, Spain*

JESÚS DEL MAZO • *Department of Cellular & Molecular Biology. Centro de Investigaciones Biológicas Margarita Salas (CIB-CSIC), Madrid, Spain*

Abhijna Krishnan Edilyam • *Indian Institute of Science Education and Research (IISER), Tirupati, Andhra Pradesh, India*

Jared M. Evatt • *Oklahoma Medical Research Foundation, Oklahoma City, OK, USA; Department of Cell Biology, University of Oklahoma Health Science Center, Oklahoma City, OK, USA*

Alfonso Fernández-Álvarez • *Instituto de Biología Funcional y Genómica (IBFG), Consejo Superior de Investigaciones Científicas (CSIC), Universidad de Salamanca (USAL), Salamanca, Spain*

John Fozard • *Department of Cell and Developmental Biology, John Innes Centre, Norwich Research Park, Norwich, UK; MRC-University of Glasgow Centre for Virus Research, University of Glasgow, Glasgow, UK*

Will H. Gittens • *Genome Damage and Stability Centre, School of Life Sciences, University of Sussex, Brighton, UK*

Rocío Gómez • *Departamento de Biología, Universidad Autónoma de Madrid (UAM), Madrid, Spain*

Silvia González-Sanz • *Department of Cellular & Molecular Biology. Centro de Investigaciones Biológicas Margarita Salas (CIB-CSIC), Madrid, Spain*

Irma Gryniuk • *Cell Cycle and Cancer Biology Research Program, Oklahoma Medical Research Foundation, Oklahoma City, OK, USA; Department of Cell Biology, University of Oklahoma Health Science Center, Oklahoma City, OK, USA*

Iván Hernández • *Gene Therapy Unit, Instituto de Investigación de Enfermedades Raras, Instituto de Salud Carlos III (ISCIII), Madrid, Spain*

Tara Hicks • *Department of Biology, The University of Iowa, Iowa City, IA, USA*

Chih-Ying Lee • *Cell Cycle and Cancer Biology Research Program, Oklahoma Medical Research Foundation, Oklahoma City, OK, USA*

Daniel León-Periñán • *Laboratory for Systems Biology of Regulatory Elements, Berlin Institute for Medical Systems Biology (BIMSB), Max Delbrück Center for Molecular Medicine in the Helmholtz Association (MDC), Berlin, Germany*

Zhen Lin • *State Key Laboratory of Molecular Biology, Shanghai Key Laboratory of Molecular Andrology, Center for Excellence in Molecular Cell Science, University of Chinese Academy of Sciences, Chinese Academy of Sciences, Shanghai, China*

Elena Llano • *Molecular Mechanisms Program, Centro de Investigación del Cáncer, Instituto de Biología Molecular y Celular del Cáncer, CSIC-Universidad de Salamanca, Salamanca, Spain; Departamento de Fisiología y Farmacología, Universidad de Salamanca, Salamanca, Spain*

Pablo López-Jiménez • *Departamento de Biología, Universidad Autónoma de Madrid (UAM), Madrid, Spain; Meiosis group, MRC Laboratory of Medical Sciences, London, UK*

Yuanlin Ma • *Reproductive Medicine Center, The First Affiliated Hospital, Sun Yat-sen University, Guangzhou, Guangdong, China; Guangdong Provincial Key Laboratory of Reproductive Medicine, Guangzhou, Guangdong, China; Guangdong Provincial Clinical Research Center for Obstetrical and Gynecological Diseases, Guangzhou, Guangdong, China*

Ana Martínez-Marchal • *Magee-Womens Research Institute, Department of Obstetrics, Gynecology & Reproductive Sciences, University of Pittsburgh, Pittsburgh, PA, USA*

Lucy McNeill • *DNA Repair & Recombination Laboratory, St Vincent's Institute of Medical Research, Fitzroy, VIC, Australia*

RODRIGO MENDOZA-MADRIGAL • *Instituto de Biología Funcional y Genómica (IBFG), Consejo Superior de Investigaciones Científicas (CSIC), Universidad de Salamanca (USAL), Salamanca, Spain*
CHRIS MORGAN • *Department of Cell and Developmental Biology, John Innes Centre, Norwich Research Park, Norwich, UK*
MATTHEW J. NEALE • *Genome Damage and Stability Centre, School of Life Sciences, University of Sussex, Brighton, UK*
ANTONY W. OLIVER • *Genome Damage and Stability Centre, School of Life Sciences, University of Sussex, Brighton, UK*
JINSHA PADMARAJAN • *Indian Institute of Science Education and Research (IISER), Tirupati, Andhra Pradesh, India*
JESÚS PAGE • *Departamento de Biología, Universidad Autónoma de Madrid (UAM), Madrid, Spain*
ALBA LÓPEZ-PALACIOS • *Department of Cellular & Molecular Biology. Centro de Investigaciones Biológicas Margarita Salas (CIB-CSIC), Madrid, Spain*
MARÍA TERESA PARRA • *Departamento de Biología, Universidad Autónoma de Madrid (UAM), Madrid, Spain*
ALBERTO M. PENDÁS • *Molecular Mechanisms Program, Centro de Investigación del Cáncer, Instituto de Biología Molecular y Celular del Cáncer, CSIC-Universidad de Salamanca, Salamanca, Spain*
IRENE PÉREZ-MORENO • *Departamento de Biología, Universidad Autónoma de Madrid (UAM), Madrid, Spain*
ROBERTO J. PEZZA • *Cell Cycle and Cancer Biology Research Program, Oklahoma Medical Research Foundation, Oklahoma City, OK, USA; Department of Cell Biology, University of Oklahoma Health Science Center, Oklahoma City, OK, USA*
JESÚS PINTO-CRUZ • *Instituto de Biología Funcional y Genómica (IBFG), Consejo Superior de Investigaciones Científicas (CSIC), Universidad de Salamanca (USAL), Salamanca, Spain*
MANUEL SÁNCHEZ-MARTÍN • *Departamento de Medicina, Universidad de Salamanca, Salamanca, Spain*
FERNANDO SÁNCHEZ-SÁEZ • *Molecular Mechanisms Program, Centro de Investigación del Cáncer, Instituto de Biología Molecular y Celular del Cáncer, CSIC-Universidad de Salamanca, Salamanca, Spain*
STEPHANIE SCHALBETTER • *Genome Damage and Stability Centre, University of Sussex, Brighton, UK*
NICOLA SILVA • *Department of Biology, Faculty of Medicine, Masaryk University, Brno, Czech Republic*
SARIT SMOLIKOVE • *Department of Biology, The University of Iowa, Iowa City, IA, USA*
VIJAYALAKSHMI V. SUBRAMANIAN • *Indian Institute of Science Education and Research (IISER), Tirupati, Andhra Pradesh, India*
MING-HAN TONG • *State Key Laboratory of Molecular Biology, Shanghai Key Laboratory of Molecular Andrology, Center for Excellence in Molecular Cell Science, University of Chinese Academy of Sciences, Chinese Academy of Sciences, Shanghai, China*
VANESSA TSUI • *DNA Repair & Recombination Laboratory, St Vincent's Institute of Medical Research, Fitzroy, VIC, Australia; The Faculty of Medicine, Dentistry and Health Science, The University of Melbourne, Parkville, VIC, Australia*
PATRICK T. WALSH • *Magee-Womens Research Institute, Department of Obstetrics, Gynecology & Reproductive Sciences, University of Pittsburgh, Pittsburgh, PA, USA*

JING WANG • *Reproductive Medicine Center, The First Affiliated Hospital, Sun Yat-sen University, Guangzhou, Guangdong, China; Guangdong Provincial Key Laboratory of Reproductive Medicine, Guangzhou, Guangdong, China; Guangdong Provincial Clinical Research Center for Obstetrical and Gynecological Diseases, Guangzhou, Guangdong, China*

ELLIE M. WRIGHT • *Genome Damage and Stability Centre, University of Sussex, Brighton, UK*

MEIXIA WU • *State Key Laboratory of Molecular Biology, Shanghai Key Laboratory of Molecular Andrology, Center for Excellence in Molecular Cell Science, University of Chinese Academy of Sciences, Chinese Academy of Sciences, Shanghai, China*

YANWEN XU • *Reproductive Medicine Center, The First Affiliated Hospital, Sun Yat-sen University, Guangzhou, Guangdong, China; Guangdong Provincial Key Laboratory of Reproductive Medicine, Guangzhou, Guangdong, China; Guangdong Provincial Clinical Research Center for Obstetrical and Gynecological Diseases, Guangzhou, Guangdong, China*

XI ZHANG • *State Key Laboratory of Molecular Biology, Shanghai Key Laboratory of Molecular Andrology, Center for Excellence in Molecular Cell Science, University of Chinese Academy of Sciences, Chinese Academy of Sciences, Shanghai, China*

Part I

Genetics and Genomics

Chapter 1

CC-seq: Nucleotide-Resolution Mapping of Spo11 DNA Double-Strand Breaks in *S. cerevisiae* Cells

George G. B. Brown, Will H. Gittens, Rachal M. Allison, Antony W. Oliver, and Matthew J. Neale

Abstract

During meiosis, Spo11 generates DNA double-strand breaks to induce recombination, becoming covalently attached to the 5′ ends on both sides of the break during this process. Such Spo11 "covalent complexes" are transient in wild-type cells, but accumulate in nuclease mutants unable to initiate repair. The CC-seq method presented here details how to map the location of these Spo11 complexes genome-wide with strand-specific nucleotide-resolution accuracy in synchronized *Saccharomyces cerevisiae* meiotic cells.

Key words Spo11, Meiosis, DNA break, DSB, Recombination, Mapping, Genome-wide, Chromosome, Next-generation sequencing, NGS, Sequencing, Single nucleotide, Covalent complex, CC, CC-seq

1 Introduction

During meiosis, homologous recombination (HR) between non-sister chromatids provides a mechanism for generating genetic diversity. HR is initiated deliberately, yet in a non-random manner, via the generation of DNA double-strand breaks (DSBs) created by the evolutionarily conserved Spo11 enzyme. Failure to initiate or properly repair Spo11 DSBs leads to chromosomal mis-segregation in meiosis creating aneuploidy. In humans, aneuploidy is a cause of genetic disorders such as Down's syndrome.

Spo11 is part of the topoisomerase family of proteins that create DNA breaks via the formation of a 5′-phosphotyrosyl-linked covalent complex (CC) [1]. The Spo11-linked DNA molecule is then endonucleolytically removed via the Mre11 complex (Mre11, Rad50, Xrs2)—an activity dependent upon the Sae2 protein [2–4], the orthologue of human CtIP [5]. Nicks created by Mre11 flanking the Spo11 DSB act as initiation sites for bidirectional resection

Jesús A. Carballo (ed.), *Meiosis: Methods and Protocols*, Methods in Molecular Biology, vol. 2818,
https://doi.org/10.1007/978-1-0716-3906-1_1,

by Mre11 and Exo1 in the 3′ to 5′ and 5′ to 3′ directions, respectively [2, 6]. The ssDNA that is revealed acts as a substrate for loading RecA-family recombinases such as Dmc1 and Rad51, which promote Spo11-DSB repair by catalyzing strand invasion between homologous chromosomes, leading to crossover and non-crossover repair outcomes.

Recombination promotes genetic variation and accurate meiotic chromosome segregation. Thus, understanding the mechanisms that govern spatial and temporal patterns of Spo11- DSB formation provides insight into patterns of genetic change arising within sexually breeding populations and the potential causes of genetic instability and aneuploidy.

Since the discovery of Spo11, several complementary methods for assessing genome-wide patterns of Spo11 DSBs have been developed—some of which have been applied to numerous experimental systems. These include:

- DNA microarray analysis of the accumulation of Spo11-bound DNA in the nuclease-deficient *rad50S* background in *S. cerevisiae* [7]
- Single-stranded DNA (ssDNA) microarray mapping using RPA ChIP-chip [8] or ssDNA enrichment [9] in DNA strand-invasion deficient *S. cerevisiae* backgrounds (*dmc1Δ* and/or *rad51Δ*) and related sequencing techniques in *Mus musculus* [10]
- High-resolution Spo11-oligo mapping in, *S. cerevisiae* [11–13], *Schizosaccharomyces pombe* [14], *M. musculus* [15], and *Arabidopsis thaliana* [16]
- High-resolution Spo11-DSB mapping methods: S1-Seq in *M. musculus* [17] *and S. cerevisiae* [18, 19] and End-Seq in *M. musculus* [20].

Here, we describe a detailed protocol for CC-seq [21, 22], an antibody-free method for determining the genome-wide locations of Spo11-DSB formation in meiotic cells with single-nucleotide accuracy (Fig. 1). The method exploits the electrostatic interaction between protein and silica to enrich Spo11-CC-linked double-stranded DNA molecules that accumulate when the Mre11-dependent nucleolytic repair pathway is prevented (i.e., in *sae2Δ*, *rad50S*, or *mre11* nuclease-dead cells). Pilot data indicate that CC-seq, while optimized for analysis of Spo11-DSBs in *S. cerevisiae* meiotic cells, is also transferable to other single-cell eukaryotes such as *S. pombe* and *Lachancea kluyveri* [23].

Briefly, DNA-containing protein-DNA covalent complexes (CC-DNA) from *S. cerevisiae* meiotic cells are prepared using an SDS and phenol chloroform extraction, fragmented by sonication, and then passed through a silica membrane in the presence of a high-salt buffer similar to prior work [1, 7]. Following a series of high-salt washes to remove background non-CC-DNA, enriched

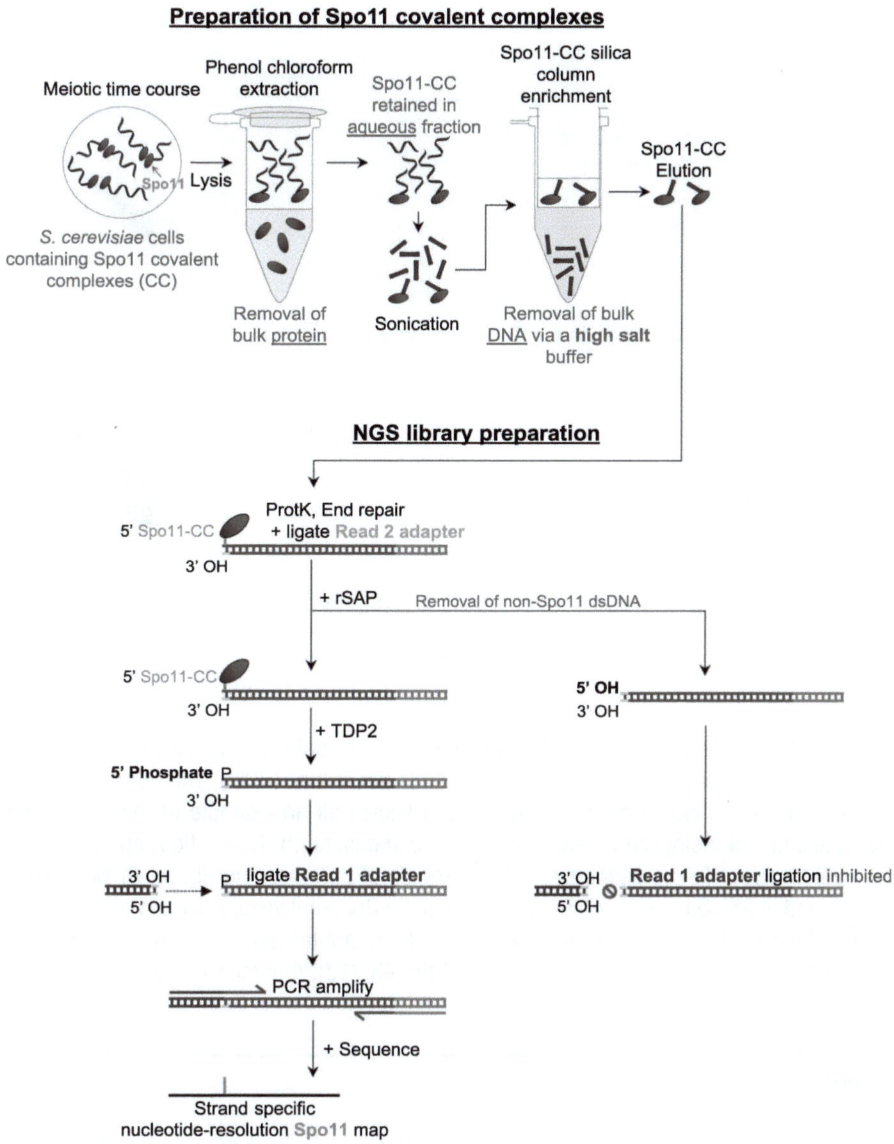

Fig. 1 Overview of the CC-seq methodology. Top: Cartoon showing the general steps of Spo11-CC enrichment via cell lysis, phenol chloroform extraction, sonication, and enrichment on silica membrane. Bottom: Cartoon showing the sequential DNA library preparation stages (see text for details). Treatment with rSAP prevents adapter ligation to non-Spo11 ends, increasing specificity

Spo11-linked DNA molecules are eluted with detergent. Adapters suitable for next-generation sequencing on the Illumina® platform are ligated to both ends of each Spo11-CC dsDNA fragment, utilizing the 5′-phosphotyrosyl-unlinking activity of recombinant vertebrate TDP2 [24] to differentiate sonicated from Spo11-linked DNA ends. Finally, resulting sequence reads are aligned to the *S. cerevisiae* reference genome generating a strand-specific nucleotide-resolution map of where Spo11-DSBs form within the genome (Fig. 2) (*see* **Note 1**).

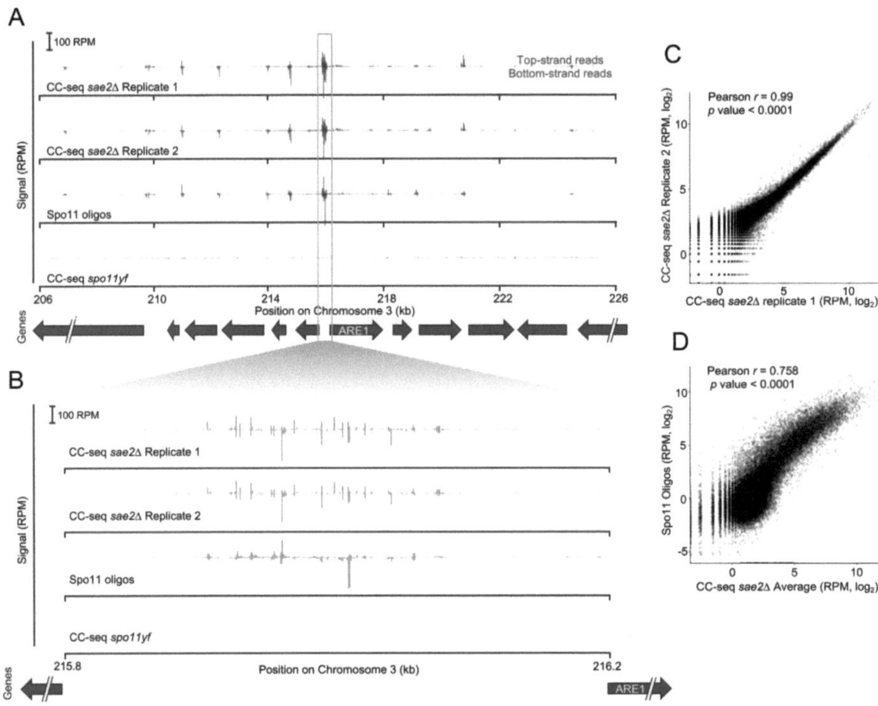

Fig. 2 Reproducibility of CC-seq maps. (**a**, **b**) Spatial reproducibility of CC-seq maps compared to Spo11-oligos (GSE84696; [13]). Mapped reads were normalized to reads per million (RPM). The bottom strand reads were offset by −1 bp. (**a**) 20 kb region of chromosome III is shown as an example of the visual reproducibility between two independent biological repeats of the *sae2Δ* datasets. (**b**) Nucleotide reproducibility of CC-seq reads at the *ARE1* hotspot (a representative 400 bp region of chromosome III). (**c**, **d**) Normalized, mapped reads were summed in 250 bp non-overlapping windows. (**c**) Quantitative comparison of CC-seq maps for the same two *sae2Δ* biological replicates (Pearson $r = 0.99$). (**d**) Quantitative comparison between the average of two *sae2Δ* biological replicates and Spo11-oligos (GSE84696; [13]) (Pearson $r = 0.758$)

2 Materials

2.1 General

1. General lab equipment, including shaking incubator, temperature-controlled floor-standing centrifuge and rotors, swinging-bucket bench-top centrifuge, vacuum aspirator, dry heat block, shaking heat block, water bath, 5 mL microcentrifuge with 5 mL tubes with suitable rotor, mini-PCR tube centrifuge, PCR tube thermocycler, 200 μL PCR tubes.
2. Qubit® 2 (Invitrogen #Q32866).
3. Qubit® dsDNA HS Assay Kit (Invitrogen #Q32854).
4. Qubit® Assay Tubes (Invitrogen #Q32856).
5. 1 × TE: 10 mM Tris–HCl, 1 mM EDTA pH 7.8.

2.2 Preparation of Spo11 Covalent Complexes

2.2.1 Meiotic Time Course and Cell Collection

1. Yeast Growth media:

   YPD: 1% yeast extract, 2% peptone, 2% D-glucose.

   BYTA: 1% yeast extract, 2% tryptone, 2% potassium (KAc), 50 mM potassium phthalate.

   SPM: 0.3% potassium (KAc), raffinose 0.02%, adenine 5 μg/mL, arginine 5 μg/mL, histidine 5 μg/mL, leucine 15 μg/mL, tryptophan 5 μg/mL, uracil 5 μg/mL.
2. Copper(II) sulfate (Merck #C1297).

2.2.2 Cell Spheroplasting and Fixing

1. Zymolyase 100 T (Amsbio #120493–1) dissolved at 50 mg/mL in 35% glycerol, 50 mM $NaHPO_4$ buffer pH 7.2, 0.5 M sucrose.
2. Beta mercapto-ethanol (βME) (Merck #M6250).
3. Spheroplasting buffer: 1 M sorbitol, 50 mM $NaHPO_4$ pH 7.2, 10 mM EDTA.

2.2.3 SDS Cell Lysis

1. STE: 2% SDS, 0.5 M Tris–HCl, pH 8.1, 10 mM EDTA, bromophenol blue 20 μg/mL.
2. Pestle for 1.5 mL tubes (VWR #431–0094).

2.2.4 Phenol Chloroform DNA Extraction

1. Phenol, chloroform, isoamyl alcohol in the ratio of 25:24:1 (v/v/v), saturated with 100 mM Tris pH 8.0, contains ~0.1% 8-hydroxyquinoline (Merck #77617).

2.2.5 DNA Sonication and DNA Quantification

1. Covaris M220 machine (Covaris #500295).
2. 1 mL milliTUBE AFA Fiber (Covaris #520135).
3. XT milliTUBE 1 mL holder (Covaris #500348).
4. 200 μL PCR tube.
5. RNase A pH 7.4 10 mg/μL (Merck #R4875) in 9 mM sodium acetate buffer pH 5.2, 100 mM Tris–HCl, pH 7.4. Heat treated to 100 °C for 15 min, and then slowly cooled.

2.2.6 Spike-in Preparation

1. Lambda DNA (~500 μg/mL) (NEB #N3011).
2. Human hTERT RPE-1 cells (ATCC #CRL-4000).
3. Dulbecco's modified Eagle's medium DMEM/F-12 (Merck #D6421–500 mL).
4. 10% Fetal calf serum (FCS).
5. Penicillin-streptomycin (10,000 U/mL) (Gibco™ #15140122).
6. Etoposide (Merck #E1383-100MG).
7. Syringe and needle (21 gauge).

2.2.7 Spo11 Covalent Complex Enrichment

1. 10 × column binding buffer: 3 M NaCl, 1% Sarkosyl, 2% TritonX100.
2. TEN: 10 mM Tris–HCl pH 7.8, 1 mM EDTA, 0.3 M NaCl.
3. TES: 10 mM Tris–HCl pH 7.8, 1 mM EDTA, 0.5% SDS.

2.2.8 Deproteinization of Spo11 Covalent Complexes

1. Proteinase K (FisherScientific #10407583) dissolved at 50 mg/mL in 50% glycerol, 20 mM Tris–HCl pH 7.8.
2. Glycogen 20 mg/μL (NEB #R0561).
3. 3 M NaOAc.

2.3 NGS Library Preparation

1. 10 mM Tris–HCl pH 8.2.
2. Ultra II DNA Library Prep Kit for Illumina® (NEB #E7645).
3. AMPure® XP magnetic beads (Beckman Coulter #A63882).
4. PCR tubes (200 μL).
5. DynaMag block (200 μL) (ThermoFisher #492025).
6. Shrimp Alkaline Phosphatase (rSAP) and rCutSmart buffer (NEB #M0371).
7. Adapter dilution buffer: 10 mM NaCl, 10 mM Tris–HCl pH 8.2.
8. 2 × Tdp2 buffer: 100 mM NaOAc, 100 mM TrisOAc, 2 mM MgOAc, 2 mM DTT (FisherScientific #10592945), 200 μg/mL Recombinant albumin (NEB #B9200S).
9. NEBNext® Multiplex Oligos for Illumina® (Index Primer Set 1) (NEB #7335S/L).
10. Bioanalyzer instrument (Agilent #G2939BA), Bioanalyzer Chip Priming Station (Agilent #5065–4401), 2100 Expert SW Laptop Bundle (Agilent #G2953CA), Agilent High Sensitivity DNA Kit (Agilent #5067–4626).

2.4 NGS Data Processing

1. terminalMapper (https://github.com/Neale-Lab).
2. CCTools (https://github.com/Neale-Lab).

3 Method

All centrifugation, incubation, and handling steps are performed at room temperature (~22 °C) unless stated otherwise.

3.1 Preparation of Spike-in DNA, Custom Adaptors, and Recombinant TDP2

To estimate relative Spo11-DSB yield between libraries, etoposide (VP16)-treated human DNA, used as a spike-in, is prepared as follows:

3.1.1 Spike-in DNA Preparation (Optional)

1. Plate hTERT RPE-1 cells (or other adherent human cells) on 4 × 15 cm plates at a density of 0.6 × 10^6 cells per plate and incubated at 37 °C in 30 mL DMEM/F12 supplemented with 10% FCS and Pen/Strep (complete media) for 72 h.

2. Cell confluency should be 60–70%, with ~5×10^6 cells per plate (20×10^6 cells total).
3. Treat each plate for 30 min with a final concentration of 100 μM etoposide in 20 mL complete media.
4. Remove treatment media and immediately add 10 mL ice-cold 70% ethanol-PBS.
5. Immediately scrape cells into this 70% ethanol-PBS buffer.
6. Pool cells from all four plates in a 50 mL falcon tube on ice.
7. Centrifuge for 5 min at 400× *g* at 4 °C.
8. Aspirate supernatant and resuspend pellet in 4 mL ice-cold 70% ethanol-PBS. Transfer to a 5 mL Eppendorf (or 10 mL falcon).
9. Centrifuge for 5 min at 400× *g* at 4 °C.
10. Aspirate supernatant.
11. Process ethanol-fixed cells according to "SDS cell lysis" and "Phenol-chloroform extraction of DNA" steps above.
12. Resuspend DNA by incubating in a thermomixer (5 min 1000 rpm, 45 °C).
13. Pipette suspended DNA into a 1 mL milliTUBE AFA Fiber.
14. Sonicate using a Covaris M220 and XT milliTUBE 1 mL holder using the following settings (*see* **Note 2**):
    - Target BP range: 200–700 bp
    - Duty Cycle: 10%
    - Intensity/peak power incidence: 75 W
    - Cycles/burst: 200
    - Time: 10 min
    - Bath temp: 7 °C
15. Pipette 12 μL sonicated VP16-treated human DNA into a PCR tube.
16. Add 3 μL RNase A (3.3 mg/mL) to a final concentration of 0.66 mg/mL.
17. Incubate at 37 °C for 20 min.
18. Measure the RNase-treated human DNA sample to calculate the DNA concentration via Qubit® using the HS dye protocol (*see* **Note 3**).
19. Using the Qubit® DNA concentration readings, dilute the RNase-treated human DNA to 80 μg/mL in 1 × TE pH 7.6 (*see* **Note 4**).

To estimate noise differences between libraries, Lambda DNA is prepared as follows:

1. Mix 2.5 mL of 500 μg/mL Lambda DNA with 2.5 mL 1 × TE pH 7.6 in a 5 mL tube and vortex for 20 s.
2. Use a syringe to pipette up and down 40x to help homogenize the Lambda DNA.
3. Aliquot Lambda DNA into 5 × 1 mL milliTUBE AFA Fiber.
4. Sonicate using a Covaris M220 and XT milliTUBE 1 mL holder using the following settings (*see* **Note 2**):
   - Target BP range: 200–700 bp
   - Duty cycle: 10%
   - Intensity/peak power incidence: 75 W
   - Cycles/burst: 200
   - Time: 10 min
   - Bath temp: 7 °C
5. Repeat the previous step for each Lambda DNA 1 mL aliquot.
6. Pool aliquots (5 mL total) in a 5 mL microcentrifuge tube and vortex heavily.
7. Pipette 2.5 μL of Lambda DNA and mix with 7.5 μL DD water in a PCR tube.
8. Measure the diluted Lambda DNA concentration after sonication via Qubit® using the HS dye protocol.
9. Using the Qubit® DNA concentration readings, dilute the Lambda DNA to 150 μg/mL in 1 × TE pH 7.6.

To estimate Spo11-DSB signal-to-noise between libraries, etoposide (VP16)-treated human DNA (signal) is mixed with Lambda DNA (noise) at a 1:50 ratio prior to "Spo11 covalent complex enrichment", as follows:

1. Mix 187.5 μL of VP16-treated human hTERT RPE-1 DNA (80 μg/mL) with 5 mL Lambda DNA (150 μg/mL) (*see* **Note 5**).
2. Prepare in advance and store 200 μL aliquots at −80 °C.

3.1.2 Custom Read 1 and Read 2 Adapter Generation

1. Custom oligonucleotides were generated using https://www.idtdna.com dissolved at a final concentration of 100 μM in 1 × TE pH 7.6:
   - Read1_Adapter_P5_top:

     ACACTCTTTCCCTACACGACGCTCTTCCGATCT
   - Read1_Adapter_5′-dephospho-P5_bottom:

     GATCGGAAGAGCGTCGTGTAGGGAAAGAGTGT /3InvdT/

- Read2_Adapter_P7_top:
  /5Phos/ GATCGGAAGAGCACACGTCTGAACTC CAGTCAC/3InvdT/
- Read2_Adapter_P7_bottom:
  GTGACTGGAGTTCAGACGTGTGCTCTTCC GATCT

2. For Read 1 and Read 2 adapter generation, mix 6 μL of top and bottom oligonucleotides.
3. Add 26 μL of 1 × TE pH 7.6.
4. Add 2 μL 1 M NaCl and mix thoroughly.
5. Anneal oligonucleotides by heating for 5 min at 100 °C and then leave to cool slowly for 2 h.
6. Final adapter concentration: 15 μM in 50 mM NaCl 1 × TE pH 7.6.
7. Prepare in advance and store 40 μL aliquots at −20 °C.

3.1.3 14M_zTDP2CAT Expression and Purification

1. To generate 14M_zTDP2CAT(240 μM), a codon-optimized synthetic gene encoding the catalytic domain of zebrafish TDP2 (amino acids 120–369)—where 14 individual single point mutations have been introduced in order to "humanize" the encoded enzyme [25]—was purchased from GeneArt (ThermoFisher Scientific, UK).
2. The coding sequence was sub-cloned into the NdeI and EcoR1 sites of pAWO-SUMO-3C, an in-house modified version of pET-17b that encodes an in-frame, human rhinovirus 3C-protease (HRV-3C) cleavable N-terminal fusion with yeast Smt3 prepended with a His6 affinity tag. The plasmid is available from AddGene: https://www.addgene.org/200512/.
3. Expression and purification of the recombinant protein followed the protocol described for m2hTDP2CAT [26], but replacing SENP1 protease with HRV-3C.
4. To generate 30 μM aliquots, dilute 120 μL of 14M_zTDP2CAT(240 μM) into 840 μL x2 Tdp2 buffer containing 50% glycerol.
5. Prepare in advance and store 100 μL aliquots at −80 °C.

3.2 Preparation of Spo11 Covalent Complexes

3.2.1 Meiotic Time Course and Cell Collection

1. From a −80 °C glycerol stock, streak out yeast cells onto a YPD plate. Incubate for 2 days at 30 °C.
2. In a glass culture tube, inoculate 4 mL YPD liquid with a single colony taken from the YPD plate. Incubate for 24 h at 30 °C with shaking (250 rpm).
3. Dilute the cell culture 1 in 50 with water to measure cell density (OD600). After 24 h, cells usually reach an OD600 of ~20.

4. In a 2 L flask, inoculate 150 mL BYTA to a cell density (OD600) of 0.3 by adding 45 OD600 units (~2.25 mL) of cell culture from the YPD overnight. Incubate for 15 h at 30 °C with shaking (250 rpm).
5. Dilute pre-sporulation BYTA culture 1 in 20 with water to measure cell density (OD600). After 15 h, cells usually reach an OD600 of ~6.5.
6. Collect 625 OD600 units (~100 mL) of pre-sporulation BYTA culture by centrifugation (5 min at 3250× *g*) in a 250 mL centrifuge bottle and drain supernatant (*see* **Note 6**).
7. Add 200 mL water, seal bottle and shake vigorously to resuspend cells.
8. Pellet cells again (5 min at 3250× *g*), drain, and resuspend in 250 mL SPM at 30 °C (final OD600 of ~2.5) by shaking vigorously.
9. Pour cell suspension into a new 2 L flask and incubate at 30 °C with shaking (250 rpm).
10. After 2 h, add 50 μL of 250 mM copper (II) sulfate (50 μM final concentration) to induce meiosis via the P_{CUP1}*IME1* promoter.
11. Monitor DNA replication timing by sampling cells at 0, 1, 1.5, 2, and 4 h after adding copper (II) sulfate by pipetting 400 μL of culture into 1 mL of 100% EtOH and processing for standard *S. cerevisiae* FACS analysis protocol [27].
12. Collect 50 mL cells at desired time points (usually between 2 and 4 h after adding copper (II) sulfate) by pouring into a 50 mL falcon tube and incubate in ice-water for 5 min.
13. Centrifuge (5 min at 3250× *g*, 4 °C) to pellet the cells. Pour off supernatant.
14. Store cell pellet in a −20 °C freezer for at least 8 h.

3.2.2 Cell Spheroplasting and Fixing

From this point, the protocol describes a single sample. In practice, four to eight samples are routinely processed in parallel.

1. Pre-aliquot 3.75 mL 100% EtOH into a 5 mL Eppendorf tube and place in the freezer (−20 °C).
2. Mix 1.2 mL spheroplast buffer, 6 μL zymo100T, 12 μL βME to a final concentration of zymo100T 250 μg/mL and βME 1% just before use.
3. Add 1.2 mL of Spheroplast solution to the frozen cell pellet.
4. Resuspend the cell pellet by vortexing.
5. Incubate the suspended cells in a water bath at 37 °C for 30 min.

6. Fix the cells by pipetting the cell suspension into the pre-aliquoted 3.75 mL EtOH (−20 °C). Shake vigorously.
7. Incubate in the freezer (−20 °C) for 10 min.
8. Collect cells by centrifugation (1 min at 1000× *g*). Aspirate supernatant.

3.2.3 SDS Cell Lysis

1. Add 800 μL STE.
2. Completely resuspend into a thick paste using a pestle.
3. Add 1600 μL STE and mix thoroughly by inversion.
4. Incubate in a dry heat block at 65 °C for 10 min, invert once during incubation.
5. Cool on ice for 1 min, then place at room temperature.

3.2.4 Phenol Chloroform DNA Extraction

1. Add 1250 μL phenol, chloroform, and isoamyl alcohol. Cap tube securely and shake vigorously for 10 s.
2. Incubate cell lysate at room temperature for 5 min.
3. Shake vigorously for 10 s.
4. Separate the cell lysate into the aqueous, interphase, and phenol phases by centrifugation (15 min at 21000× *g*).
5. Carefully remove 2000 μL of supernatant, using a P1000 tip with ~5 mm cut off the tip to increase the bore diameter, and pipette into a clean 5 mL tube containing 3.3 mL 100% EtOH.
6. Mix by inversion 20x. A large stringy DNA precipitate will form.
7. Collect DNA by gentle centrifugation (1 min at 21000× *g*), and aspirate supernatant.
8. Wash the DNA pellet with 5 mL 70% EtOH carefully.
9. Collect DNA by centrifugation (1 min at 21000× *g*), and aspirate supernatant.
10. Dry at room temperature for 10 min.
11. Add 1 mL 1 × TE (pH 7.8) to the DNA pellet and leave at 4 °C to soak overnight.
12. (Optional) Break point (*see* **Note 7**).

3.2.5 DNA Sonication and DNA Quantification

1. Resuspend DNA by incubating in a thermomixer (5 min, 1000 rpm, 45 °C).
2. Pipette suspended DNA into a 1 mL milliTUBE AFA Fiber.
3. Sonicate using a Covaris M220 and XT milliTUBE 1 mL holder using the following settings (*see* **Note 2**):
   - Target BP range: 200–700 bp
   - Duty cycle: 10%

- Intensity/peak power incidence: 75 W
- Cycles/burst: 200
- Time: 15 min
- Bath temp: 7 °C

4. Pipette 12 μL sonicated DNA into a PCR tube.
5. Add 3 μL RNase A (3.3 mg/mL) to a final concentration of 0.66 mg/mL.
6. Incubate at 37 °C for 20 min.
7. Analyze 10 μL of RNase-treated DNA using a 1% agarose gel with ethidium bromide concentration at 0.5 μg/mL to determine sonication efficiency (Fig. 3) (*see* **Note 2**).
8. Measure the RNase-treated DNA sample to calculate the DNA concentration via Qubit® using the HS dye protocol (*see* **Note 8**).

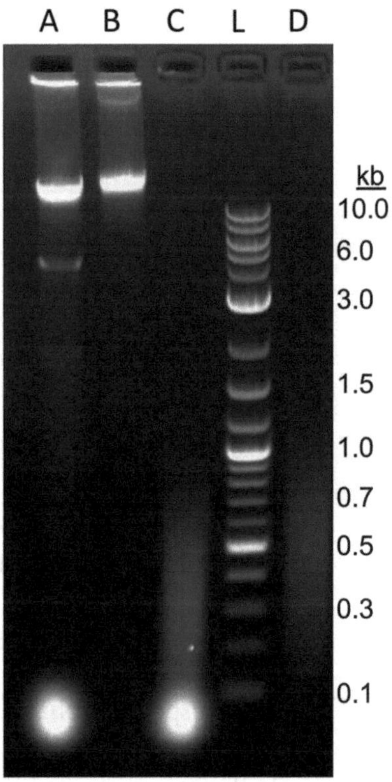

Fig. 3 DNA size distribution for a typical CC-seq library. Samples are separated on a 1% agarose gel containing 0.5 μg/mL EtBr for 1 h at 100 V. Library samples were treated as follows: Lane A, untreated. Lane B, RNase treated only. Lane C, sonicated only. Lane L, 1 μg of $\log_2$ DNA ladder (NEB #N3200). Lane D, RNase treated and sonicated

3.2.6 Spo11 Covalent Complex Enrichment

1. Pipette 900 μL of sonicated DNA into a 1.5 mL microcentrifuge tube.
2. Add 110 μL of the column-binding buffer.
3. (Optional) Add a DNA spike control to a final % DNA concentration of 0.1% VP16-treated human DNA and 5% Lambda DNA (see Subheading 3.1.1).
4. Top up the column-binding buffer and sonicated DNA solution with 1 × TE pH 7.6 to 1.1 mL and mix thoroughly.
5. Load two Qiaprep silica columns each with 550 μL of sonicated DNA solution and incubate for 5 min.
6. Centrifuge (1 min at 21000× *g*).
7. Reload each column by pipetting the flow-through back onto the same column and incubate for 5 min.
8. Centrifuge (1 min at 21000× *g*).
9. Wash column with six sequential washes with 600 μL TEN. Centrifuge (1 min at 21000× *g*) each time and discard flowthrough.
10. Perform a final dry centrifugation (1 min at 21000× *g*) to remove any residual wash buffer.
11. Add 125 μL TES (warmed to 60 °C) to column and incubate at room temperature for 5 min to elute Spo11-CC DNA.
12. Centrifuge (1 min at 21000× *g*).
13. Repeat steps 11–12 with a second aliquot of 125 μL TES.
14. Pool eluates (500 μL total) in a 2 mL microcentrifuge tube.

3.2.7 Deproteinization of Spo11 Covalent Complexes

1. Add 10 μL Proteinase K (50 mg/mL) to a final concentration of ~1 mg/mL.
2. Incubate in a thermomixer (1 h, 800 rpm, 50 °C).
3. Allow samples to cool to room temperature.
4. Add 5 μL of glycogen (20 mg/mL) to a final concentration of ~0.2 μg/μL and vortex.
5. Add 50 μL of 3 M NaOAc (265.5 mM final) and vortex.
6. Add 1.41 mL of 100% EtOH and vortex.
7. Incubate at −20 °C overnight to precipitate the DNA.

Caution: For **steps 8** and **10**, the DNA pellet is translucent and difficult to see, so proceed with care.

8. Collect precipitated DNA via centrifugation (60 min at 21000× *g*, 4 °C) and aspirate supernatant.
9. Wash the DNA with 70% EtOH (−20 °C) slowly and do not vortex.

10. Collect precipitated DNA via centrifugation (15 min at 21000× *g*, 4 °C) and aspirate supernatant.
11. Air-dry at room temperature for 10 min.
12. Add 25 μL of 10 mM Tris–HCl pH 8.2.
13. Resuspend DNA by thermomixing (5 min, 1000 rpm, 45 °C).
14. Remove 25 μL to a new PCR tube.
15. (Optional) Measure the DNA concentration via Qubit® using the HS dye protocol (*see* **Note 9**).
16. (Optional) Break point (*see* **Note** 7).

3.3 NGS Library Preparation

This protocol describes one library but normally four to eight libraries can be processed in parallel.

3.3.1 DNA End Repair and a Tailing

To 25 μL of CC-enriched sample (thawed at room temperature if previously frozen):

1. Add 3.5 μL NEBNext® Ultra II End Prep Reaction Buffer.
2. Add 1.5 μL NEBNext® Ultra II End Prep Reaction.
3. Vortex to mix and quickly spin down in mini-PCR centrifuge.
4. Place in a thermocycler and run the following program: 30 min at 20 °C, 30 min at 65 °C.

3.3.2 Read 2 Adapter Ligation

1. During the 60 min incubation, prepare a mastermix of the following components:

   15 μL NEBNext® Ultra II Ligation Master and 0.5 μL NEBNext® Ligation Enhancer and keep on ice (*see* **Note 11**).
2. In a separate tube, mix 1 μL of **Read 2 P7** adapter (15 μM) with 9 μL of adapter dilution buffer and mix thoroughly. This tenfold diluted stock is sufficient for eight libraries. Do **not** add adapters to the master mix.
3. Remove sample from thermocycler and cool to room temperature. Add 15.5 μL of the NEBNext® master mix.
4. Add 1.25 μL of diluted Read 2 P7 adapter (1.5 μM).
5. Mix thoroughly by vortexing for 2 s and quickly spin down.
6. Incubate at 20 °C overnight in a thermocycler with the heated lid off.

3.3.3 Generic AMPure® XP Bead Cleanup Protocol

Refer to main protocol sections for specific volumes and buffers to use at each purification stage.

1. Take AMPure® XP beads out of the fridge at least 30 mins before use and shake vigorously 30–40 times to completely suspend.

   Add ***specified volume*** of AMPure® XP beads (0.83 × final concentration) and vortex for 2 s.

2. Invert the PCR tubes until the beads are completely suspended.
3. Incubate samples on the bench for 5 min (*see* **Note 10**).
4. Place samples on a PCR tube DynaMag block (200 μL size).
5. After 5 min, pipette off and discard the supernatant being careful not to disturb the beads on the sides of the tube.
6. Add 200 μL of 80% freshly prepared EtOH to the beads while in the magnetic stand. Incubate at room temperature for 30 s, and then carefully remove and discard the supernatant. Be careful not to disturb the beads that contain DNA targets.
7. Repeat **Step 6** for a total of two washes. Be sure to remove all visible liquid after the second wash. If necessary, briefly spin the tube/plate, place back on the magnet, and remove all traces of ethanol with a P10 pipette tip.
8. Air-dry the beads for 5 min while the PCR tubes are on the magnetic stand with the lid open.
9. To elute the DNA targets add ***specified volume*** of ***specified buffer*** and vortex.
10. Incubate samples on the bench for 5 min at room temperature.
11. Place samples on a PCR tube (200 μL) DynaMag block.
12. Transfer eluate, now containing the DNA targets, to a new PCR tube using a pipette. Be careful not to disturb the beads.

3.3.4 rSAP (Phosphatase) Treatment

1. Remove sample from thermocycler and add 46.75 μL of 10 mM Tris–HCl pH 8.2 to double the volume to 93.5 μL.
2. Perform an AMPure® XP bead cleanup by adding 78 μL beads and eluting with 85 μL DD water into a clean PCR tube (see Subheading 3.3.3).
3. Add 10 μL 10 × rCutsmart buffer, mix, then add 5 μL rSAP enzyme.
4. Incubate for 30 min at 37 °C.
5. Incubate for 15 min at 65 °C.

3.3.5 Removal of 5′-Linked Spo11 Peptide Using Recombinant TDP2

1. Perform an AMPure® XP bead cleanup by adding 83 μL beads and eluting with 50 μL of 10 mM Tris–HCl pH 8.2 into a clean PCR tube (see Subheading 3.3.3).
2. Add 50 μL 2 × Tdp2 buffer and vortex for 5 s.
3. Add 3 μL of 30 μM 14M_zTDP2CAT (see Subheading 3.1.3) and vortex for 5 s.
4. Incubate at 37 °C for 1 h.

3.3.6 Read 1 Adapter Ligation

1. During the 60 min incubation, prepare a mastermix of the following components:
   15 μL NEBNext® Ultra II Ligation Master and 0.5 μL NEBNext® Ligation Enhancer and keep on ice (*see* **Note 11**).
2. In a separate tube, mix 1 μL of **Read 1 P5** adapter (15 μM) with 9 μL of adapter dilution buffer and mix thoroughly. This tenfold diluted stock is sufficient for eight libraries. Do not add adapters to the master mix.
3. Remove sample from thermocycler and cool to room temperature.
4. Perform an AMPure® XP bead cleanup by adding 85.5 μL beads and eluting with 26.5 μL of 10 mM Tris–HCl pH 8.2 into a clean PCR tube (see Subheading 3.3.3).
5. Add 3.5 μL NEBNext® Ultra II End Prep Reaction Buffer.
6. Add 15.5 μL of the NEBNext® master mix.
7. Add 1.25 μL of diluted Read 1 P5 adapter (1.5 μM).
8. Mix thoroughly by vortexing and quickly spin down.
9. Incubate at 20 °C overnight in a thermocycler with the heated lid off.

3.3.7 PCR Amplification

1. Perform an AMPure® XP bead clean up adding 78 μL beads and eluting with 15 μL of 10 mM Tris–HCl pH 8.2 into a clean PCR tube.
2. To 15 μL elution, add 25 μL NEBNext® Ultra II Q5 Master Mix, 5 μL Index Primer (use a unique index primer for each sample), and 5 μL Universal PCR Primer. Vortex for 2 s to mix.
3. Perform a PCR reaction with the following steps:

Temperature	Time	Cycles
98 °C	30 s	1
98 °C 65 °C	10 s 75 s	15
65 °C	5 min	1
4 °C	Hold	

4. After the PCR reaction, add 50 μL of 10 mM Tris–HCl pH 8.2.
5. Make up 1 mL of 0.1 × TE pH 7.6 by mixing 100 μL 1 × TE pH 7.6 with 900 μL DD water.
6. Perform an AMPure® XP bead cleanup by adding 83 μL beads and eluting with 30 μL 0.1 × TE pH 7.6 into a clean PCR tube (see Subheading 3.3.3).

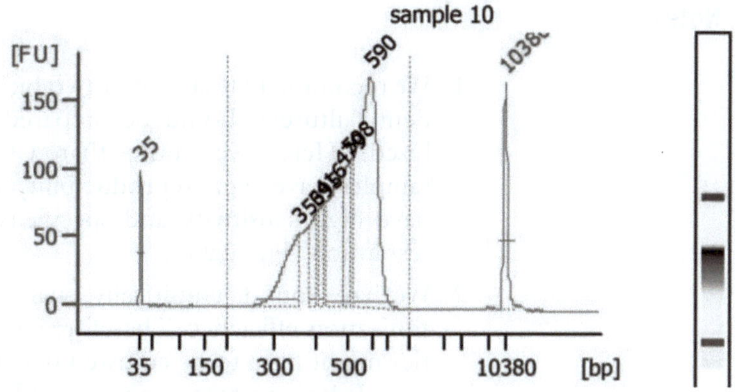

Fig. 4 A typical CC-seq library bioanalyzer 2100 trace. Samples were diluted 1 in 10 after PCR and AMPure® XP bead cleanup

3.3.8 Library Quantification and Quality Control

1. Prepare a high-sensitivity Bioanalyzer 2100 DNA chip following manufacturer's protocol.
2. Pipette 1 μL of each library into a clean tube containing 9 μL DD water and mix thoroughly (1 in 10 dilution).
3. Carefully load 1 μL of each diluted library onto a Bioanalyzer chip.
4. Load and run the chip on the Bioanalyzer machine. A typical Bioanalyzer 2100 trace is shown in Fig. 4 (*see* **Note 12**).

3.3.9 Library Sequencing

1. Dilute library in DD water as recommended by the NGS manufacturer.
2. Libraries are compatible with any Illumina®-compatible DNA sequencer (e.g., MiSeq™, NextSeq™ or HiSeq™). Multiple libraries (each with unique indexes, added at the PCR stage) can be multiplexed together (*see* **Note 13**).

3.4 NGS Data Processing

1. Download the demultiplexed FASTQ files.
2. Align the FASTQ files to the genome of interest, for example "sacCer3," using terminalMapper (https://github.com/Neale-Lab/terminalMapper), a data processing pipeline that aligns via Bowtie2 and then calls the read 1 (Spo11-CC coordinate) for each molecule, generating "FullMap" files.
3. Use R or other analysis scripts to plot and analyze the data to explore where Spo11-CCs form (e.g., using functions within the CCTools package, https://github.com/Neale-Lab/CCTools, see Fig. 2).

4 Notes

1. We recommend that at least two biological replicates (independent cultures) should be prepared per genotype being analyzed. Here, we show that CC-seq biological replicate samples have high reproducibility (Pearson's $r = 0.99$) and have high sensitivity and specificity for Spo11 at nucleotide resolution (Fig. 2a–c).
2. We have found, empirically, that DNA concentration affects sonication efficiency. Therefore, sonication timing or intensities might have to be adjusted depending on DNA concentration. It is important to aim for all DNA fragments to be below 1 kb, and the majority in the range between 200 and 700 bp.
3. To maintain EtBr staining in the gel, use a running buffer containing 1 × TAE with 0.5 μg/mL EtBr. We have also, empirically, found that the concentration of EtBr matters and that low concentrations of EtBr affect the DNA running speeds. We believe that the RNase binds to the DNA and impedes movement of DNA through agarose and that high concentrations of EtBr reduce ionic interactions between proteins and DNA.
4. Excess human CC-enriched DNA can be stored at −80 °C for future preparation of additional human:lambda spike material.
5. The final concentration of the spike-in can in theory be lower but must be greater than 50 μg/mL in order to spike in libraries up to a DNA concentration of 100 μg/mL.
6. Final OD600 value of ~6.5 is usually very robust. However, if OD600 values differ significantly between cultures processed on the same day, sample volumes can be normalized with DD water to balance the centrifuge.
7. DNA samples can be stored in the freezer (−20 °C) for up to 2 weeks.
8. DNA concentrations should be between 15 μg/mL and 100 μg/mL.
9. DNA concentrations should be between 15 ng/mL and 100 ng/mL.
10. It is important to not overdry the AMPure® XP beads because this affects DNA recovery. Bead drying times may have to be adjusted based on the number of samples or pipetting time. The beads should appear "glossy" after the drying stage; if the beads appear "cracked," then they have been overdried.
11. If required, the NEBNext® Master Mix can be made up to 8 h before use.

12. Library concentrations obtained should be between 20 and 200 nM.
13. Aim for at least two million reads per library, although most library-to-library differences can be detected with much less reads of ~0.5 million.

Acknowledgments

This work was supported by Wellcome Trust Investigator (200843/Z/16/Z) and Discovery (225852/Z/22/Z) Awards to MJN.

References

1. Keeney S, Giroux CN, Kleckner N (1997) Meiosis-specific DNA double-strand breaks are catalyzed by Spo11, a member of a widely conserved protein family. Cell 88:375–384. https://doi.org/10.1016/s0092-8674(00)81876-0
2. Neale MJ, Pan J, Keeney S (2005) Endonucleolytic processing of covalent protein-linked DNA double-strand breaks. Nature 436:1053–1057. https://doi.org/10.1038/nature03872
3. Keeney S, Kleckner N (1995) Covalent protein-DNA complexes at the 5′ strand termini of meiosis-specific double-strand breaks in yeast. Proc Natl Acad Sci USA 92:11274–11278. https://doi.org/10.1073/pnas.92.24.11274
4. Prinz S, Amon A, Klein F (1997) Isolation of COM1, a new gene required to complete meiotic double-strand break-induced recombination in Saccharomyces cerevisiae. Genetics 146: 781–795. https://doi.org/10.1093/genetics/146.3.781
5. Sartori AA, Lukas C, Coates J, Mistrik M, Fu S, Bartek J, Baer R, Lukas J, Jackson SP (2007) Human CtIP promotes DNA end resection. Nature 450:509–514. https://doi.org/10.1038/nature06337
6. Garcia V, Gray S, Allison RM, Cooper TJ, Neale MJ (2015) Tel1(ATM)-mediated interference suppresses clustered meiotic double-strand-break formation. Nature 520:114–118. https://doi.org/10.1038/nature13993
7. Gerton JL, DeRisi J, Shroff R, Lichten M, Brown PO, Petes TD (2000) Global mapping of meiotic recombination hotspots and coldspots in the yeast Saccharomyces cerevisiae. Proc Natl Acad Sci USA 97:11383–11390. https://doi.org/10.1073/pnas.97.21.11383
8. Borde V, Robine N, Lin W, Bonfils S, Géli V, Nicolas A (2009) Histone H3 lysine 4 trimethylation marks meiotic recombination initiation sites. EMBO J 28:99–111. https://doi.org/10.1038/emboj.2008.257
9. Buhler C, Borde V, Lichten M (2007) Mapping meiotic single-strand DNA reveals a new landscape of DNA double-strand breaks in Saccharomyces cerevisiae. PLoS Biol 5:e324. https://doi.org/10.1371/journal.pbio.0050324
10. Khil PP, Smagulova F, Brick KM, Camerini-Otero RD, Petukhova GV (2012) Sensitive mapping of recombination hotspots using sequencing-based detection of ssDNA. Genome Res 22:957–965. https://doi.org/10.1101/gr.130583.111
11. Pan J, Sasaki M, Kniewel R, Murakami H, Blitzblau HG, Tischfield SE, Zhu X, Neale MJ, Jasin M, Socci ND, Hochwagen A, Keeney S (2011) A hierarchical combination of factors shapes the genome-wide topography of yeast meiotic recombination initiation. Cell 144: 719–731. https://doi.org/10.1016/j.cell.2011.02.009
12. Thacker D, Mohibullah N, Zhu X, Keeney S (2014) Homologue engagement controls meiotic DNA break number and distribution. Nature 510:241–246. https://doi.org/10.1038/nature13120
13. Mohibullah N, Keeney S (2017) Numerical and spatial patterning of yeast meiotic DNA breaks by Tel1. Genome Res 27:278–288. https://doi.org/10.1101/gr.213587.116
14. Fowler KR, Sasaki M, Milman N, Keeney S, Smith GR (2014) Evolutionarily diverse determinants of meiotic DNA break and recombination landscapes across the genome. Genome

Res 24:1650–1664. https://doi.org/10.1101/gr.172122.114

15. Lange J, Yamada S, Tischfield SE, Pan J, Kim S, Zhu X, Socci ND, Jasin M, Keeney S (2016) The landscape of mouse meiotic double-strand break formation, processing, and repair. Cell 167:695–708.e16. https://doi.org/10.1016/j.cell.2016.09.035
16. Choi K, Zhao X, Tock AJ, Lambing C, Underwood CJ, Hardcastle TJ, Serra H, Kim J, Cho HS, Kim J, Ziolkowski PA, Yelina NE, Hwang I, Martienssen RA, Henderson IR (2018) Nucleosomes and DNA methylation shape meiotic DSB frequency in Arabidopsis thaliana transposons and gene regulatory regions. Genome Res 28:532–546. https://doi.org/10.1101/gr.225599.117
17. Yamada S, Hinch AG, Kamido H, Zhang Y, Edelmann W, Keeney S (2020) Molecular structures and mechanisms of DNA break processing in mouse meiosis. Genes Dev 34:806–818. https://doi.org/10.1101/gad.336032.119
18. Mimitou EP, Keeney S (2018) S1-seq assay for mapping processed DNA ends. Meth Enzymol 601:309–330. https://doi.org/10.1016/bs.mie.2017.11.031
19. Mimitou EP, Yamada S, Keeney S (2017) A global view of meiotic double-strand break end resection. Science 355:40–45. https://doi.org/10.1126/science.aak9704
20. Paiano J, Wu W, Yamada S, Sciascia N, Callen E, Paola Cotrim A, Deshpande RA, Maman Y, Day A, Paull TT, Nussenzweig A (2020) ATM and PRDM9 regulate SPO11-bound recombination intermediates during meiosis. Nat Commun 11:857. https://doi.org/10.1038/s41467-020-14654-w
21. Gittens WH, Johnson DJ, Allison RM, Cooper TJ, Thomas H, Neale MJ (2019) A nucleotide resolution map of Top2-linked DNA breaks in the yeast and human genome. Nat Commun 10:4846. https://doi.org/10.1038/s41467-019-12802-5
22. Ruiz LML, Johnson D, Gittens WH, Brown GGB, Allison RM, Neale MJ (2024) Meiotic prophase length modulates Tel1-dependent DNA double-strand break interference. PLoS Genet 20:e1011140. https://doi.org/10.1371/journal.pgen.1011140
23. https://pubmed.ncbi.nlm.nih.gov/38478689/
24. Cortes Ledesma F, El Khamisy SF, Zuma MC, Osborn K, Caldecott KW (2009) A human 5′--tyrosyl DNA phosphodiesterase that repairs topoisomerase-mediated DNA damage. Nature 461:674–678. https://doi.org/10.1038/nature08444
25. Ribeiro CJA, Kankanala J, Shi K, Kurahashi K, Kiselev E, Ravji A, Pommier Y, Aihara H, Wang Z (2018) New fluorescence-based high-throughput screening assay for small molecule inhibitors of tyrosyl-DNA phosphodiesterase 2 (TDP2). Eur J Pharm Sci 118:67–79. https://doi.org/10.1016/j.ejps.2018.03.021
26. Hornyak P, Askwith T, Walker S, Komulainen E, Paradowski M, Pennicott LE, Bartlett EJ, Brissett NC, Raoof A, Watson M, Jordan AM, Ogilvie DJ, Ward SE, Atack JR, Pearl LH, Caldecott KW, Oliver AW (2016) Mode of action of DNA-competitive small molecule inhibitors of tyrosyl DNA phosphodiesterase 2. Biochem J 473:1869–1879. https://doi.org/10.1042/BCJ20160180
27. Rosebrock AP (2017) Analysis of the budding yeast cell cycle by flow cytometry. Cold Spring Harb Protoc 2017:pdb.prot088740. https://doi.org/10.1101/pdb.prot088740

Chapter 2

SPO-Seq: An Accessible Method for Efficient Evaluation of Spo11 Catalytic Activity and Profiling Meiotic DSB Hotspots in *Saccharomyces cerevisiae*

Iván Hernández, David Álvarez-Melo, Mario García de Lacoba, and Jesús A. Carballo

Abstract

Meiotic recombination is a key process facilitating the formation of crossovers and the exchange of genetic material between homologous chromosomes in early meiosis. This involves controlled double-strand breaks (DSBs) formation catalyzed by Spo11. DSBs exhibit a preferential location in specific genomic regions referred to as hotspots, and their variability is tied to varying Spo11 activity levels. We have refined a ChIP-Seq technique, called SPO-Seq, to map Spo11-specific DSB formation in *Saccharomyces cerevisiae*. The chapter describes our streamlined approach and the developed bioinformatic tools for processing data and comparing with existing DSB hotspot maps. Through this combined experimental and computational approach, we aim to enhance our understanding of meiotic recombination and genetic exchange processes in budding yeast, with the potential to expand this methodology to other organisms by applying a few modifications.

Key words Tyrosyl group, Spo11, DSBs, Hotspots, Transposase, Tagmentation, ChIP-Seq

1 Introduction

Meiosis is a highly regulated cell division process, involving two divisions, meiosis I and meiosis II, which result in the formation of four-cell products with half the initial chromosome endowment [1]. A pivotal event during meiosis I is prophase I, where meiotic recombination occurs, facilitating the exchange of genetic material and the formation of crossovers between homologous chromosomes. This process promotes genetic diversity and ensures accurate chromosome segregation during metaphase I [2].

The initiation of meiotic recombination is marked by the introduction of double-strand breaks (DSBs) in DNA, catalyzed by the conserved topoisomerase Spo11 [3, 4]. This event leads to the formation of nicks at a distance from the Spo11 cleavage site,

Jesús A. Carballo (ed.), *Meiosis: Methods and Protocols*, Methods in Molecular Biology, vol. 2818,
https://doi.org/10.1007/978-1-0716-3906-1_2,

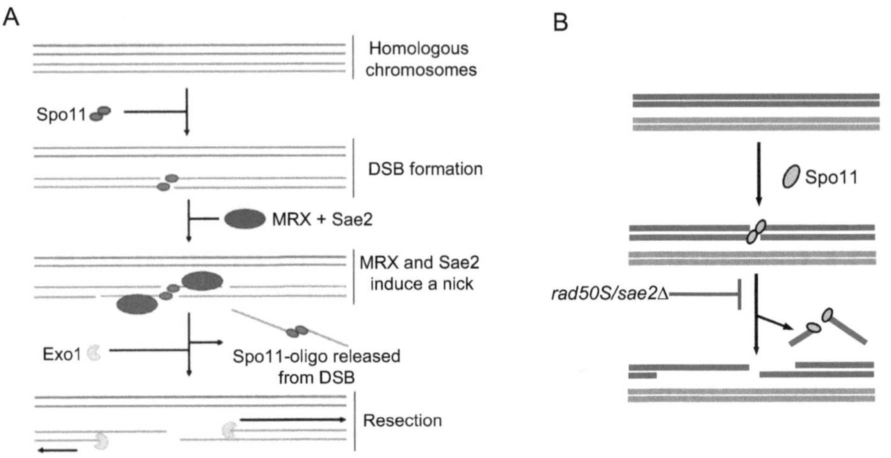

Fig. 1 (**a**) Diagram illustrating the sequential events that culminate in the generation of a double-strand break (DSB), involving Spo11 and various nucleases in the process. (**b**) The mutations in *rad50S/mre11S/sae2Δ* impact a specific step, resulting in Spo11 remaining covalently attached to the 5′ ends of DSBs

followed by the creation of protruding 3′ ends through resection [5, 6] (see Fig. 1a). The single-stranded DNA generated is processed by recombinases [7], ultimately leading to joint molecule formation and the progression towards crossover or non-crossover for repair [8, 9].

Spo11, a member of the type II topoisomerase family, is evolutionarily conserved and plays a central role in initiating recombination. It functions as a homodimer, catalyzing a transesterification reaction that results in DSBs [10]. Once a strand receives a break, it remains covalently bound to the enzyme, preventing further cleavage, a mechanism facilitated by the catalytic tyrosine Tyr135 [4]. After these initial events, accessory proteins process the nicks, generating oligonucleotides of specific size ranges [11]. These fragments serve as by-products of recombination initiation and can be detected through several molecular techniques.

Besides Spo11, at least ten other accessory proteins are required for DSB formation in *S. cerevisiae*, including Ski8/Rec103, Rec102, Rec104, Mre11, Rad50, Xrs2, Sae2, Mer2, Mei4, and Rec114. Deletion of genes encoding these proteins leads to defects in the early stages of meiosis [12, 13].

In terms of structure and function, the MRX complex, comprising Mre11, Rad50, and Xrs2 in combination with Sae2, plays a vital role in the initial processing of DSBs [14]. Xrs2, while not as conserved as its mammalian counterpart NBS1, serves an adaptor and chaperone function, allowing the localization of Mre11 in the nucleus and interacting with various regulatory proteins, including Tel1/ATM [15]. The Mre11 protein exhibits enzymatic activity, including 3′→ 5′ single-stranded exonucleolytic and endonucleolytic enzymatic activity. This enzyme is responsible for performing the nick in the DNA at some distance from the original Spo11 cleavage, enabling the release of Spo11 bound to oligonucleotides. It also participates in resecting the break in the 3′→ 5′ direction, while resection in the 5′→ 3′ direction is mediated by Exo1 [16]. Rad50 shares protein motifs with the SMC family and regulates the nuclease activity of Mre11 when bound to ATP. It is also responsible for binding to chromosomes susceptible to breakage [17].

Mutations in the genes encoding these proteins can have significant effects on somatic development and the initiation of meiotic recombination. Null alleles of the Mre11 (*mre11Δ*) and Rad50 (*rad50Δ*) genes lead to the complete abolition of recombination, resulting in somatic phenotypes such as hypersensitivity to induced DNA damage and slow growth rate in culture [18, 19]. On the other hand, point mutations, such as *rad50S* and *mre11S*, allow normal somatic development but hinder the processing of double-strand breaks, preventing resection and the release of Spo11-oligos [18]. *sae2Δ* mutants exhibit a phenotype similar to that of *S*-type mutations (Fig. 1b).

DSBs in organisms like *Saccharomyces cerevisiae* are not randomly distributed but localized to specific regions called recombination hotspots. These hotspots are associated with epigenetic signals such as histone H3 lysine 4 trimethylation. Some organisms, like *Mus musculus* and *Homo sapiens*, utilize proteins like PRDM9 to define hotspot regions by promoting DSBs in specific genomic sequences [20–22].

In *Saccharomyces cerevisiae*, recombination hotspots are primarily located in interstitial regions of chromosome arms, with fewer occurrences in telomeric and pericentromeric regions, suggesting a level of control over large-scale double-strand break formation [23]. Examining smaller spatial scales, hotspots exhibit a strong negative correlation with the meiosis-specific cohesin subunit Rec8, indicating their preference for DNA loops emanating from chromosomal axes [24]. While hotspots are predominantly located in promoter regions, some are found within open reading frames, and they do not seem to be associated with specific sequences [25, 26]. Nucleosome-depleted regions (NDRs) are positively correlated with recombination hotspots and their activation frequency, with the "strongest" hotspots being larger and having larger NDR regions [26], displaying variations in their strength.

Until the early years of this century, determining the prevalence of certain hotspots over others relied on the fragmentation analysis of genomic regions containing hotspots, detected through traditional techniques such as Southern blotting [11, 15]. While effective for routine experiments, this method had limitations in quantitatively assessing Spo11 activity at individual hotspots. More recent approaches employing massive sequencing have validated previous findings [26–29]. Nevertheless, these techniques, while valuable, have proven labor-intensive and costly, limiting their widespread application.

To address these challenges and create a more accessible, quantitative, and sensitive method, we combined and optimized existing techniques. Leveraging mutations that enhance the covalent binding of Spo11 to DSBs, we enabled the immunoprecipitation of the hotspot-bound protein (ChIP) after cleavage [23, 25]. This modification eliminates the need for formaldehyde-based crosslinking, a common step in ChIP-based techniques [30, 15]. Our aim is to develop a simpler, cost-effective technology for mapping meiotic DSB formation in budding yeast. A second methodological breakthrough addresses the challenge of sequencing blocked ends, inherent to the covalent binding of Spo11 at the 5′ end of the DSB. A transposon DNA tagging approach overcomes this hurdle, enabling the tagging of DNA with adaptors for sequencing. Complementing these techniques is the development of a bioinformatic workflow, enhancing our comprehension of budding yeast meiotic recombination and genetic exchange processes.

2 Materials

2.1 Strains and Culture Media

1. The strain used for this study is a diploid strain of SKl genetic background of rapid and synchronous sporulation [31], JCY1667. The genotype of the strain employed in the current study is as follows:

   MAT a/α ho::LYS2/ ho::LYS2, lys2/ lys2, hisG::leu2/ hisG::leu2, ura3/ura3, trp1/trp1, SPO11-MYC18::TRP1/ SPO11-MYC18::TRP1, rad50S::URA3/ rad50S::URA3
2. YPG plates: 1% w/v yeast extract, 2% w/v bacto-peptone, 3% w/v glycerol, and 2% bacto-agar.
3. YPD: 1% w/v yeast extract, 2% w/v bacto-peptone, 2% w/v D-glucose. For YPD plates, 2% bacto-agar was added to liquid YPD.
4. YPA medium: 2% bacto-peptone, 1% yeast extract, 1% potassium acetate, 100 μL of 1% defoamer per 100 mL of YPA.
5. SPM: 1% potassium acetate, and 100 μL 1% defoamer per 100 mL SPM.

2.2 Meiotic Synchronization and Cell Harvesting

1. Two-liter Pyrex flasks.
2. TBS: Tris–HCl 50 mM, 150 mM NaCl, pH 7.4.
3. Polypropylene microcentrifuge tubes (2 mL).
4. Liquid nitrogen.

2.2.1 Monitoring Meiotic S-phase Progression

1. Fixative: ethanol 40%, sorbitol 0.1 M.
2. Ethanol 70%.
3. RNAse solution: 50 mM Tris–HCl pH 7.5; 1 mg/mL RNAse A.
4. Pepsin solution: Pepsin 5 mg/mL, HCl 50 mM.
5. Sytox green solution: 5 μg/mL Sytox Green, 50 mM Tris–HCl pH 7.5.
6. Ultrasonic cell disruptor.
7. Round-bottom polypropylene tubes (5 mL).

2.3 Whole Cell Extract (WCE)

1. Complete lysis buffer (CLB): 50 mM Hepes-KOH pH 7.5, 140 mM NaCl; 1 mM EDTA, 1% Triton X-100, 0.1% sodium deoxycholate, 1X protease inhibitor cocktail, 1 mM PMSF, 0.06 mg/mL aprotinin.
2. Polypropylene tubes with screw cap and O-ring seal (1.5 mL).
3. Glass beads 0.5 mm in diameter.
4. Bead beating grinder and lysis system (*see* **Note 1**).
5. Conical polypropylene tubes with screw cap (15 mL).
6. Syringe Needles (26G).

2.4 Chromatin Immunoprecipitation (ChIP)

1. TE-SDS buffer: 10 mM Tris–HCl pH 8, 1 mM EDTA, 1% SDS.
2. PBS/BSA blocking solution: 5 mg/mL of BSA dissolved in PBS.
3. Chip-grade mouse monoclonal anti-Myc antibody clone 9E11 (*see* **Note 2**).
4. Sterile phosphate-buffer saline, 1X (PBS).
5. Dynabeads Protein A (*see* **Note 3**).

2.5 DNA Isolation and Purification

1. TE buffer: 10 mM Tris–HCl pH 8; 1 mM EDTA.
2. TE-SDS buffer: 10 mM Tris–HCl pH 8; 1 mM EDTA, 1% SDS.
3. Complete lysis buffer supplemented (CLB-S): 50 mM Hepes-KOH pH 7.5, 360 mM NaCl; 1 mM EDTA, 1% Triton X-100, 0.1% sodium deoxycholate, 1X protease inhibitor cocktail, 1 mM PMSF, 0.06 mg/mL aprotinin.
4. Washing buffer: 10 mM Tris–HCl pH 8, 250 mM LiCl, 1 mM EDTA, 0.5% sodium deoxycholate, 0.5% NP-40.
5. Elution buffer: 1% SDS, 10 mM EDTA, 50 mM Tris–HCl pH 8.
6. Glycogen.
7. Proteinase K (10 mg/mL).

8. PCIA: phenol–chloroform–isoamyl alcohol (25:24:1).
9. Top-bench refrigerated centrifuge for 1.5–2 mL tubes.
10. Ethanol 96% and 70% (stored at −20 °C).
11. RNAse A (DNAse free).
12. 3 M sodium acetate.
13. Nuclease-free deionized filtered autoclaved water.

2.6 Library Preparation

1. Fluorimeter and fluorescent probes for DNA concentration calculation (*see* **Note 4**).
2. Nextera XT library preparation kit (Illumina®; FC-131-1024):
   - Amplicon Tagment Mix (ATM).
   - Tagment DNA Buffer (TD).
   - Neutralize Tagment buffer (NT).
   - Nextera PCR Master Mix (NPM).
   - Resuspension Buffer (RSB).
3. Other consumables:
   - Nextera XT Index Kit (FC-131-1001).
   - 0.2 mL PCR tubes.
   - Centrifuge tube adapters for 0.2 mL PCR tubes.
   - Freshly prepared 80% ethanol.
   - AMPure XP (Purification Beads).
   - Nuclease-free water.

3 Methods

3.1 Meiotic Time-Courses

Time course experiments were conducted in duplicate to obtain cultures with synchronous meiosis. The process involved several steps to achieve meiotic synchrony.

1. Isolate single colonies allowing them to grow for up to three days on YPED-agar plates.
2. Inoculate individual colonies into 5 mL liquid YPED culture, incubating for an additional 24 h with gentle agitation at 30 °C.
3. The following day, centrifuge the cells (2880× *g*, 30 °C, 3 min), wash them with 5 mL of YPA medium, and inoculate them into 220 mL of presporulation medium (YPA) in 2-Liter PYREX flasks at an optical density in between $O.D._{600} = 0.2 - 0.25$.
4. Grow the cultures for 13.5 h at 30 °C in the incubator with agitation (250 RPM) until the desired optical density of the cultures at 600 nm is between 1.4 and 1.8.

5. Centrifuge each culture in 2× 125 mL centrifuge tubes (2000× *g*, 30 °C, 4 min). After centrifugation, rinse the pellets with 25 mL of sporulation medium (SPM). Combine the separated cultures and resuspend the pooled pellets with 220 mL of fresh SPM in newly prepared 2-Liter flasks, previously tempered at 30 °C. Keep the cells in sporulation medium for seven and a half hours.
6. Collect samples for the ChIP analysis by extracting 25 mL from each 220 mL culture at designated timepoints, including 0, 1.5, 2.5, 3.5, 4.5, 5.5, 6.5, and 7.5 h throughout the time course. Place these samples in 50 mL Falcon tubes. Keep the collected samples on ice until further processing.
7. Simultaneously, gather 1 mL samples for all eight timepoints in separate 1.5 mL Eppendorf tubes to analyze meiotic progression and synchrony. Keep the collected samples on ice until further processing.
8. Perform two washes with ice-cooled TBS on the 25 mL samples collected for ChIP (**step 6**, this subheading) and centrifuge at 2880× *g*, 4 °C, 4 min. Resuspend the cells in 1 mL TBS in 2 mL polypropylene microcentrifuge tubes and centrifuge at 3220× *g*, 4 °C, 4 min. Eliminate the supernatant and place samples at −80 °C until use.
9. On the 1 mL samples used to monitor meiotic progression and synchrony (**step** 7, this subheading), spin down the cells and wash the pellets twice with 1 mL of ethanol-sorbitol fixative solution and keep in this solution at −20 °C until use.

Optimization of these time course experiments is crucial for obtaining cultures of cells with highly synchronized meiosis, providing with consistent and reliable samples for further study.

3.2 Fluorescence-Activated Cell Sorting (FACS)

FACS was used to observe the progression and synchrony during meiotic genome duplication, and to choose the most synchronized cultures for conducting SPO-Seq analyses.

1. Pipette 500 μL of each of the eight samples obtained to monitor the status of meiotic synchronization and collect the sedimented cells by centrifugation (3220× *g*, 4 °C, 3 min). Immediately wash the cells with 500 μL of 70% ethanol.
2. Treat the cells with 250 μL of RNAse solution for 1.5 h. Incubate the cells for 2.5 min with 500 μL of Pepsin solution at RT.
3. Stain the cells with 1 mL Sytox Green solution and allow to incorporate the dye into the DNA for half an hour.
4. The next day, disperse the cell clusters by sonication at 3.5 power level for 10 s (Branson Sonifier 150, Emerson Industrial Automation).

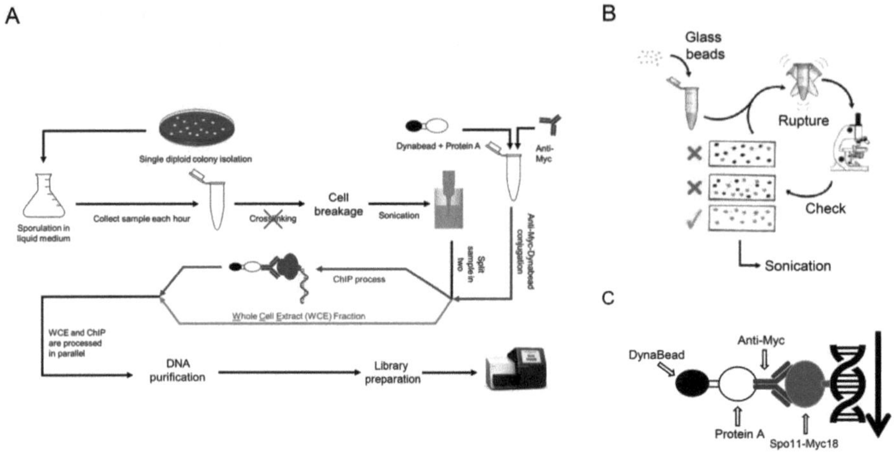

Fig. 2 (**a**) Overview of the experimental workflow for SPO-Seq, detailing the key steps and procedures involved in the tagmented Spo11-bound DNA sequencing process. (**b**) Summary of the steps preceding sonication, encompassing both cell breakage and confirmation processes. If insufficient, the rupture step needs to be repeated. (**c**) To immunoprecipitatet Spo11-Myc18, we utilized a multi-step precipitation method. This involved Dynabeads, magnetic particles coated with protein A, separated by a magnet. The precipitation utilized an anti-myc mouse monoclonal antibody, specific to an antigen in 18 copies attached to Spo11. Importantly, Spo11 is covalently bonded to DNA through the catalytic tyrosine (Y135)

5. Transfer 250 μL of the preparation to special tubes for use in the cytometer. Proceed to analyze the samples using a flow cytometer such as Coulter Epics XL cytometer (Beckman Coulter). Count a minimum of 25000 cells on each of the samples.
6. Analyze the results from the flow cytometer using a software package such as FlowLogic™.

3.3 Chromatin Immunoprecipitation (ChIP)

To perform Chromatin Immunoprecipitation (ChIP) the methodologies described in [32] were adopted with several modifications. The provided protocol is specifically tailored for handling one timepoint or sample in each iteration. Repeat the procedure as needed to accommodate the total number of samples intended for analysis. The overall protocol is summarized in Fig. 2a.

3.3.1 Antibody and Magnetic Beads Conjugation

- Prepare at least three ice boxes; one for the samples, another to put all the necessary solutions of the day to be cooled, and a third one for the prelabelled empty tubes to be used in sequential steps.
- In addition, for each sample to process, label four 15 mL polypropylene conical bottom tubes, eight 1.7 mL polypropylene low binding tubes, and four screw capped 1.5 mL tubes. Keep all tubes constantly chilled on ice.

- Set up and prechill a centrifuge at 4 °C with the compatible rotors for the pre-labelled tubes.
- Prepare a water bath at 65 °C.

1. Add 60 μL of Dynabeads to a 1.7 mL low-binding polypropylene tube. Add 500 μL of PBS/BSA solution to the beads to remove the DMSO in which they are dissolved. Collect the beads to the tube with the magnetic rack and remove the liquid. Repeat the washing (*see* **Note 5**).
2. In a new 1.7 mL microcentrifuge tube, prepare 282 μL of PBS/BSA and add 18 μL of anti-Myc antibody. Then add the 300 μL to the beads.
3. Incubate in the cold chamber in the rotator for 4 h to generate the conjugation between antibodies and Dynabeads (*see* **Note 6**).

3.3.2 Whole Cell Extract Preparation for ChIP

- All the following steps should be handled on ice.

1. Prepare a 15 mL polypropylene conical bottom tube containing 6 mL of Complete Lysis Buffer (CLB) and keep it on ice.
2. Add 1.6 mL of CLB to the sampled cells (Subheading 3.1, **step 8**) and resuspend the pellet.
3. Split 400 μL of the cell suspension into each of the four prelabelled and prechilled screw-capped 1.5 mL tubes and add glass beads until the liquid phase is no longer observed. Once is covered, do not pour more glass beads on top of that.
4. Using a Bead Beating system such as FastPrep Instruments, ideally located in a cold room, place the tubes in the instrument's rack, lock the tubes and the lid properly, and turn on the instrument using a program with the following parameters:
   - Speed: 4.5 m/s
   - Time: 60 s

   Immediately after the bead-beating stops, quickly place the four tubes on ice and leave them to rest for 1 min. Perform three repeats of this cycle. If insufficient, the rupture step needs to be repeated (Fig. 2b; *see* **Note 7**).
5. To separate the extract from the glass beads, heat a needle using a Bunsen burner and insert it perpendicularly, while still incandescent, on the top flat side of the screw cap. Make a hole as small as you can (inserting the end of the tip is enough), and place it upside down in the 15 mL polypropylene conical tubes. Do this for all four tubes. Centrifuge in a prerefrigerated centrifuge at 2880× *g* for 3 min at 4 °C.

6. Remove the screw cap tubes from the conical bottom 15 mL tubes and collect with a P1000 micropipette all the extract recovered in the centrifugation (including the pellet), avoiding pipetting any glass beads that were not retained in the screw cap tube. Transfer the extract into the prechilled 1.7 mL polypropylene low-binding tubes. Keep the tubes on ice until the next step.

3.3.3 Fragmentation of Chromatin

1. After elution of the whole cell extract (**step 6**, Subheading 3.3.2), the prechilled samples are subjected to the sonication process. This process is divided into 5 cycles of 15 s at 20% power in a sonicator model Branson Ultrasonic Sonifier S-250D (Emerson Industrial Automation), centrifuging before the first and between each cycle at 1960× *g*, 4 °C for 1 min.
2. After the last cycle of sonication, perform centrifugation at 11270× *g*, 4 °C for 5 min, in order to remove macroscopic cellular debris (*see* **Note 8**).
3. Transfer the supernatant into the remaining pre-chilled and empty 1.7 mL polypropylene low-binding tubes.
4. Add 20 μL from the sonicated sample in 1.5 mL labeled tubes containing 380 μL of 1% TE/SDS for analysis of the whole cell extract (WCE) fraction. Keep these tubes on ice stored inside the cold room or a refrigerator (*see* **Note 9**).

3.3.4 Immunoprecipitation

Following the WCE fraction collection, immunoprecipitation is conducted with the remaining sample as follows:

1. Retrieve the tubes from the rotator, which contain the Dynabead-Protein A-antibody conjugate (as detailed in **step 3**, Subheading 3.3.1). Wash the conjugate twice using 500 μL of PBS/BSA, employing the magnetic rack.
2. Reconstitute the Dynabead-Protein A-antibody conjugate in 60 μL of PBS/BSA.
3. Dispense 15 μL of the antibody-bead slurry onto each of the four 1.7 mL low-binding tubes containing the samples with the sheared chromatin (**step 3**, Subheading 3.3.3.). Mix carefully (refer to Fig. 2a).
4. To perform the immunoprecipitation, incubate the tubes at 4 ° C for 3 h on a rotator.

3.3.5 DNA Purification

1. After incubation with the antibody-Dynabeads conjugate, perform sequential washes of 4 min long with different solutions, using the magnetic rack between washes to pack the magnetic Dynabeads to one side of the tube (Fig. 2c).

   The washing buffers must be added in the following order:

(a) CLB (repeat once)
(b) CLB-S (repeat once)
(c) Washing Buffer (repeat once)
(d) TE (final wash)

2. Once all the washes are done and the wash solution removed, add to the beads 40 μL of Elution Buffer. Seal the tubes with Parafilm® and incubate at 65 °C in a water bath/Thermoblock with agitation for 15 min.
3. During the 15-min incubation period, every 4 min take them out of the bath, tap them at least 20 times with your finger, and return them to the bath/Thermoblock (*see* **Note 10**).
4. After 15 min remove the 1.7 mL low-binding tubes from the bath and spin down the Dynabeads. Transfer the 40 μL to 4 × 1.5 mL tubes preloaded with 160 μL of TE, 1% SDS.
5. Add 132.5 μL TE, 3 μL glycogen, and 15 μL proteinase K to each tube (to both ChIP and WCE tubes). Incubate at 37 °C for 2 h. During this time, prepare and label two sets of eight 1.5 mL tubes. Add 300 μL of PCIA to the first set of eight tubes.
6. After the 2-hour incubation with proteinase K, add 300 μL of PCIA to the tubes. Vortex the eight tubes at full power for at least 1 min per tube. Centrifuge the tubes in a benchtop centrifuge at top speed and room temperature for 7 min.
7. Using a micropipette, recover the upper (aqueous) phase, avoiding the organic remains (the cloudy fraction left at the interface) and the lower phase (phenolic or organic phase). Transfer the pipetted aqueous phase to the prelabelled tube already containing 300 μL of PCIA (**step 5**, this subheading).
8. Repeat the Phenol-Chloroform-Isoamyl alcohol phase separation step (vortex and centrifugation) and this time transfer the aqueous phase to the second set of empty 1.5 mL polypropylene tubes. Add twice the recovered volume of 96% ethanol to the tubes and from 6 to 12 μL of 200 mM NaCl (*see* **Note 11**).
9. Invert the tubes to mix the contents and leave them in the freezer overnight.
10. The next day, centrifuge all tubes at top speed for 30 min and 4 °C. Remove the supernatant and wash by adding 700 μL of 70% ethanol to the pellet containing the DNA followed by centrifugation at top speed for 5 min at 4 °C. Remove the ethanol and centrifuge again at top speed and 4 °C for 1 min in order to remove any remnants of the ethanol.
11. Place the tubes with the lid open in the laminar flow cabinet for at least 20 min until the pellets are dry.

12. Add 30 μL of RNAse-DNAse Free solution to each tube and incubate at 37 °C for 1 h.
13. Around the 30-min mark of the 1-h incubation period, consolidate tubes from the ChIP sample and the WCE sample by combining each set of four tubes into two separate single tubes. One tube will contain the contents from the same ChIP sample, and the other tube will include the contents from the WCE aliquots. This process results in a total reduction of eight tubes to two.
14. Add 480 μL of 100% ethanol and 24 μL of sodium acetate to the WCE samples. To the ChIP samples add 240 μL 100% ethanol and 12 μL 3 M sodium acetate. Incubate in the freezer at −20 °C for at least 1 h.
15. Centrifuge the tubes for 30 min, at top speed and 4 °C. Perform a wash with 70% ethanol, and after decanting the liquid following the 1-min spin, allow the pellets to air-dry in the cabinet.
16. Add 15 μL of sterile filtered water and allow DNA resuspension in the refrigerator at 4 °C overnight. Keep at −80 °C until use.

3.4 DNA Library Preparation

- The Nextera XT kit from Illumina was employed to attach Illumina-compatible adaptors, addressing the hindrance in adaptor ligation caused by the tyrosyl group left by the covalent bonding of the Spo11 protein to the 5′-ends of the double-strand breaks (DSBs). The Nextera XT library preparation method is based on its capability to carry out in vitro transposition reaction of the target DNA where transposons tag the DNA in order to ligate the adaptors (Fig. 3).
- The protocol shown here was modified to handle a limited number of samples. If you need to prepare a larger number of samples, you can find a comprehensive protocol provided by the supplier at this web address:

  https://support-docs.illumina.com/LP/NexteraXTRef/Content/LP/Nextera/XT/Protocol.htm
- A summary of the Library preparation workflow is depicted in Fig. 4a.

3.4.1 Transposon DNA Tagging

1. Measure DNA concentration using Qubit3® and the High Sensitivity dye provided by the manufacturer. The desired DNA concentration from ChIP samples was 1–30 ng/μL. For input samples (WCEs), the concentration was 100–500 ng/μL.
2. For each sample, use 1 ng for library preparation. Dilute your DNA when necessary using appropriate volumes of nuclease-free deionized sterile water.

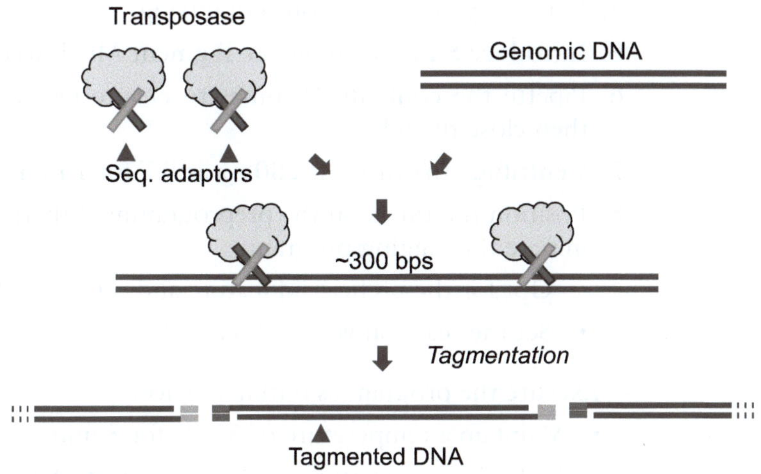

Fig. 3 Schematic representation of the tagmentation process, illustrating the simultaneous fragmentation and tagging of DNA molecules. The process involves the integration of DNA fragments with unique molecular tags, facilitating subsequent analysis. (The image is modified from the Nextera XT DNA Library Prep Kit Reference Guide, copyright 2017 Illumina, Inc.)

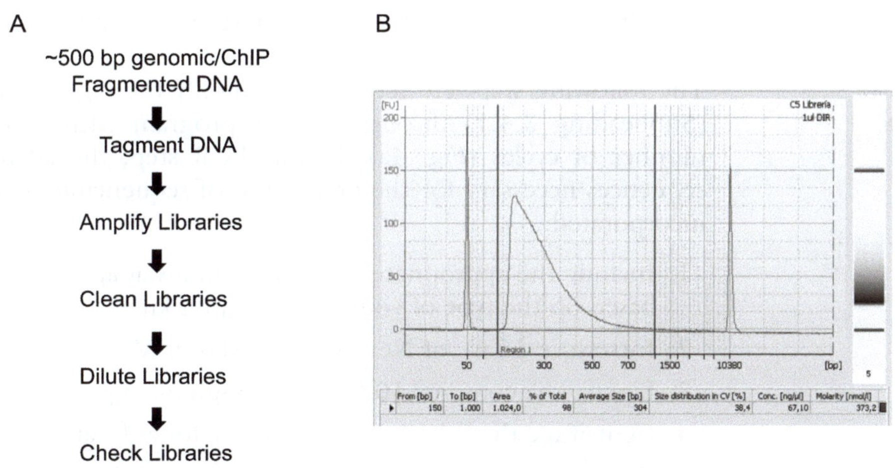

Fig. 4 (**a**) Nextera XT Kit workflow for library preparation. The process includes initial DNA fragmentation and simultaneous tagging, followed by limited-cycle PCR amplification incorporating unique dual indices. Subsequent purification ensures the generation of ready-to-sequence libraries, facilitating efficient and streamlined sample preparation for Illumina sequencing platforms. (**b**) Bioanalyzer results for library size and quality assessment. The graph displays the distribution of DNA fragment sizes, offering insights into the integrity and suitability for subsequent sequencing

3. To initiate genomic DNA tagmentation, dispense the specified volumes in the order provided into a 0.2 mL PCR tube per sample to be analyzed:
   - Tagment DNA Buffer (10 μL)
   - 1 ng DNA (5 μL)

4. Utilize a pipette for thorough mixing.
5. Introduce 5 μL of Amplicon Tagment Mix into each tube.
6. Pipette the contents 10 times to ensure proper mixing, and then close the lid.
7. Centrifuge the tubes at 280× *g* at RT for 1 min.
8. Position the tubes on the preprogrammed thermal cycler, and initiate the tagging program:
   - Opt for the preheat lid feature and set it to 100 °C.
   - Set the reaction volume to 50 μL.

– Execute the program as outlined below:
  - Maintain a temperature of 55 °C for 5 min.
  - Hold the temperature at 10 °C (*see* **Note 12**).

9. Add to each tube 5 μL of Neutralize Tagment buffer.
10. Pipette the contents 10x to achieve thorough mixing.
11. Centrifuge the tubes at 280× *g* at RT for 1 min using the centrifuge adaptors for 0.2 mL tubes.
12. Allow incubation at room temperature for 5 min.

3.4.2 Library Amplification

The following steps are required to amplify the tagmented DNA (Subheading 3.4.1) through a PCR program with a restricted number of cycles (Fig. 4a). In this PCR step, the adapter and sequences necessary for the generation of sequencing clusters are incorporated.

1. Include the appropriate volumes of index adapters per sample based on the type of your index adapter kit.
2. Introduce 15 μL of Nextera PCR Master Mix into each tube.
3. Pipette the contents 10X for thorough mixing.
4. Centrifuge the tubes at 280× *g* at RT for a duration of 1 min.
5. Position the tubes on the pre-programmed thermal cycler and execute the following PCR program:
   – Choose the preheat lid option and set to 100 °C.
   – Set the reaction volume to 50 μL.
     - 72 °C for 3 min
     - 95 °C for 30 s
   – 15 cycles of:
     - 95 °C for 10 s
     - 55 °C for 30 s
     - 72 °C for 30 s

- 1 final cycle of:
  - 72 °C for 5 min
  - Hold at 10 °C

3.4.3 Library Cleaning

1. Spin the samples in a centrifuge at 280× *g* and RT for 1 min.
2. Take 50 μL of supernatant from each PCR tube and transfer them to new 0.2 mL tubes.
3. For small PCR amplicon sample input, add the AMPure XP beads volume based on your input size (*see* **Note 13**). In this case, we added 90 μL of AMPure XP beads to reach 1.8x bead to sample ratio.
4. Gently mix the tubes in a shaker for 2 min.
5. Incubate at room temperature for 5 min.
6. Place the tubes on a magnetic stand for PCR tubes and wait until the liquid is clear (approximately 2 min).
7. Remove and discard all supernatant without disturbing the beads.
8. Wash the beads twice with 200 μL of 80% EtOH as follows:
   (i) With the tubes on the magnetic stand, add 200 μL of fresh 80% EtOH without mixing.
   (ii) Incubate for 30 s.
   (iii) Without disturbing the beads, remove and discard all supernatant.
9. Use a 20 μL pipette to remove and discard any residual EtOH.
10. Air-dry on the magnetic stand for 15 min.
11. Remove from the magnetic stand.
12. Add 17.5 μL of Resuspension Buffer to the beads.
13. Close the tubes and mix the content on an orbital shaker at 1800 rpm for 2 min.
14. Incubate at room temperature for 2 min.
15. Place the tubes on the magnetic stand and wait until the liquid is clear (approximately 2 min).
16. Transfer 15 μL of the supernatants to new 1.5 mL screw cap microtubes and store them at −80 °C until use.

3.4.4 Library Quality Check and Library Dilution

1. Add 1 μL of each undiluted library to analyze on an Agilent Technology 2100 Bioanalyzer using a DNA 7500 or a High Sensitivity DNA chip.
2. Assess the library's size by conducting a Bioanalyzer trace with heightened sensitivity. The estimated average library size following tagmentation ranges from 100 base pairs to 1 kilobase (Fig. 4b).

3. Quantify the library concentration using a high-sensitivity dsDNA Qubit after the library cleanup, and determine the library size using a Bioanalyzer to calculate molarity.
4. Determine the molarity of the library or combined libraries using the provided formula:

$$\text{Molarity (nM)} = \frac{\text{concentration in } \frac{\text{ng}}{\mu\text{L}}}{660\ \frac{\text{g}}{\text{mol}}\ \text{x average library size in bp}} \times 10^6$$

5. Utilize the molarity value to calculate the volumes of Resuspension Buffer and library required for diluting libraries to reach the concentration suitable for your system.
6. Dilute the libraries using Resuspension Buffer to reach the starting concentration of the sequencing system to be used.
7. Follow the denaturation and dilution instructions provided by the guide of your system to achieve the desired final loading concentration.
8. Sequence your samples (*see* **Note 14**).

3.5 Bioinformatic Processing

3.5.1 Workflow for the Mapping and Alignment of NGS Reads

1. Following the completion of the paired-end 75 bp sequencing run, obtain the reads in fastq format from all your samples. Ensure that you retrieve the relevant data for further analysis.
2. To map your reads onto the reference genome, such as *S. cerevisiae* (PacBioSK1) or SacCer2/3 (S288c), utilize a bioinformatic tool like BWA-MEM [33].
3. If your samples involve strain backgrounds different from those in the reference genomes, configure the alignment options to accommodate a limited number of mismatches (*see* **Note 15**).

3.5.2 Peak Calling

To create a list of DSB hotspots using the sequencing data, which are genomic areas with a high concentration of clustered meiotic double-strand breaks, and to facilitate comparison with existing DSB hotspot maps, a peak-calling algorithm was developed ab initio. The step-by-step workflow for generating this hotspot map is outlined as follows:

1. Convert your datasets, which contain raw counts, into normalized units, such as Reads Per Million Mapped Reads (RPM).
2. Smooth the high-resolution RPM-normalized Spo11-break map using a sliding window approach with a Hann window of 201 base pairs.
3. Implement a cut-off threshold of 8 RPM.
4. Merge the coordinates obtained from your ChIP datasets where the DSB hotspots have a signal strength greater than

8 RPM. To create a comprehensive hotspot map, utilize at least two datasets, preferably from different timepoints, to capture a broader range of early and late DSBs. In our case, we utilized four datasets, consisting of two duplicates from two distinct timepoints (*see* **Note 16**):

$$C_{\mathrm{Ref}} = C_1 \cap C_2 \cap C_3 \cap \ldots C_{n+1}$$

5. Combine the datasets derived from the input samples (Whole Cell Extracts, WCE) corresponding to each of the aforementioned ChIP samples using the same procedure:

$$W_{\mathrm{Ref}} = W_{C1} \cap W_{C2} \cap W_{C3} \cap \ldots W_{Cn+1}$$

6. To remove background sequencing noise, exclude the coordinates of hotspots in C_{Ref} that coincide with regions containing a high abundance of non-specific reads in W_{Ref} (>8 RPM).
7. After generating the map, remove hotspots that are shorter than 75 base pairs, as these are produced since coordinates from *Cs* and *Ws* do not usually match perfectly.
8. In this step, use genomic coordinates from a known high-resolution DSB hotspot map (e.g., [26–28]), as a reference. Merge matching hotspots from this map with those in your new map, adjusting coordinates to fit the transposon adaptor insertion method and high-resolution maps. The resulting map includes high-resolution data, featuring hotspots overlapping (partially or fully) with your newly mapped DSB hotspots. This adaptation is advantageous for future analyses involving comparisons with maps published by other laboratories.
9. Cluster the newly listed hotspots by combining the coordinates of those that are less than 200 base pairs apart, but ensure they do not exceed a size of 4 kilobases. Establish a criterion for splitting mega-hotspots into two separate hotspots. Following these steps, the described workflow mapped 4543 hotspots.
10. Now, populate your resulting 1D map (BED file) with sequencing coverage, expressed either in raw counts or in normalized units such as RPMs or any other arbitrary unit obtained after normalization.

3.5.3 Dataset Normalization

For the normalization of datasets to enable inter-dataset comparisons, various methods documented in the literature can be applied. In this study, we employed the following method:

1. For each sequenced sample calculate the total of reads mapping exclusively within the coordinates of all your assigned hotspots.
2. Represent each hotspot as a fractional proportion of the total for each sample.

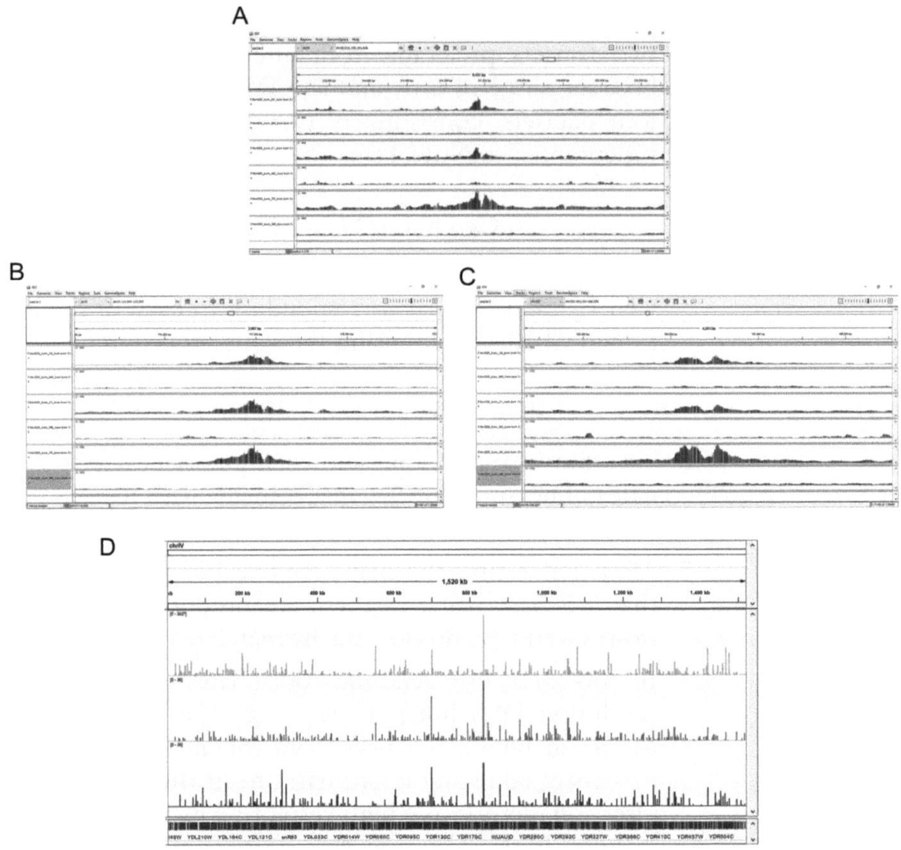

Fig. 5 (**a**) The outcomes following the sequencing and alignment after performing SPO-Seq (peaks; the genomic tracks displaying a prominent increase in read coverage) and input DNA (background; the genomic tracks displaying a lack of enrichment in read coverage) are presented for established hotspots on chromosome III. (**b**) Same as (A) but for chromosome VI. (**c**) Same as (A-B) but for chromosome VIII. (**d**) An inter-dataset visual comparison featuring two SPO-Seq samples (in red and blue) alongside a previously published hotspot map in green [28]

3. Scale the depicted coverage values by using an appropriate positive power of ten, ensuring proper visualization in the majority of genome viewers and browsers.
4. Analyze your DSB maps, for example, for comparison between different mutants or treatments (Fig. 5).

4 Notes

1. Tubes must fit in the model employed, so check with the manufacturer to avoid compatibility issues.
2. A good antibody is critical for your signal/noise ratio in your ChIP-seq datasets. Make sure you choose one that has been previously validated by other labs or that you have extensively tested.

3. For other solid support matrixes to immobilize your antibody alternative consider also Sepharose™ or agarose slurry conjugated with protein A or G.
4. Qubit dsDNA assay was the preferred method for DNA concentration calculation. Picogreen assay can also employed for High Sensitivity using a fluorescence spectrophotometer or a dedicated fluorometer.
5. It is better to prepare two pipettes: one with 1 mL and another with 500 μL for the whole protocol, to avoid losing time changing the suction volumes.
6. One and a half hours to go before the end of the 4-h antibody-beads conjugation period, start the chromatin purification protocol (Subheading 3.3.2).
7. If a different instrument is used, set up conditions equivalent to those listed for FastPrep Instruments.
8. Alternative instruments for chromatin fragmentation or shearing, such as the Bioruptor or similar devices capable of processing multiple tubes simultaneously, may be employed. However, it is imperative to establish specific conditions for each distinct instrument within each laboratory to ensure the achievement of the desired range for average DNA fragment size.
9. To avoid keeping the sample unattended for a long period, label and preload the solution in the 1.5 mL tubes while the samples are being centrifuged.
10. This is crucial; the mechanical tapping aids in separating the DNA from the precipitate pool.
11. If you recovered 300 μL add 12 μL and if 150 μL add 6 μL, in between you have to calculate and estimate the volume to add.
12. As soon as the program reaches 10 °C, proceed directly to **step 9** since the transposome remains active.
13. For input size 300–500 bp, the AMPure beads recommendation are 1.8x AMPure beads using a volume of 90 μL. For input size >500 bp the recommended AMPure bead size is 0.6x and a volume of 30 μL.
14. Due to the existence of several sequencing systems, the ensuing procedure must adhere to the guide for your system of choice.
15. If you have mapped your reads to a reference genome distinct from other publicly available datasets, employ the LiftOver tool (https://genome.ucsc.edu/cgi-bin/hgLiftOver) to realign your read coordinates to the reference genome of interest.
16. To enhance specificity, merge only the coordinates of matching hotspots from each dataset. When encountering less-

represented hotspots or those not detected in one of your samples, consider makingexceptions for hotspots that appear in three out of four datasets. Perform permutations to encompass all possible combinations. However, with an increasing number of samples, the analysis becomes more complex.

Acknowledgments

We would like to thank the Unidad de Genómica Parque Científico de Madrid for their valuable support and advice in helping design and execute the SPO-Seq technique. We would like to thank Matt Neale for helpful advice in the bioinformatic analysis and with the normalization of the datasets. This work has been funded by the grant numbers: BFU2015-64361-P, PID2021-125830NB-I00, and RyC-2013-13950. IH was funded by the JAE Intro (2017) program.

References

1. Gray S, Cohen PE (2016) Control of meiotic crossovers: from double-strand break formation to designation. Annu Rev Genet 50:175–210
2. Hunter N (2015) Meiotic recombination: the essence of heredity. Cold Spring Harb Perspect Biol 7(12)
3. Bergerat A, de Massy B, Gadelle D, Varoutas PC, Nicolas A, Forterre P (1997) An atypical topoisomerase II from Archaea with implications for meiotic recombination. Nature 386(6623):414–417
4. Keeney S, Giroux CN, Kleckner N (1997) Meiosis-specific DNA double-strand breaks are catalyzed by Spo11, a member of a widely conserved protein family. Cell 88(3):375–384
5. Longhese MP, Bonetti D, Guerini I, Manfrini N, Clerici M (2009) DNA double-strand breaks in meiosis: checking their formation, processing and repair. DNA Repair (Amst) 8(9):1127–1138
6. Manfrini N, Guerini I, Citterio A, Lucchini G, Longhese MP (2010) Processing of meiotic DNA double strand breaks requires cyclin-dependent kinase and multiple nucleases. J Biol Chem 285(15):11628–11637
7. Shinohara M, Shinohara A (2013) Multiple pathways suppress non-allelic homologous recombination during meiosis in Saccharomyces cerevisiae. PLoS One 8(4):e63144
8. Zakharyevich K, Ma Y, Tang S, Hwang PY, Boiteux S, Hunter N (2010) Temporally and biochemically distinct activities of Exo1 during meiosis: double-strand break resection and resolution of double Holliday junctions. Mol Cell 40(6):1001–1015
9. Alonso-Ramos P, Álvarez-Melo D, Strouhalova K, Pascual-Silva C, Garside GB, Arter M, Bermejo T, Grigaitis R, Wettstein R, Fernández-Díaz M, Matos J, Geymonat M, San-Segundo PA, Carballo JA (2021) The Cdc14 phosphatase controls resolution of recombination intermediates and crossover formation during meiosis. Int J Mol Sci 22(18)
10. Robert T, Nore A, Brun C, Maffre C, Crimi B, Bourbon HM, de Massy B (2016) The TopoVIB-like protein family is required for meiotic DNA double-strand break formation. Science (New York, NY) 351(6276):943–949
11. Neale MJ, Pan J, Keeney S (2005) Endonucleolytic processing of covalent protein-linked DNA double-strand breaks. Nature 436(7053):1053–1057
12. Keeney S (2008) Spo11 and the formation of DNA double-strand breaks in meiosis. Genome Dyn Stab 2:81–123
13. Yadav VK, Claeys Bouuaert C (2021) Mechanism and control of meiotic DNA double-strand break formation in S. cerevisiae. Fro Cell Dev Biol 9:642737
14. Gobbini E, Cassani C, Villa M, Bonetti D, Longhese MP (2016) Functions and regulation of the MRX complex at DNA double-strand breaks. Microbial cell (Graz, Austria) 3(8):329–337

15. Carballo JA, Panizza S, Serrentino ME, Johnson AL, Geymonat M, Borde V, Klein F, Cha RS (2013) Budding yeast ATM/ATR control meiotic double-strand break (DSB) levels by down-regulating Rec114, an essential component of the DSB-machinery. PLoS Genet 9(6): e1003545
16. Garcia V, Phelps SE, Gray S, Neale MJ (2011) Bidirectional resection of DNA double-strand breaks by Mre11 and Exo1. Nature 479(7372):241–244
17. Lim HS, Kim JS, Park YB, Gwon GH, Cho Y (2011) Crystal structure of the Mre11-Rad50-ATPγS complex: understanding the interplay between Mre11 and Rad50. Genes Dev 25(10):1091–1104
18. Alani E, Padmore R, Kleckner N (1990) Analysis of wild-type and rad50 mutants of yeast suggests an intimate relationship between meiotic chromosome synapsis and recombination. Cell 61(3):419–436
19. Usui T, Ohta T, Oshiumi H, Tomizawa J, Ogawa H, Ogawa T (1998) Complex formation and functional versatility of Mre11 of budding yeast in recombination. Cell 95(5): 705–716
20. Smagulova F, Gregoretti IV, Brick K, Khil P, Camerini-Otero RD, Petukhova GV (2011) Genome-wide analysis reveals novel molecular features of mouse recombination hotspots. Nature 472(7343):375–378
21. Brick K, Smagulova F, Khil P, Camerini-Otero RD, Petukhova GV (2012) Genetic recombination is directed away from functional genomic elements in mice. Nature 485(7400): 642–645
22. Myers S, Freeman C, Auton A, Donnelly P, McVean G (2008) A common sequence motif associated with recombination hot spots and genome instability in humans. Nat Genet 40(9):1124–1129
23. Blitzblau HG, Bell GW, Rodriguez J, Bell SP, Hochwagen A (2007) Mapping of meiotic single-stranded DNA reveals double-stranded-break hotspots near centromeres and telomeres. Curr Biol 17(23):2003–2012
24. Blat Y, Protacio RU, Hunter N, Kleckner N (2002) Physical and functional interactions among basic chromosome organizational features govern early steps of meiotic chiasma formation. Cell 111(6):791–802
25. Buhler C, Borde V, Lichten M (2007) Mapping meiotic single-strand DNA reveals a new landscape of DNA double-strand breaks in Saccharomyces cerevisiae. PLoS Biol 5(12): e324
26. Pan J, Sasaki M, Kniewel R, Murakami H, Blitzblau HG, Tischfield SE, Zhu X, Neale MJ, Jasin M, Socci ND, Hochwagen A, Keeney S (2011) A hierarchical combination of factors shapes the genome-wide topography of yeast meiotic recombination initiation. Cell 144(5): 719–731
27. Thacker D, Mohibullah N, Zhu X, Keeney S (2014) Homologue engagement controls meiotic DNA break number and distribution. Nature 510:241–246
28. Mohibullah N, Keeney S (2017) Numerical and spatial patterning of yeast meiotic DNA breaks by Tel1. Genome Res 27(2):278–288
29. Lange J, Yamada S, Tischfield SE, Pan J, Kim S, Zhu X, Socci ND, Jasin M, Keeney S (2016) The landscape of mouse meiotic double-strand break formation, processing, and repair. Cell 167(3):695–708 e616
30. Panizza S, Mendoza MA, Berlinger M, Huang L, Nicolas A, Shirahige K, Klein F (2011) Spo11-accessory proteins link double-strand break sites to the chromosome axis in early meiotic recombination. Cell 146(3): 372–383
31. Kane SM, Roth R (1974) Carbohydrate metabolism during ascospore development in yeast. J Bacteriol 118(1):8–14
32. Mendoza MA, Panizza S, Klein F (2009) Analysis of protein-DNA interactions during meiosis by quantitative chromatin immunoprecipitation (qChIP). Methods Mol Biol 557:267–283

Chapter 3

Hi-C2B: Optimized Detection of Chromosomal Contacts Within Synchronized Meiotic *S. cerevisiae* Cells

Ellie M. Wright, Stephanie Schalbetter, and Matthew J. Neale

Abstract

Hi-C, a genome-wide chromosome conformation capture assay, is a powerful tool used to study three-dimensional genome organization by converting physical pairwise interactions into counts of pairwise interactions. To study the many temporally regulated facets of meiotic recombination in *S. cerevisiae*, the Hi-C assay must be robust such that fine- and wide-scale comparisons between genetic datasets can be made. Here we describe an updated protocol for Hi-C (Hi-C2B) that generates reproducible libraries of interaction data with low noise and for a relatively low cost.

Key words Meiosis, Hi-C, Chromosome conformation capture, Paired-end sequencing, Genome stability

1 Introduction

Chromosome conformation capture methodologies based on "3C"—a "one-to-one" approach for studying contact frequencies between known genomic sequences—have advanced our understanding of chromosome and nuclear organization in many organisms [1–10, 14]. Prior to the evolution of 3C, chromosome structure was studied via a combination of electron microscopy, immunofluorescence microscopy, and the global, spatial patterning of proteins of interest determined by chromatin immunoprecipitation (ChIP). Now, chromosome conformation can also be analyzed with respect to the underlying genomic sequence, unveiling a wealth of information about the structural properties and spatial conformations of chromosomes within individual or populations of cells.

Early discoveries attributed to the 3C assay include identification of "flexible" chromosomes within *S. cerevisiae* [1]; observations of enhancer–promoter interactions for gene regulation at the β-globin locus [2], furthered by studies that showed enhancer–

Jesús A. Carballo (ed.), *Meiosis: Methods and Protocols*, Methods in Molecular Biology, vol. 2818,
https://doi.org/10.1007/978-1-0716-3906-1_3,

promoter interactions, require tissue-specific transcription factors [3, 4]. Architectural chromatin loops have been uncovered as stable chromosome structures mediated, in mammals, by a partnership between ubiquitously expressed CTCF protein and the cohesin complex [5–9]. On the back of such discoveries, and a greater accessibility and availability of high-throughput DNA sequencing technologies, development of many 3C-related techniques—including Hi-C—have transpired. Each 3C-based method customizes the technique to provide independent, detailed documentation of the spatial organization of chromosomes [10].

To capture chromosome conformation via 3C-related methodologies, the same initial approach is employed. Briefly, intact cells are isolated and subjected to treatment with formaldehyde in order to cross link physical DNA–protein and protein–protein interactions occurring in three-dimensional space [1]. Cross-linked material is digested with a 6- or 4-bp cutting restriction enzyme, the latter increasing the spatial resolution of resulting data. Subsequent adaptations of the Hi-C method have also employed the use of MNase (Micro-C) to generate Hi-C interaction maps at nucleosome resolution [11]. In all methods, fragmented DNA is subjected to in situ ligation prior to cross-link reversal. At this stage, a 3C "template" is obtained which contains both linear and circular polymeric concatemers made from genomic segments reshuffled with respect to their spatial proximity at the time of fixation [10]. Following generation of a 3C "template," 3C-based techniques differ in their detection and quantification of 3C ligation junctions [10].

Like classic 3C, Hi-C employs one of a number of restriction enzymes to digest cross-linked DNA to generate 5′ overhangs that are subsequently filled with a biotin-labeled nucleotide (Fig. 1) [12]. Blunt-end fragments are then ligated to produce products where the ligation junction is marked with the biotin label, referred to as the 3C template [12]. A Hi-C library is then created by mechanical shearing followed by enrichment steps before paired-end high-throughput sequencing, and mapping reads to a reference genome [12]. The result is a two-dimensional matrix of the sequence-specific interactions occurring within the three-dimensional nuclear space [12].

Hi-C, the genome-wide adaptation of 3C, was introduced more than a decade ago (Hi-C1) [1]. Hi-C2, described as an optimized method for the production of high-resolution chromosome conformation, includes a number of changes to the Hi-C1 protocol, particularly to the blunt-end ligation reaction [13]: (1) removal of additional chromatin solubilization by SDS prior to blunt-end ligation, with an aim to preserve nuclear structure, thus reducing the occurrence of random ligation events; (2) reduced reaction length; (3) abolition of large-scale dilution during ligation which was included in the original Hi-C protocols

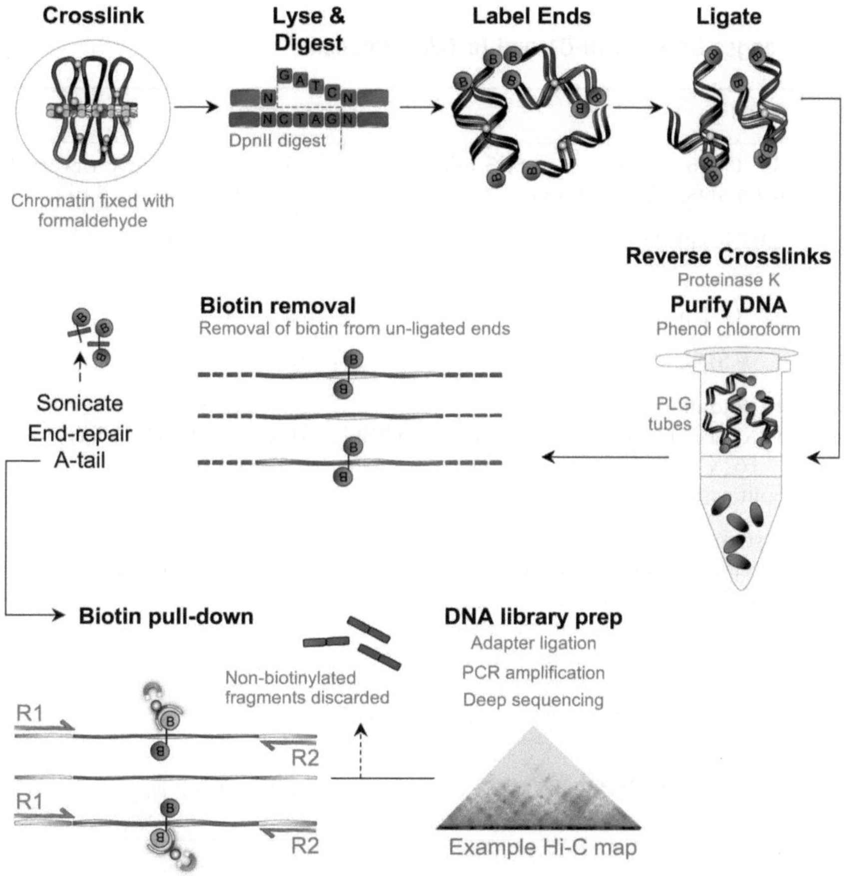

Fig. 1 Schematic outlining of the various steps of Hi-C library preparation
Chromatin is fixed with formaldehyde to capture DNA–DNA and DNA–protein interactions. Cells are lysed and chromatin solubilized before the DNA is digested with DpnII to generate 5′-GATC overhangs. Overhangs are filled with nucleotides including biotin-14-dCTP (green filled circles). Blunt ends are ligated together, and cross-links are reversed via proteinase K treatment. DNA is extracted with phenol:chloroform. Biotin is removed from unligated ends, chimeric DNA fragments are sheared, sonicated ends are repaired and subjected to poly(A)-tailing. Ligation junctions are enriched via streptavidin-coated beads which have a high affinity for the incorporated biotin. Non-biotinylated fragments are discarded before adapter ligation, PCR amplification, deep sequencing, and mapping (see Methods, Subheading 3 for full details)

to avoid random intermolecular ligation events. Hi-C1 *in situ* ligation reactions are diluted 15-fold, whereas the Hi-C2 protocol documents just a two-fold dilution; finally, (4) Hi-C2 includes a molecular crowding agent, polyethylene glycol (PEG), during ligation.

The methodological adaptations of Hi-C2 yielded a protocol with smaller processing and reaction volumes, thereby dramatically reducing the overall cost of Hi-C library preparation [13]. In addition, smaller processing volumes make handling of samples much easier such that more samples can be processed in parallel. Performing Hi-C1 restricted handling to a maximum of two

Table 1
Summary of changes between Hi-C1 and Hi-C2B protocols

Protocol	Hi-C1	Hi-C2B
Spheroplast & lyse	35 °C, 15 min, invert every 5 min; 0.1% SDS, 65 °C, 10 min	35 °C, 10 min, invert every 2–3 min; 0.01% SDS, 65 °C, 5 min
Digest enzyme	2.07 U/μL DpnII	0.92 U/μL DpnII
Label	37 °C, 2 h	23 °C, 4 h
Solubilize	1.5% SDS, 65 °C, 20 min	NA
Blunt end ligation	15-fold dilution 10X NEB T4 DNA ligase buffer (w/o PEG), 16 °C, 8 h Split across 6x 2 mL tubes	2-fold dilution 10X NEB T4 DNA ligase buffer (w/o PEG), 16 °C, 4 h Split across 16x PCR tubes
Extract DNA	2-step DNA extraction Precipitate whole ligation reaction in EtOH, 4 °C, overnight Spin down and dissolve the pellet in TE. Proteinase K cross-link reversal, 65 °C, 2 h PCI DNA extraction and precipitation, 4 °C, 2 h	PCI DNA extraction and precipitation of whole ligation reaction, 4 °C, overnight
Remove biotin	dATP, dTTP, dCTP, dGTP used in reaction	dATP, dGTP *only* used in reaction

samples per assay for *S. cerevisiae* cells. However, approximately eight samples can be processed in a single run of Hi-C2. Despite these benefits, we found that employing a Hi-C2-like protocol in *S. cerevisiae* generally yielded libraries with high levels of inter-chromosomal (trans) Hi-C interactions, indicative of random ligation.

Here we present a composite chromosome conformation capture assay that combines beneficial aspects of both the Hi-C1 and Hi-C2 protocols in order to generate high-quality, reproducible *S. cerevisiae* Hi-C interaction maps with high yield, throughput, and relatively low cost. For a summary of changes, see Table 1.

2 Materials

Media solutions are prepared using double distilled water, dH_2O, and autoclaved. Prepare and store all reagents at room temperature (unless indicated otherwise). Materials are listed in the first section in which they are used. Product codes are denoted by # followed by a number.

2.1 Meiotic Cell Cultures

1. Amino acid supplement (AAHLTU) 200X: Weigh 2 g of the adenine hemisulfate salt (#A3159), L-arginine monohydrochloride (#A5131), L-histidine monohydrochloride monohydrate (#H8125), L-leucine (#L8000), L-tryptophan (#T0245), and uracil (#U0750).
2. YPD (2%) media: 1% (w/v) yeast extract, 2% (w/v) peptone, 2% (v/v) glucose, 20%, 0.0137% (w/v) adenine + uracil (AU) supplement.
3. BYTA media: 1% (w/v) yeast extract, 2% (w/v) tryptone, 1% (w/v) potassium acetate, 50 mM potassium phthalate, 0.001% (v/v) antifoam 204 (Sigma #A8311).
4. Sporulation media (SPM): 0.3% (w/v) potassium acetate, 0.02% (w/v) raffinose, 0.004% (v/v) AAHLTU, 200X, 0.001% (v/v) antifoam solution.
5. Copper (II) sulfate (250 mM): Weigh 0.39 g $CuSO_4$ powder (Sigma #C1297) and dissolve in 10 mL dH_2O.

2.2 Fixation

1. 37% Formaldehyde (FA): formaldehyde solution (Sigma #F8775).
2. 2.5 M Glycine: Weigh 75 g glycine powder (Sigma #G5417) and dissolve in 1 L dH_2O. Filter to sterilize.
3. Dry ice.

2.3 Spheroplast, Lyse, Digest

1. Spheroplast buffer: 50 mM Tris–HCl (pH 7.5), 1 M sorbitol. Store at 4 °C.
2. 2-Mercaptoethanol (Sigma Aldrich #M6250).
3. 100T Zymolyase stock (50 mg/mL): Weigh 0.1 g Zymolyase 100 T powder (#120493-1) and dissolve in 1.4 mL 50% glycerol, 0.1 mL 1 M $NaHPO_4$ buffer (pH 7.2), 0.5 mL 2 M sucrose. Vortex to dissolve, store at −20 °C. Dilute to 5 mg/mL by taking 200 μL of the 50 mg/mL stock and diluting in 800 μL 50% glycerol. Store at −20 °C.
4. NEBuffer™ 1X r3.1: Dilute 10X NEBuffer™ r3.1 (#B6003S) in dH_2O. Keep on ice.
5. 1% SDS: Dilute 10% SDS in dH_2O.
6. 10% Triton X-100: Dilute Triton X-100 in dH_2O.
7. NEB™ DpnII (#R0543L). Store at −20 °C.

2.4 Fill in, Blunt-End Ligate, Reverse Cross-Links

1. NEBuffer™ 10X r3.1 (#B6003S). Store at −20 °C.
2. Thermo Scientific™ 10 mM dATP, dTTP, dGTP: Individually dilute 10 μL of 100 mM dNTPs (#10520651) with 90 μL of Tris–EDTA (TE). Store at −20 °C.
3. Invitrogen™ Biotin-14-dCTP: 0.4 mM biotin-14-dCTP (#10022582), 100 mM Tris–HCl (pH 7.5), 0.1 mM EDTA.

4. NEB™ Klenow DNA Polymerase I (5 U/μL) (M0210L). Store at −20 °C.
5. Eppendorf™ ThermoMixer™ (#5382000031).
6. NEB™ T4 DNA Ligase Buffer (10X): 50 mM Tris–HCl (pH 7.5), 10 mM $MgCl_2$, 1 mM ATP, 10 mM DTT (#B0202S).
7. Invitrogen™ T4 DNA Ligase Buffer (5X): 250 mM Tris–HCl (pH 7.6), 50 mM $MgCl_2$, 5 mM ATP, 5 mM DTT, 25% (w/v) polyethylene glycol-8000 (included with: #10443242).
8. Bovine serum albumin (BSA) 10 mg/mL: Weigh 10 mg BSA (#B9001S) and dissolve in 1 mL dH_2O to make 10 mg/mL.
9. Invitrogen™ T4 DNA Ligase (5 U/μL) (#10443242). Store at −20 °C.
10. Eppendorf™ 2 mL safe-lock tubes (#10430423).
11. Proteinase K (10 mg/mL): Weigh 0.1 g of proteinase K powder (Fisher Scientific #10407583) and dissolve in 10 mL of 20 mM Tris–HCl 50% glycerol solution. Store at −20 °C.

2.5 Extract, Precipitate, and Quantify DNA

1. Phenol–chloroform–isoamyl alcohol mixture (#77617). Store at 4 °C.
2. 5PRIME Phase Lock Gel (PLG) Light 2 mL tubes (#2302820).
3. 3 M sodium acetate, pH 5.2: Weigh 246.1 g of sodium acetate and dissolve in 800 mL of dH_2O. Adjust the pH to 5.2 with glacial acetic acid. Allow the solution to cool to RT before use.
4. Millipore Amicon® Ultra 0.5 mL centrifugal filters (#UFC500396).
5. Tris Low-EDTA (TLE): 10 mM Tris–HCl (pH 8.0), 0.1 mM EDTA.
6. Ribonuclease A (RNase A) 10 mg/mL: Weigh 0.1 g of RNase A powder (Sigma Aldrich #R4875) and dissolve in 9 mL of 10 mM sodium acetate buffer (pH 5.2). Heat to 100 °C for 15 min and allow to cool to room temp before use. Adjust to pH 7.4 by adding 0.1 mL of 1 M Tris–HCl (pH 7.4). Store at −20 °C.
7. Qubit™ Fluorometer (purchased from Thermo Scientific™).
8. Invitrogen™ Qubit™ dsDNA high sensitivity assay kit (#Q32851).

2.6 Remove Dangling Ends

1. NEBuffer™ 10X r2.1 (#B6002S). Store at −20 °C.
2. 10 mM dATP, dGTP: Prepared as described in 2.4.
3. NEB™ T4 DNA Polymerase I (3 U/μL) (#M0203L). Store at −20 °C.

2.7 Fragment and Repair Broken Ends

1. Covaris microTUBE: AFA® fiber pre-slit snap-cap (130 μL) (#520045).
2. Covaris instrument: We use a low-throughput focused-ultra-sonicator, M220 (#500295).
3. 100 mM dNTP mix: In a single Eppendorf, mix 100 μL of each dNTP (#10520651). Store at −20 °C.
4. NEB™ T4 Polynucleotide Kinase (PNK): Store at −20 °C (#M0201S).

2.8 Enrich, Ligate, PCR-Amplify

1. DNA LoBind Tube 1.5 mL (#022431021).
2. Invitrogen™ Dynabeads® MyOne™ Streptavidin C1 beads (#65001).
3. Tween Wash Buffer (TWB): 50 mM Tris–HCl (pH 8.0), 0.5 mM EDTA, 1 M NaCl, 0.05% (v/v) Tween.
4. Invitrogen™ Dynal MPC™: Magnetic particle concentrator.
5. Binding Buffer (2X): 10 mM Tris–HCl (pH 8.0), 1 mM EDTA, 2 M NaCl.
6. Binding Buffer (1X): Dilute 5 mL 2X BB in 5 mL dH_2O. Make fresh for each experiment and store on ice.
7. Klenow Fragment (3′–5′ exo-) (#M0212S). Store at −20 °C.
8. T4 DNA Ligase Buffer (1X): Dilute Invitrogen™ T4 DNA Ligase Buffer (5X) in dH_2O. Make fresh for each experiment and store on ice.
9. NEXTFLEX® DNA Barcodes (#NOVA-514103).

2.9 Clean, Quantify, Sequence, Map

1. Beckman Coulter™ Agencourt AMPure® XP (#A63881). Store at 4 °C.
2. Bioanalyzer instrument: We use Agilent 2100 Bioanalyzer Instrument (#G2939BA).
3. Agilent High Sensitivity DNA Kit (#5067–4626).
4. Illumina NexSeq 500/550 High Output Kit v2.5 (75 cycles) (#20024906).
5. Linux workstation running Hi-C Pro [15] (version 3.1.0).

3 Methods

The protocol presented here is optimized for the detection of intra- and inter-chromosomal contacts from highly synchronous *S. cerevisiae* cells. Both cycling and sporulating budding yeast cells can be investigated using the Hi-C2B assay. We combine methodological steps from Hi-C1 and Hi-C2 in order to improve the recovery and resolution of *S. cerevisiae* Hi-C libraries. A schematic

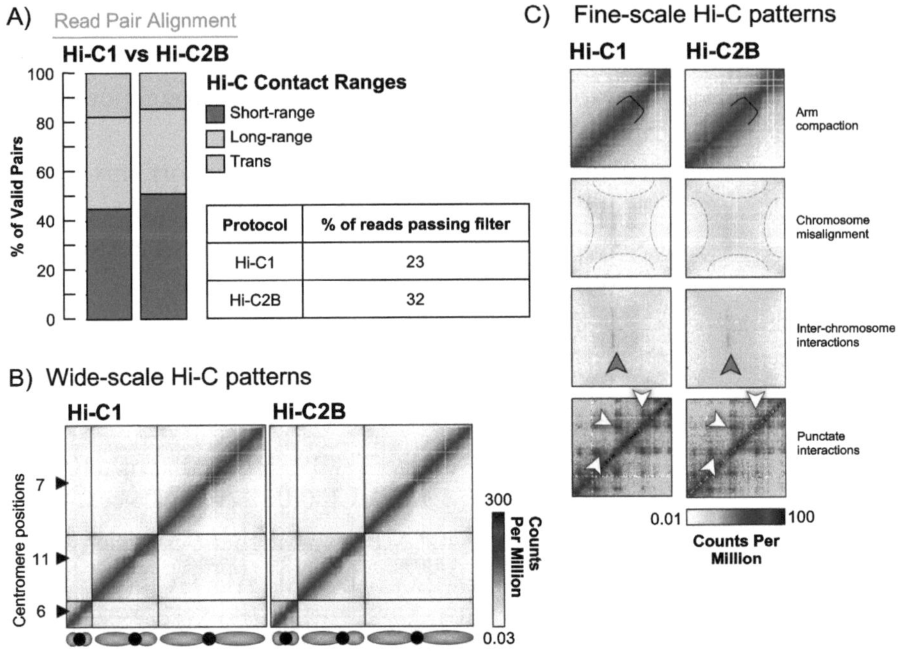

Fig. 2 Comparisons between Hi-C1 and Hi-C2B data output
(**a**) Fragment size distribution of short- and long-range intra-chromosomal (cis) contacts versus long-range inter-chromosomal (trans) contacts as reported by the Hi-C Pro pipeline [14] for libraries assayed by Hi-C1 or Hi-C2B methodologies. The table shows the percentage of reads that pass all Hi-C Pro filters as a fraction of the starting read value (all pairs). (**b**) Hi-C interaction maps from meiotic *S. cerevisiae* cells prepared by Hi-C1 or Hi-C2B. Assessment of wide-scale patterns of Hi-C interaction highlight the similarity in output data from either methodology. Hi-C contact maps of chromosome 6, 11, and 7 are plotted at 5 Kb resolution. The black arrows indicate positions of the centromeres. (**c**) Assessment of known chromosome conformations demonstrates similarities in the fine-scale Hi-C interaction patterns from libraries prepared by Hi-C1 or Hi-C2B. Arm compaction: axial compaction is indicated by the width of the main diagonal relative to the fixed-width black clamp. Chromosome misalignment: dashed pinked lines outline the inter-chromosomal cross-shape structures that manifest due to chromosome misalignment (present in meiotic recombination mutants for example). Inter-chromosome interactions: enhanced domains of inter-chromosomal contacts indicated by green arrowheads. Punctate interactions: focal grid-like patterns that arise from loops generated by meiotic cohesin (Rec8) [18]; prominent examples are indicated by white arrowheads. Hi-C contact maps from meiotic *S. cerevisiae* cells are plotted at 2 Kb resolution

outlining the various steps of the Hi-C library preparation is shown in Fig. 1, with representative comparative examples described in Fig. 2.

3.1 Meiotic Cell Cultures

1. From a −80 °C glycerol stock, streak out yeast cells onto a YPD plate. Incubate for 2 days at 30 °C.
2. In a glass culture tube, inoculate 4 mL YPD liquid with a single colony taken from the YPD plate. Incubate 24 h at 30 °C with shaking (250 rpm), until the culture is saturated (~OD 20).

3. In a conical flask, inoculate the saturated culture in BYTA pre-sporulation media to a final OD of 0.3. Incubate for 16–18 h at 30 °C with shaking (250 rpm) (*see* **Note 1**).
4. Collect cells by centrifugation (4 min, 4000 *g*, room temperature) and drain supernatant. Wash with water by sealing the tube and shaking vigorously (*see* **Note 2**). Pellet cells again, drain, and resuspend with SPM. Return to a conical flask and incubate at 30 °C with shaking (250 rpm) for the duration of the time course (*see* **Note 3**).

3.2 Fixation

1. Transfer 8 mL (20–30 OD_{600} units) of synchronously sporulating *S. cerevisiae* cells to a 50 mL Falcon tube.
2. Add 648 μL 37% formaldehyde (~3% final concentration). Incubate for 20 min at 30 °C with shaking (250 rpm).
3. Quench formaldehyde by adding 1.3 mL of 2.5 M glycine (~0.35 M final) and incubate for a further 5 min at 30 °C with shaking (250 rpm) (*see* **Note 4**).
4. Collect cells by centrifugation (2 min, 2500 *g*, 4 °C) and aspirate supernatant. Wash with 1.2 mL water by resuspending material using a pipette; do this slowly to avoid making bubbles.
5. Split the suspension into four separate 1.7 mL Eppendorf tubes (~300 μL per tube) (*see* **Note 5**).
6. Collect cells again by centrifugation using a table-top centrifuge (1 min, 14,000 *g*, room temperature). Carefully remove the supernatant using a pipette and snap-freeze the pellets on dry ice (*see* **Note 6**).

3.3 Spheroplast, Lyse, Digest

1. Thaw cell pellets on ice for 1–2 h.
2. Wash cells in 1 mL ice-cold spheroplasting buffer by pipetting up and down (do not vortex, the cells are fragile at this stage).
3. Collect cells by centrifugation using a table-top centrifuge (2 min, 2500 *g*, room temperature). Carefully remove supernatant using a pipette (*see* **Note** 7).
4. Resuspend cells in 1 mL spheroplasting buffer. Add 2 μL 50 mg/mL zymolyase 100T (100 μg/mL final) and 10 μL 2-mercaptoethanol and mix thoroughly.
5. Incubate for 10 min at 35 °C, inverting slowly after 5 min.
6. Collect cells by centrifugation (2 min, 2500 *g*, room temperature) and then resuspend the pellet in 500 μL ice-cold 1X NEB 3.1 restriction enzyme buffer (*see* **Note 8**).
7. Collect cells as described in **step 6**. Resuspend the pellet in 360 μL ice-cold 1X NEB 3.1 restriction enzyme buffer.
8. Solubilize chromatin by adding 3.8 μL 1% SDS and mixing gently by pipetting up and down to avoid making bubbles. Incubate at 65 °C for 5 min.

9. Cool mixture by placing it on ice for 5 min immediately after incubation with SDS.
10. When the mixture has cooled, add 43 μL of 10% Triton X-100 and invert gently two to three times to quench SDS. Incubate for 15 min at 37 °C, 300 rpm (*see* **Note 9**).
11. Add 8 μL DpnII, mix thoroughly by inversion, and incubate overnight in a thermomixer at 37 °C, 300 rpm. The final volume at this stage will be ~460 μL.

3.4 Fill in, Blunt-End Ligate, Reverse Cross-Links

1. Incubate at 65 °C for 20 min in order to deactivate DpnII.
2. Fill in DNA ends by adding 2 μL water, 6 μL 10X NEB 3.1 restriction enzyme buffer, 1.5 μL 10 mM dATP, dGTP, and dTTP, 37.5 μL 0.4 mM biotin-14-dCTP, and 5 U/μL Klenow fragment DNA polymerase I. Mix by carefully pipetting and gentle inversion (3–4 times).
3. Incubate mixture at 23 °C for 4 h with interval agitation in a ThermoMixer (*see* **Note 10**).
4. For blunt-end ligation, make a reaction mixture containing 120 μL 10% Triton X-100, 120 μL 10X ligation buffer (NEB), 12 μL 10 mg/mL BSA, 363 μL dH_2O, 50 μL 5 U/μ L T4 DNA Ligase (Invitrogen). Total will be 665 μL (*see* **Note 11**).
5. Dilute the sample twofold by adding the ligation reaction mixture to the fill-in reaction tube and mixing fully by inverting the tube three to four times.
6. Split volume across sixteen 0.2 mL PCR tubes (~70 μL per tube) (*see* **Note 12**).
7. Incubate at 16 °C for 4 h inverting slowly every 30–45 min.
8. Pool the ligation reaction into a single 2 mL safe-lock tube (pulse spin to recover material that may be in the lid).
9. Add 100 μL of 10 mg/mL proteinase K and incubate at 65 °C overnight to reverse cross-links.

3.5 Extract, Precipitate, and Quantify DNA

1. Cool mixture by placing it on ice for ~20 min.
2. Transfer half of the mixture (~600 μL) to a second 2 mL safe-lock tube (two tubes per sample).
3. Add an equal volume of PCI (25:24:1) to each tube and vortex heavily. Transfer the white, milky solution to a pre-spun 2 mL phase-lock tube (2 per sample).
4. Centrifuge at room temperature for 5 min, 14,000 *g*.
5. Transfer the aqueous phase containing DNA to pre-chilled 1.7 mL Eppendorf tubes (2 per sample) containing 50 μL 3 M NaOAc and 1 mL 100% EtOH (pre-chilled).

6. Store on dry ice for a few minutes or at −20 °C for at least 2 h to precipitate DNA (*see* **Note 13**).
7. Collect the DNA precipitate by centrifugation at 4 °C for 30 min, 16,000 *g*, and aspirate supernatant (pulse spin to remove residual EtOH).
8. Wash the pellet with 1 mL ice-cold 70% EtOH.
9. Repeat **step 7** (*see* **Note 14**).
10. Air-dry the pellet for ~20 min.
11. Dissolve the pellet in 250 μL 1X TLE (pH = 8.0) for ~30 min.
12. Pool into a single Amicon 30 kDa column and centrifuge at room temperature, 14,000 *g*, until volume reaches ~100 μL (*see* **Note 15**).
13. Wash with 400 μL 1X TLE (pH = 8.0) and repeat spin until volume reaches ~100 μL.
14. Flip the Amicon 30 kDa column containing the concentrated sample into a new collection tube, spin at room temperature for 2 min, 14,000 *g*.
15. Transfer sample to a 1.7 mL Eppendorf tube and determine the volume (usually ~100–130 μL).
16. Add 1:100 μL of 1 mg/mL RNase A (typically ~1–1.3 μL) and incubate at 37 °C for 30 min.
17. Measure the DNA concentration of the Hi-C sample using the Qubit dsDNA HS Assay Kit and proceed to the removal of "dangling ends" (*see* **Note 16**).

3.6 Remove Dangling Ends

1. Remove biotin-14-dCTP from unligated ends by adding 12 μL 10X NEB 2.1 restriction enzyme buffer, 0.3 μL 10 mM dATP, 0.3 μL 10 mM dGTP, 12 μL 300 U/μL T4 DNA polymerase (*see* **Note 17**).
2. Incubate the mixture at 20 °C for 4 h and inactivate the enzyme by incubating at 75 °C for 20 min.
3. Transfer the mixture into a single Amincon 30 kDa column and centrifuge at room temperature, 14,000 *g*, until volume reaches ~100 μL.
4. Wash column with 400 μL 1X TLE (pH = 8.0) and repeat spin until volume reaches ~130 μL.
5. Flip the Amicon 30 kDa column containing the concentrated sample into a new collection tube, spin at room temperature for 2 min, 14,000 *g*. Determine the volume in the column.
6. Bring the Hi-C sample to 130 μL with 1X TLE and load into a Covaris microTUBE.

3.7 Fragment, Repair Broken Ends

1. Load the microTUBE containing the Hi-C sample into a Covaris M220.
2. Run the following program (*see* **Note 18**):
   1. Duty factor: 20%
   2. Peak power: 50 W
   3. Cycles per burst: 200
   4. Temperature: 20 °C
   5. Run time: 240 s
3. Transfer the sonicated Hi-C sample to a 0.2 mL PCR tube.
4. To repair broken DNA ends, add to the sample 15 μL 10X ligation buffer (NEB), 1.5 μL 100 mM dNTP mix (25 mM each), 5.4 μL 3 U/μL T4 DNA polymerase, 5.4 μL 10 U/μL T4 polynucleotide kinase, 1.1 μL 5 U/μL Klenow DNA polymerase I. Make the volume up to 30 μL with water and mix.
5. Incubate at 20 °C for 30 min and inactivate enzyme activity by incubating at 75 °C for 20 min. Maintain at 4 °C.

3.8 Enrich, Ligate, PCR-Amplify

1. Vortex Streptavidin C1 beads and transfer 10 μL into a 1.7 mL DNA LoBind tube (*see* **Note 19**).
2. Wash beads with 100 μL of Tween Wash Buffer (TWB) by pipetting up and down.
3. Incubate the beads with the buffer at room temperature for 3 min with rotation.
4. Collect the beads against a magnetic particle concentrator at room temperature for 1 min and remove and discard the supernatant.
5. Resuspend the beads in 100 μL of TWB and transfer to a new LoBind tube.
6. Repeat **step 4**.
7. Resuspend the beads in 150 μL of 2X Binding Buffer (BB) and add to the end repair reaction. Pipette up and down to mix and to avoid making bubbles.
8. Incubate at room temperature for 1 h with rotation.
9. Repeat **step 4**.
10. Resuspend beads in 400 μL 1X BB and transfer to a new LoBind tube. Gently invert to mix.
11. Repeat **step 4**.
12. Resuspend beads with 100 μL 1X TLE (pH 8.0) and transfer to a new LoBind tube.
13. Repeat **step 4**.

14. Resuspend beads in 41 μL 1X TLE (pH 8.0).
15. Digest unligated biotinylated dangling ends by adding 5 μL 10X NEB 2.1 restriction enzyme buffer, 1 μL 10 mM dATP, 3 μL Klenow DNA Polymerase (3′–5′ exo-).
16. Incubate at 37 °C for 30 min and inactivate the enzyme activity by incubating at 65 °C for 20 min. Cool mixture by incubating at 4 °C.
17. Transfer the reactions to a new LoBind tube, then repeat **step 4**.
18. Resuspend beads in 400 μL 1X T4 DNA ligation buffer.
19. Repeat **step 4**.
20. Resuspend beads in 40 μL 1X T4 DNA ligation buffer.
21. Ligate a unique NEXTFLEX® barcoded adaptor by adding 2 μL 5X T4 DNA ligation buffer (Invitrogen), 3 μL 25 μM NextFlex barcoded adaptor, 3 μL 5X T4 DNA ligase (Invitrogen), 2 μL water and mix.
22. Incubate at 22 °C for 2 h (*see* **Note 20**).
23. Repeat **step 4**.
24. Wash beads twice with 400 μL TWB by repetition of **step 4** in between washes. Incubate at room temperature for 5 min with rotation.
25. Repeat **step 4**.
26. Resuspend beads in 200 μL 1X BB and transfer to a new LoBind tube.
27. Repeat **step 4**.
28. Resuspend beads in 200 μL 1X TE and transfer to a new LoBind tube.
29. Repeat **step 4**.
30. Resuspend beads in 10 μL 1X TE and transfer to a new LoBind tube.
31. Minimally amplify the Hi-C library by adding 40 μL 2X NEXTFLEX® DNA PCR Master Mix, 6.7 μL NEXTFLEX® Primer Pair Mix, and 110 μL water to the DNA on beads and mix.
32. Divide the reaction between 4 PCR tubes (~41 μL per tube) and run the following program:
    (a) 98 °C, 2 min
    (b) 10–14 cycles of:
        (i) 98 °C, 30 s
        (ii) 65 °C, 30 s
        (iii) 72 °C, 1 min
    (c) 72 °C, 4 min

33. Collect the beads against a magnetic particle concentrator at room temperature for 1 min and **retain** the supernatant by pooling into a single 1.7 mL Eppendorf tube. The total volume will be ~150 μL. **This is the Hi-C sample**. Store at −20 °C before proceeding to the next step.

3.9 Clean, Quantify, Sequence, Map

To remove primer dimers, purify the amplified Hi-C library from the supernatant using Ampure XP beads. Allow the Ampure XP mixture to come to room temperature (for ~30 min) and vortex heavily prior to use. Pre-warm 130 μL (per sample) 1X TLE buffer to 65 °C.

1. Add 1.1x Ampure XP beads (~165 μL) to the Hi-C sample.
2. Vortex and lightly spin to remove mixture from the lid.
3. Incubate at room temperature for 10 min.
4. Collect the beads against a magnetic particle concentrator at room temperature for 5 min and remove and discard supernatant.
5. Wash beads by adding 1 mL 70% EtOH to the tube, incubate at room temperature for 1 min, then aspirate. Repeat this step (*see* **Note 21**).
6. Elute DNA by resuspending the DNA/beads in 100 μL 1X TLE preheated to 65 °C (see above).
7. Incubate at room temperature for 10 min.
8. Add 1.1x Ampure XP beads (110 μL) and repeat **steps 2–5** with the eluent/bead mixture.
9. Repeat **step 6**, but resuspend in 30 μL 1X TLE preheated to 65 °C.
10. Incubate at room temperature for 10 min.
11. Repeat **step 4**; however, this time *keep* the supernatant (Hi-C library) and transfer to a new 1.7 mL Eppendorf tube labeled with the library name/number.
12. Assess the size and concentration of DNA using the Agilent 2100 Bioanalyzer system.
13. Sequence Hi-C libraries using paired-end reads of at least 42 bp (using, for example, a NextSeq sequencing kit) (*see* **Note 22**).
14. Sequenced Hi-C data can be mapped using Hi-C Pro using standard parameters (https://github.com/nservant/HiC-Pro) (*see* **Note 23**).

4 Notes

1. Under the control of the *CUP1* promoter, transcriptional induction of *IME1* depends solely on the presence of copper ions [15]. As such, whereas acetate-based pre-sporulation media has been shown to boost sporulation compared to glucose-based in non-inducible cells, acetate-based pre-growth is dispensable for sporulation induced from the *CUP1* promoter [15–17]. An alternative pre-sporulation medium that is compatible with the Hi-C2B assay is low-glucose YP (as described in [15]). Cells are prepared as described in Subheading 3.1, **step 1** with a few additional steps. Following **step 2**, the saturated culture is transferred to fresh YPD media in either a glass culture tube or conical flask (depending on the final volume) to a final OD of 0.1 and grown to exponential phase for a few hours. In place of **step 3**, the exponential culture should be inoculated in reduced glucose YP pre-sporulation media (we use 1%) to a final OD of 0.05. As above, incubate for 16–18 h at 30 °C with shaking (250 rpm), until the culture reaches OD 10–12. Proceed to **step 4**. For a more detailed protocol, please ref. to 15.
2. We find that heavily vortexing the cellular material for a full resuspension at this stage improves synchronous release into meiosis. Water should be stored at room temperature.
3. For strains where *IME1* is placed under the control of the *CUP1* promoter, after 2 h in sporulation medium, copper (II) sulfate (50 μM final) is added to induce *IME1* expression and initiate sporulation synchronously.
4. Holding the mixture on ice for ~10 min helps to further quench fixed cells resulting, in our experience, in an improved library yield.
5. Hi-C2B features a twofold reduction in starting material. While the total volume remains the same, the concentration of reagents in proportion to the cellular material is higher in Hi-C2B. Reducing the amount of starting material appears to improve the resolution of Hi-C libraries.
6. Snap frozen, fixed cell pellets should be stored at −80 °C.
7. Cell pellets are relatively unstable at this point, so care should be taken when removing the supernatant. The same applies for **step 6**.
8. Dilute 10X NEB 3.1 restriction enzyme buffer in Milli-Q water and chill on ice.

9. Short incubation of the solubilized material with Triton X-100 prior to the addition of the enzyme improves DpnII efficiency—inferred by improved resolution of final Hi-C libraries—likely due to a reduction in SDS-associated inhibition of enzyme activity.
10. Eppendorf ThermoMixer Settings: 23 °C, 4 h; 900 rpm for 10 s every 5 min.
11. Despite using T4 DNA Ligase from Invitrogen, it is important to substitute the 5X Ligation Buffer for the 10X NEB variety which does not contain polyethylene glycol (PEG). While previous methods have reported the use of ligase buffer containing PEG for generation of Hi-C libraries [16], replacement of PEG-containing (Invitrogen) with non-PEG-containing (NEB) ligation buffer minimizes non-specific intermolecular ligation efficiency, thereby favoring intramolecular ligation. This alteration is crucial for the generation of high-resolution Hi-C contact maps that are not swamped with artifactual intermolecular interactions (see Fig. 3 for further details).
12. High levels of intermolecular ligation yield Hi-C libraries saturated with "noise." To reduce the efficiency of the ligation reaction, the mixture is split across 16 PCR tubes, further reducing the relative frequency of non-specific intermolecular ligation.
13. Precipitated DNA should show evidence of white threads, or clouds; samples should be viscous, not solid.
14. A shorter spin can be used for the wash step. We use 5–10 min.
15. There should be two pellets per sample; the total volume will be ~500 μL when pooled.
16. In our experience, between 1 and 5 μg of DNA per sample is sufficient to generate high-resolution Hi-C libraries. Samples with <1 μg can be processed however; the library complexity is likely to be low. We recommend discarding samples with <1 μg.
17. For samples with >5 μg of DNA, we suggest splitting the Hi-C sample between two separate reaction tubes for the dangling end removal step.
18. Hold any additional samples on ice during sonication.
19. To avoid cross contamination during NGS library preparation, filter tips should be used for all following steps.
20. The adapter ligation reaction can be run for more than 2 h. We have performed the incubation for several hours or overnight without any effect on the final Hi-C library.

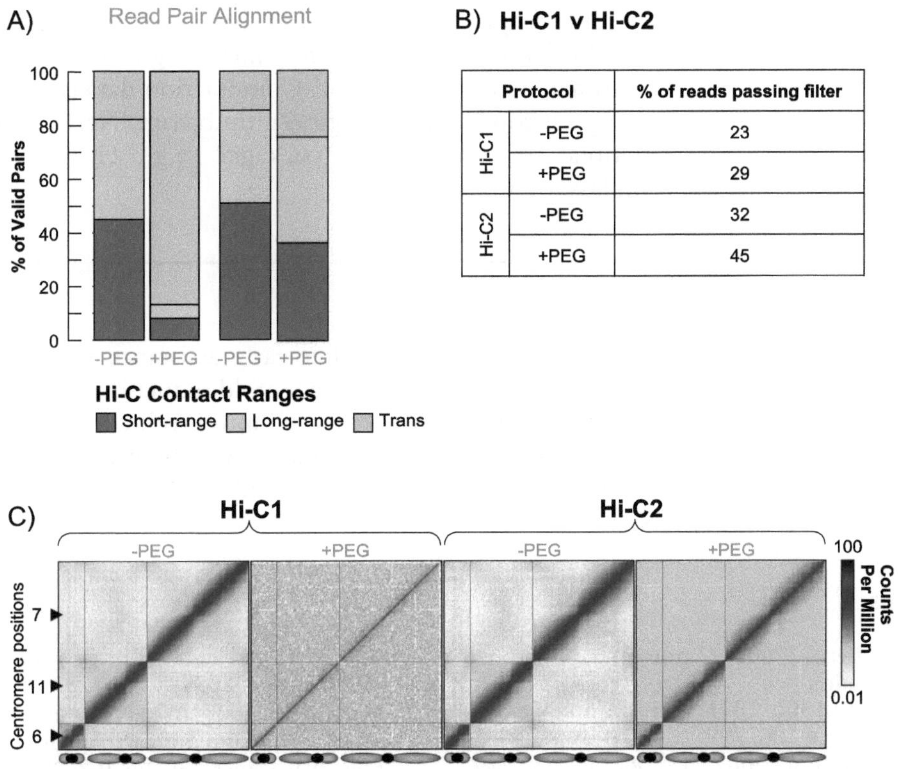

Fig. 3 The effect of crowding agents during ligation stage
(**a**) Fragment size distribution of short- and long-range intra-chromosomal (cis) contacts versus long-range inter-chromosomal (trans) contacts as reported by the Hi-C Pro pipeline [14] for libraries assayed by Hi-C1 ±PEG or Hi-C2 ±PEG (where Hi-C2 -PEG is equivalent to Hi-C2B, reported here). Addition of PEG-8000 to either protocol increases the rate of artifactual intermolecular ligation (detectable as increased trans contacts and higher matrix background (C), which is unwanted in Hi-C libraries). (**b**) Table showing the percentage of reads that pass all Hi-C Pro filters as a fraction of the starting read value (all pairs) for Hi-C libraries assayed by Hi-C1 or Hi-C2 as described above. Increasing the rate of intermolecular ligation through the addition of PEG in either protocol generates an increase in the percentage of reads that pass all Hi-C Pro filters, but this increase in ligation efficiency is disproportionately made up of artifactual inter-molecular ligation products (detectable as increased trans contacts) at the expense of true intra-molecular proximity ligation events. (**c**) Hi-C interaction maps from meiotic *S. cerevisiae* cells prepared by Hi-C1 or Hi-C2 as described above. Hi-C contact maps of chromosome 6, 11, and 7 are plotted at 10 Kb resolution. The black arrows indicate positions of the centromeres

21. Between each wash, aspirate the ethanol while keeping the tubes on the magnetic particle collector. Ensure the DNA/-beads are completely dry before resuspending (eluting) in TLE following the second ethanol wash. Residual ethanol may interfere with downstream reactions.
22. In order to obtain high-resolution Hi-C interaction maps, libraries should be sequenced to a depth of 10–30 million reads after passing all quality control filters. This usually equates to 50–100 million starting/unfiltered reads per Hi-C library. If wanting to quality control prior to deep sequencing, load small amounts of each library (aim for <1 million reads).

23. Default output generates ICED (balanced) and RAW matrices in addition to several other files including .bed files relating to each mapped resolution. Hi-C interaction data can be plotted using custom scripts (https://github.com/Neale-Lab/Hi-C-Scripts), or via standard packages (e.g., Hi-Glass: http://higlass.io/)

Acknowledgments

We thank S. Schalbetter for her early assistance with this work, specifically, for implementing the tenfold reduction of SDS concentration during the cell lysis step and production of "Hi-C1" libraries. This work was supported by Wellcome Trust Investigator (200843/Z/16/Z) and Discovery (225852/Z/22/Z) Awards to MJN.

References

1. Dekker J et al (2002) Capturing chromosome conformation. Science 295(5558):1306–1311. https://doi.org/10.1126/science.1067799
2. Palstra RJ et al (2003) The β-globin nuclear compartment in development and erythroid differentiation. Nat Genet 35(2):190–194. https://doi.org/10.1038/ng1244
3. Drissen R et al (2004) The active spatial organization of the β-globin locus requires the transcription factor EKLF. Genes Dev 18(20):2485–2490. https://doi.org/10.1101/GAD.317004
4. Vakoc CR et al (2005) Proximity among distant regulatory elements at the β-globin locus requires GATA-1 and FOG-1. Mol Cell 17(3):453–462. https://doi.org/10.1016/J.MOLCEL.2004.12.028
5. Splinter E et al (2006) CTCF mediates long-range chromatin looping and local histone modification in the β-globin locus. Genes Dev 20(17):2349–2354. https://doi.org/10.1101/GAD.399506
6. Parelho V et al (2008) Cohesins functionally associate with CTCF on mammalian chromosome arms. Cell 132(3):422–433. https://doi.org/10.1016/J.CELL.2008.01.011
7. Wendt KS et al (2008) Cohesin mediates transcriptional insulation by CCCTC-binding factor. Nature 451(7180):796–801. https://doi.org/10.1038/nature06634
8. Hadjur S et al (2009) Cohesins form chromosomal cis-interactions at the developmentally regulated IFNG locus. Nature 460(7253):410–413. https://doi.org/10.1038/nature08079
9. Handoko L et al (2011) CTCF-mediated functional chromatin interactome in pluripotent cells. Nat Genet 43(7):630–638. https://doi.org/10.1038/ng.857
10. Denker A, de Laat W (2016) The second decade of 3C technologies: detailed insights into nuclear organization. Genes Dev 30(12):1357–1382. https://doi.org/10.1101/gad.281964.116
11. Hsieh THS et al (2016) Micro-C XL: assaying chromosome conformation from the nucleosome to the entire genome. Nat Methods 13(12):1009–1011. https://doi.org/10.1038/nmeth.4025
12. Lieberman-Aiden E et al (2009) Comprehensive mapping of long-range interactions reveals folding principles of the human genome. Science 326(5950):289. https://doi.org/10.1126/SCIENCE.1181369
13. Belaghzal H, Dekker J, Gibcus JH (2017) Hi-C 2.0: an optimized hi-C procedure for high-resolution genome-wide mapping of chromosome conformation. Methods 123:56–65. https://doi.org/10.1016/j.ymeth.2017.04.004
14. Servant N et al (2015) HiC-Pro: an optimized and flexible pipeline for Hi-C data processing. Genome Biol 16(1):1–11. https://doi.org/10.1186/s13059-015-0831-x
15. Chia M, van Werven FJ (2016) Temporal expression of a master regulator drives synchronous sporulation in budding yeast. Genes

Genomes Genetics 6(11):3553–3560. https://doi.org/10.1534/g3.116.034983
16. Kassir Y, Granot D, Simchen G (1988) IME1, a positive regulator gene of meiosis in *S. cerevisiae*. Cell 52(6):853–862. https://doi.org/10.1016/0092-8674(88)90427-8
17. Elrod SL et al (2009) Optimizing sporulation conditions for different saccharomyces cerevisiae strain backgrounds. Methods Mol Biol 557:21–25. https://doi.org/10.1007/978-1-59745-527-5_2/TABLES/4
18. Schalbetter SA et al (2019) Principles of meiotic chromosome assembly revealed in *S. cerevisiae*. Nat Commun 10(1):4795. https://doi.org/10.1038/s41467-019-12629-0

Chapter 4

Efficient Enrichment of Synchronized Mouse Spermatocytes Suitable for Genome-Wide Analysis

Agustin Carbajal, Irma Gryniuk, Rodrigo O. de Castro, and Roberto J. Pezza

Abstract

Chromatin undergoes extensive remodeling during meiosis, leading to specific patterns of gene expression and chromosome organization, which ultimately controls fundamental meiotic processes such as recombination and homologous chromosome associations. Recent game-changing advances have been made by analysis of chromatin binding sites of meiotic specific proteins genome-wide in mouse spermatocytes. However, further progress is still highly dependent on the reliable isolation of sufficient quantities of spermatocytes at specific stages of prophase I. Here, we describe a combination of methodologies we adapted for rapid and reliable isolation of synchronized fixed mouse spermatocytes. We show that chromatin isolated from these cells can be used to study chromatin-binding sites by ChIP-seq. High-quality data we obtained from INO80 ChIP-seq in zygotene cells was used for functional analysis of chromatin-binding sites.

Key words Mouse spermatocytes, ChIP-seq, Genomic analysis, Spermatocyte fraction enrichment

1 Introduction

Meiosis involves highly regulated cell developmental transitions, which are coordinated at chromatin level and ultimately control proper reproductive tissue development and sustained germ cell production in adults. For example, during the meiotic prophase I, spermatocytes undergo extensive changes in the patterns of gene expression supporting a series of unique events including pairing, synapsis, and recombination of chromosomes to ultimately ensure their accurate segregation. These processes are coordinated and

---

Supplementary Information: The online version contains supplementary material available at https://doi.org/10.1007/978-1-0716-3906-1_4.

Jesús A. Carballo (ed.), *Meiosis: Methods and Protocols*, Methods in Molecular Biology, vol. 2818,
https://doi.org/10.1007/978-1-0716-3906-1_4,

controlled at the chromatin level by regulators such as chromatin remodeling complexes, which meiotic functions in any organism are still poorly understood.

A key task for researchers working in various aspects of meiosis at the genome-wide level is obtaining enough high-quality spermatocytes enriched at any of the different stages of prophase I for use in assays such as ChIP-seq and RNA-seq. An important limitation in this endeavor is that meiotic cells are asynchronous—adult testis germ cell population can be found as a combination of all stages, from pre-meiotic spermatogonia to spermatozoa. Importantly, not all stages of maturation are equally represented within the adult testis population, some cell stages like leptotene or zygotene are of lower abundance and hence difficult to purify. Classical methods for spermatocyte enrichment such as STA-PUT velocity sedimentation [1, 2] and elutriation [3] work well for spermatocyte stages such as pachytene because of its relative high abundance, but they are not suitable for spermatocytes with low abundance as those in early stages of prophase I. On the other hand, FAC sorting of Hoechst 3334 dyed cells obtained from adult mice can enrich early prophase spermatocytes [4], but it has a relatively low yield and the methodology setup is labor intensive.

Recent breakthrough methodology [5, 6] allows synchronization of spermatocytes in testis. In this approach, spermatocyte maturation is arrested at the stage of undifferentiated spermatogonia on the newborn mouse by administration of WIN18,446, a compound that inhibits the synthesis of retinoic acid. Following this treatment, an injection of retinoic acid induces gametogenesis progression, resulting in the generation of synchronously maturing population of spermatocytes. The time encompassed between retinoic acid administration and cell harvest determines the stage of prophase I in which most spermatocytes are found (Fig. 1a). An important challenge after spermatocyte synchronization is the isolation of the spermatocytes from other cell populations such as spermatogonia and somatic cells. FACS sorting has been used for enrichment of synchronized spermatocytes [5, 7, 8]. However, as described above, this FACS technique imposes several disadvantages, including longer experimentation times when a substantial number of cells are required, which is often the case for genome-wide analysis approaches. Here, we describe a rapid, simple, and robust approach to further enrich fraction of synchronized mouse spermatocytes at different stages of prophase I. In our approach, we employ centrifugation of PFA-fixed single cell suspension on a discontinuous Percoll gradient (Fig. 1b), from which we obtain high yield of enriched spermatocytes in a relatively short time (Fig. 1c, d, and Table 1) (*See* **Note 2**). As a proof of concept, we performed INO80 ChIP-seq on early zygotene cells obtained using this method (Fig. 2). INO80 is a chromatin remodeler shown to be indispensable for meiocytes to pass beyond the zygotene stage [9]. Hence, knowing the genome-wide location of INO80 during

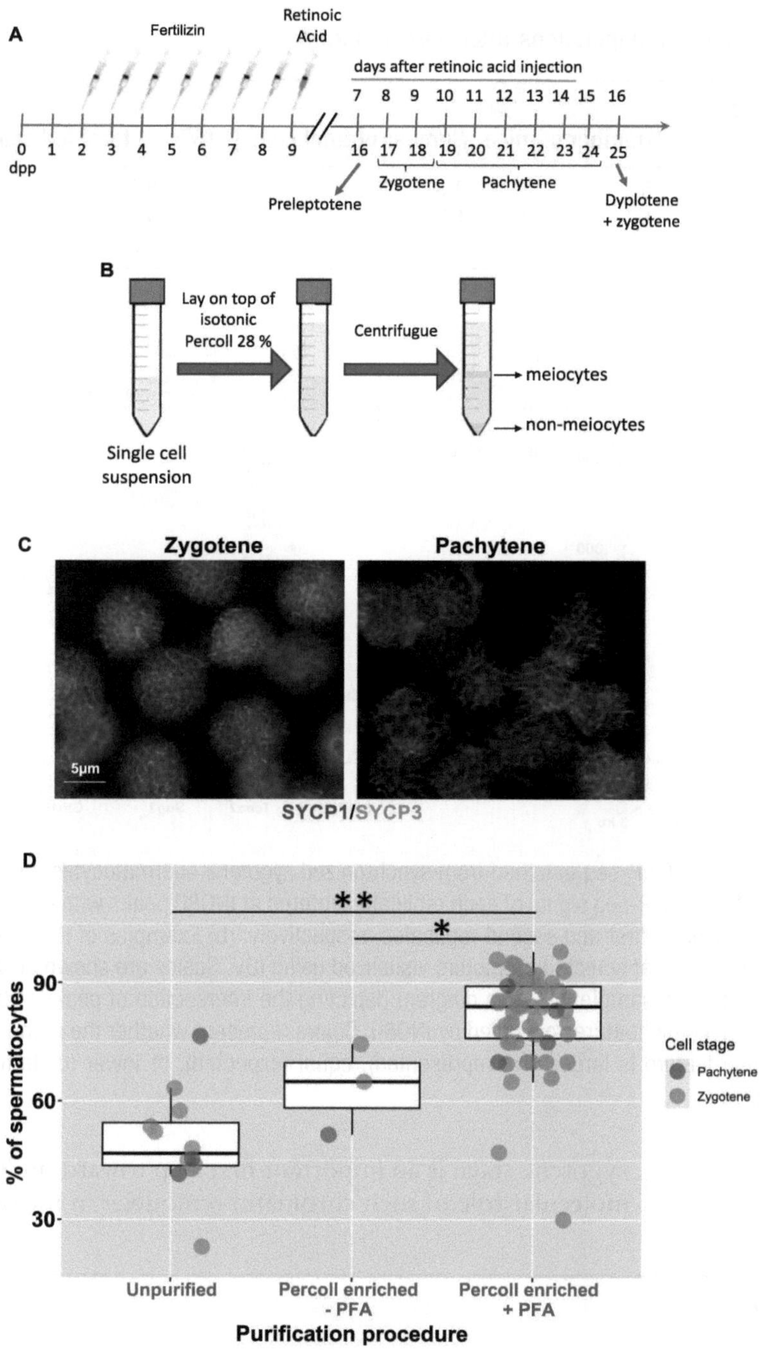

Fig. 1 Enrichment of synchronized mouse spermatocytes. (**a**) Schematic showing neonate mice treatment for synchronization of spermatocytes and collection of cells at different stages of prophase I. (**b**) Diagram representing PFA-fixed spermatocyte centrifugation on a discontinuous Percoll gradient. (**c**) Representative image showing zygotene and pachytene spermatocytes after Percoll gradient. (**d**) Percentage of spermatocytes before Percoll gradient (mean $\pm$ standard deviation, 49.4 $\pm$ 13.1%, n $=$ 12), purified using a discontinuous Percoll gradient without cell fixation ("- PFA," 63.7 $\pm$ 11.6%, n $=$ 3) or purified after fixing cells ("+PFA," 81.3 $\pm$ 11.8%, n $=$ 55) (*See* **Note 1**). Tukey's significance test, $p < 0.05$ (*) or < 0.001 (**)

Table 1
Analysis of zygotene cell populations after Percoll purification

Targeted meiotic stage	Number of mice	Number of total cells ($\times 10^6$)	% targeted cells after percoll
Zygonema	5	9.3	87
	6	12	96
	4	8.7	91

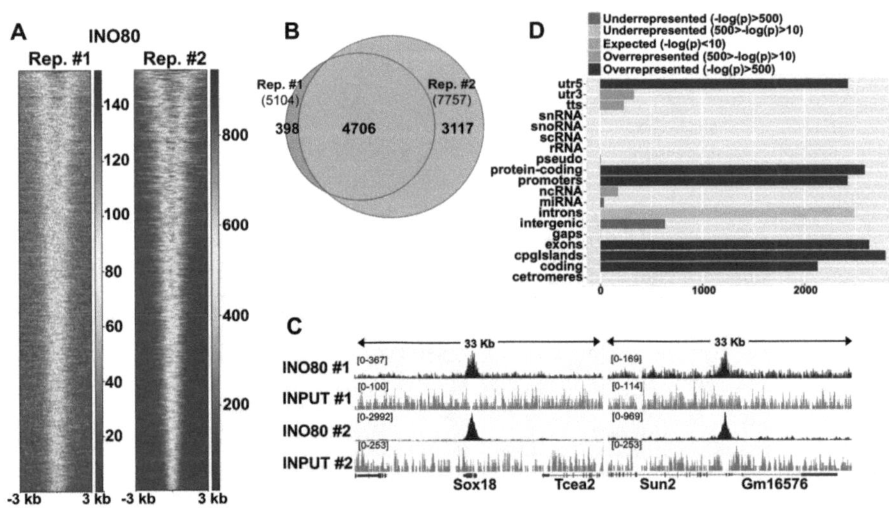

Fig. 2 Analysis of INO80 ChIP-seq obtained from synchronized zygotene spermatocytes. (**a**) Heatmap profiles corresponding to INO80 ChIP-seq signal of each replicate centered at INO80 peaks with IDR =< 0.05. Rep. #1 and Rep. #2 correspond to first and second replicates, respectively. (**b**) Examples of INO80 ChIP-seq signals and their respective inputs at selected promoters visualized using IGV. Scales are shown between brackets at the top left corner of each sample. (**c**) Venn diagram depicting the intersection of peaks of both INO80 ChIP-seq replicates. (**d**) Genomic features occupied by INO80. Colors represent whether the number of peaks found at a given genomic feature is larger (overrepresented), equal (expected), or lower (underrepresented) than expected by chance

zygotene stage is an important first step toward understanding the molecular role of such chromatin remodeler in meiosis.

2 Materials

2.1 Synchronization of Spermatocyte in Mouse Testis

1. 5 mg solid WIN18,446 (also known as N,N′- -Octamethylenebis (2,2-Dichloroacetamide), MP Biomedicals, cat. # 158050) is suspended in 500 μL of Tragacanth powder in water (1%). Suspension can be prepared 1 h before injection.
2. Retinoic acid *stock solution*: dissolve 50 mg of retinoic acid (Sigma, cat. # R2625) in 1664 mL of DMSO. This solution can be stored for 1 week at −80 °C. Cover from light and use glass containers and pipettes when possible.

3. Retinoic acid *suspension*: mix 83 μL of retinoic acid stock in 1917 μL of sterile milliQ water. Cover from light and use glass containers and pipettes when possible. Prepare right before use.
4. 26-gauge needles, 3/8 in. long (or less).

2.2 Retrieving Testis and Preparing Single Cell Suspension

1. Testis are harvested from synchronized male mouse. All experiments conformed to relevant regulatory standards guidelines and were approved by the Oklahoma Medical Research Foundation—IACUC (Institutional Animal Care and Use Committee, protocol 20.33).
2. Scissors and fine tips forceps.
3. Cell culture dishes 60 ×15 mm (Cellstar, cat. # 628160).
4. Micropipettes and tips 1–10 μL, 20–200 μL, and 100–1000 μL.
5. 1.5 mL plastic microtubes.
6. 40 μm (Fisherbrand, cat. # 22363547) and 70 μm (Fisherbrand, cat. # 22363548) mesh cell strainer.
7. Hanks' Balanced salt solution: (0.14 mM NaCl, 5 mM KCl, 1 mM $CaCl_2$, 0.4 mM $MgSO_4.7H_2O$, 0.5 mM $MgCl_2.6H_2O$, 0.3 mM $Na_2HPO_42H_2O$, 0.4 mM KH_2PO_4, 6 mM D-Glucose (Dextrose), 4 mM sodium bicarbonate; Sigma, cat. # H8264). Store at 4 °C.
8. Fetal bovine serum (Atlanta Biologicals).
9. Type I collagenase (Worthington, cat. # LS004196).
10. Turbo Nuclease (Nacalai, cat. # NU0103M).
11. Trypsin (Sigma, cat. # T0303).
12. Protease inhibitor (Pierce, cat. # A32965, 1 tablet in 1 mL of water).

2.3 Cell Fixation and Percoll Treatment

1. Ice cold 1 X PBS.
2. 16% PFA solution (Electron Microscopy Sciences, cat. # 15710).
3. 1.25 M glycine solution.
4. 28% v/v Percoll solution (28 mL Percoll (Sigma-Aldrich, cat. # P1644) + 10 mL 10 X PBS solution, completed to a total volume of 100 mL using milliQ water).

2.4 ChIP-seq

1. ChIP buffer containing 5 μL of protease inhibitor cocktail: (CalbioChem, cat. # 539134-1) (50 mM Tris–HCl pH 8, 1 mM EDTA, 150 mM NaCl, 1% Triton X-100, 0.1% sodium deoxycholate, 0.1% SDS).

2. Antibody (INO80 antibodies, Abcam cat. # ab105451, lot GR3206182-2). INO80 antibody specificity was confirmed by Western blot analysis of INO80 knockout and wild-type control MEFs [10].
3. ChIP-grade protein A/G magnetic beads (Thermo Fisher, cat # 26162).
4. ChIP washing solution: (20 mM Tris–HCl pH 8, 2 mM EDTA, 150–500 mM NaCl, 1% Triton X-100, 0.1% sodium deoxycholate).
5. Lithium buffer: (10 mM Tris–HCl pH 8, 1 mM EDTA, 250 mM LiCl, 1% Igepal, 0.7% sodium deoxycholate).
6. TE buffer pH 7.4: (10 mM Tris–HCl, 1 mM EDTA).
7. Elution solution: (1% SDS, 100 mM $NaHCO_3$ pH 9, made freshly).
8. PFA cross-link reverse solution: 5 M NaCl.
9. 0.5 M EDTA.
10. 1 M Tris–HCl (pH 6.5).
11. RNAse A (Qiagen cat # 19101).
12. Proteinase K (approximately 3 U, Qiagen, cat. # 19131).
13. minElute PCR purification kit (Qiagen, cat. # 28004).

2.5 Spermatocyte Chromosome Spread

1. Microscope slides (Epredia, cat. # 6776214).
2. Chromosome spread fix solution (paraformaldehyde 1% (Electron Microscopy Sciences, cat. # 15710), 0.2% Triton X-100, cat. # 1002268324; pH, 10.2 adjust with 20 mM Na_2B4O_7 solution).
3. Primary antibodies, mouse SYCP3 antibody (chicken polyclonal [11, 12]); SYCP1 antibody (Novus, cat. # NB300-229).
4. Photo-Flo 200 solution (0.4% in water) (Kodak, cat. # 74257).
5. Fluoro-Gel II with DAPI (Electron Microscopy Sciences, cat. # 17985-51).

2.6 Library Preparation and Sequencing

1. Library quantification kit (Kapa, cat. # KK4854).
2. End-repair mix: (0.66 mM dNTP mix (NEB, cat. # N0447S), 100 U/mL T4 DNA polymerase (NEB, cat. # M0203L), 33 U/mL Klenow fragment (NEB, cat. # M0210S), 333 U/mL T4 PNK (NEB, cat. # M0201L), and 1.67 x T4 PNK buffer).
3. A-tailing mix (0.25 mM dATP, 125 U/mL Klenow fragment (NEB, cat. # M02105S)).
4. Truseq single index adaptors (Illumina, cat. # 20015960 or 20015961).

5. Ligation mix (30 U/μL T4 DNA ligase (NEB, cat. # M0202L) and 1.58 X ligase buffer).
6. HiFi Hot Start library amplification kit (Kapa, cat. # KK2621).
7. ACCEL-NGS® 1S PLUS DNA LIBRARY KIT (NEB, cat. # 10024).
8. Unique dual indexing kit (Swift, cat. # X9096).

2.7 Equipment

1. Orbital Shaker.
2. Dry bath at 32 °C.
3. Epifluorescent microscope (Zeiss, Inc.)
4. Qubit (Life Technologies).
5. Tape station 4150 (Agilent).

3 Methods

3.1 Synchronization of Spermatocytes in Mouse Testis

Synchronize spermatocytes at prophase I as described in [5]. Briefly, starting at 2 dpp, mice are fed for seven consecutive days with WIN18,446 solid suspension on Tragacanth gum per gram of body weight to arrest gametogenesisGametogenesis at the spermatogonia stage. The next day (9 dpp), mice are injected with 10 μL of retinoic acid suspension per gram of body weight to induce their coordinated maturation (Fig. 1a). Collect zygotene and pachytene cells, euthanizing mice at 8 and 10 days after retinoic acid injection, respectively. The stage of cells maturation at a particular time after retinoic acid injection could vary for different mice breed and should be explored first. Synchronization efficiency and meiotic cell stages can be quickly evaluated by SYCP3 antibodies immunostaining of spermatocyte chromosome spreads.

3.2 Enrichment of Primary Spermatocytes by Centrifugation on Discontinuous Percoll Gradient

In the following protocol, indicated amounts have been adjusted to two mice, four testes.

1. After detunication, place testes in a 15 mL conical tube containing 2 mL of prewarmed (35 °C) collagenase solution (1 mg/mL type I collagenase (Worthington, cat. # LS004196) and 60 U/mL Turbo Nuclease (Nacalai, cat. # NU0103M) in Hank's balanced salt solution) for 7 min at 35 °C with gentle horizontal shaking (60 rpm).
2. Keep disaggregated seminiferous tubules in vertical position at room temperature for 2 min to allow tubules to decant.
3. Discard supernatant and add 2 mL of prewarmed (35 °C) collagenase-trypsin solution (0.55 mg/mL collagenase, 0.37 mg/mL trypsin (Sigma, cat. # T0303), and 60 U/mL TurboNuclease in Hank's balanced salt solution).

4. Disintegrate tubules by gently pipetting 3–5 times with a transfer pipette and incubate them for 20 min at 35 °C with gentle horizontal shaking (60 rpm). Then, further process tubules by gently pipetting 3–5 times with a pipette.
5. Add 800 μL of serum and 50 μL of protease inhibitor (Pierce, cat. # A32965, 1 tablet in 1 mL of water), mix by inversion, and incubate on ice for 2 min to allow remaining tubules to decant.
6. Filter cells through a 40 μm cell strainer (avoid taking intact tubules to the strainer as they will reduce their filtering capability). At this point, a sample of 50 μL is taken to determine spermatocyte staging by immunostaining of spermatocyte chromosome spread (γH2AX, SYCP1 and SYCP3 are used as staging markers).
7. Complete volume to 8 mL with 1 X PBS (room temperature) and add 500 μL of 16% PFA solution (Electron Microscopy Sciences, cat. # 15710). Incubate cells at room temperature with rotation for 5 min.
8. Add 1 mL of 1.25 M glycine solution to quench PFA and incubate cells for 4 min at room temperature with rotation.
9. Pellet cells by centrifugation at 800× *g* for 10 min 4 °C on a swinging bucket rotor.
10. Discard the supernatant and resuspend the pellet containing the cells with 1 mL of ice-cold 1X PBS by gently pipetting until no cell clumps are detected.
11. Add ice cold 1X PBS to complete a total volume of 8 mL. Then, add 80 μL of 1.25 M glycine.
12. Use a trimmed plastic 1 mL pipette tip to gently place the cell suspension on top of 4 mL of isotonic 28% v/v Percoll solution in a 15 mL tube on ice.
13. Centrifuge at 800× *g* (4 °C) for 20 min on a swinging bucket rotor with the brake set on "slow" or no-brake" so that the interphase is not disrupted by the braking process.
14. Collect spermatocytes from the 1X PBS: 28% Percoll interphase (Fig. 1b) and place them on a new 15 mL tube. Do not collect more than 3 mL.
15. Add ice-cold 1 X PBS to complete a total volume of 14 mL and mix by inversion. Take 300–600 μL sample to assess cell purity by immunostaining of spermatocyte chromosome spreads (γH2AX, SYCP1, and SYCP3 are used as markers, Fig. 1c).
16. Centrifuge cells at 800× *g* (4 °C) for 10 min and discard the supernatant.

We have observed that treatment of spermatocytes (zygotene/pachytene) obtained from synchronized testis with 1% PFA resulted in higher cell enrichment (81.3 ± 11.8%, n = 55) when compared to samples before purification (49.4 ± 13.1%, n = 12) and those obtained after Percoll gradient using non-fixed spermatocytes (Fig. 1d and Table 1). See **Notes 1** and **2**.

3.3 ChIP-seq

1. Resuspend PFA-fixed Percoll-enriched spermatocytes in 1 mL of ChIP buffer containing 5 μL of protease inhibitor cocktail (CalbioChem, cat. # 539134-1) (50 mM Tris–HCl pH 8, 1 mM EDTA, 150 mM NaCl, 1% Triton X-100, 0.1% sodium deoxycholate, 0.1% SDS).
2. Sonicate for 24 min using the Covaris E220 evolution (peak power 140, duty factor 5, 200 cycles per burst). Optimal sonication time can vary for different proteins and hence should be fine-tuned using an empirical approach.
3. Centrifuge the sample at 12,000× *g* for 10 min at 4 °C. Take 10 μL of supernatant as input control and incubate the resulting supernatant (containing chromatin) with 20 μg of antibody overnight at 4 °C. In our case we ChIP INO80 on early purified zygotene spermatocytes (Abcam cat. # ab105451, lot GR3206182-2) to test cell quality and to evaluate suitability of the material for genome wide analysis. We chose INO80 because it is a chromatin remodeler required for normal meiotic progression [9]. We reason that INO80 chromatin-binding profiles may help explain the observed phenotype at the molecular level, specifically by identifying INO80 target genes or other genomic structures.
4. Add ChIP-grade protein A/G magnetic beads (Thermo Fisher, cat # 26162) to capture antibody/chromatin complexes for 2 h at 4 °C.
5. Wash 2 times with increasing salt concentrations (20 mM Tris–HCl pH 8, 2 mM EDTA, 150–500 mM NaCl, 1% Triton X-100, 0.1% sodium deoxycholate), and once with lithium buffer (10 mM Tris–HCl pH 8, 1 mM EDTA, 250 mM LiCl, 1% Igepal, 0.7% sodium deoxycholate).
6. Wash beads twice with TE buffer pH 7.4 (10 mM Tris–HCl, 1 mM EDTA) and elute DNA twice by incubation at 65 °C for 30 min using 150 μL elution solution (1% SDS, 100 mM $NaHCO_3$ pH 9 made freshly).
7. Reverse PFA cross-link by adding 12 μL of 5 M NaCl and incubating at 65 °C for at least 18 h.
8. Next day, add 6 μL of 0.5 M EDTA, 12 μL of 1 M Tris–HCl (pH 6.5), and 1 μL RNAse A (Qiagen cat. # 19101) to the sample and incubate for 30 min at 37 °C.

Table 2
Library preparation and sequencing methods

Sample	Library preparation method	Sequencing
INO80 replicate 1	In-house	Paired end, 50 bp
INPUT 1	In-house	Paired end, 50 bp
INO80 replicate 2	Adaptase kit	Paired end, 150 bp
INPUT 2	In-house	Paired end, 150pb

All samples were sequenced at OMRF clinical genomics core, with an Illumina Novaseq 6000 sequencer

9. Add 5 μL of proteinase K (approximately 3 U, Qiagen, cat. # 19131) and incubate for 1 h at 56 °C.
10. Purify DNA using the minElute PCR purification kit (Qiagen cat. # 28004) using 7 volumes of PB buffer (provided with the kit) and eluting twice in 12 μL of elution buffer each time.
11. Quantify eluted DNA using Qubit (Life Technologies) before library preparation.

3.4 ChIP-seq Library Preparation and Sequencing

Library preparation and sequencing methods are detailed in Table 2.

Library Preparation Method (In House)

1. Incubate 20 μL of the eluted DNA with 30 μL of end-repair mix (0.66 mM dNTP mix (NEB, cat. # N0447S), 100 U/mL T4 DNA polymerase (NEB, cat. # M0203L), 33 U/mL Klenow fragment (NEB, cat. # M0210S), 333 U/mL T4 PNK (cat. # M0201L), and 1.67 x T4 PNK buffer) at 20 °C for 30 min.
2. Purify DNA using Qiagen Minelute kit using 7 volumes of PB buffer and elute in 12 μL of EB buffer.
3. Mix 10 μL of eluted DNA with 40 μL of A-tailing mix (0.25 mM dATP, 125 U/mL Klenow fragment (cat. # NEB M02105S), and Klenow buffer 1.25 X) and incubate at 37 °C for 30 min.
4. Purify DNA again using Qiagen Minelute kit using 7 volumes of PB buffer and elute in 12 μL of EB buffer.
5. Dilute truseq single index adaptors (Illumina, cat. # 20015960 or 20015961) according to the DNA concentration (insert: adaptor molar ratio of 1:2, with maximal dilution of 1:50) and add 1 μL of a specific diluted adaptor to each sample. Add 20 μL of ligation mix (30 U/μL T4 DNA ligase (cat. # NEB M0202L) and 1.58 X ligase buffer) to 10 μL of the insert: adaptor mix and incubate for 30 min at 20 °C.

6. Purify DNA using Qiagen Minelute kit using 7 volumes of PB buffer and elute in 12 μL of EB buffer.
7. Finally, amplify DNA using the Kapa HiFi Hot Start library amplification kit (cat. # KK2621) according to manufacturer's instructions.
8. Quantify library by qPCR using Kapa library quantification kit (cat. # KK4854) and assess size and quality of DNA using Agilent Tape station.

Commercial Adaptase Library Preparation Method. As an alternative for library preparation (used for INO80 replicate 2 (Table 2)), ACCEL-NGS® 1S PLUS DNA LIBRARY KIT (cat. # 10024) can be used in conjunction with Swift unique dual indexing kit (cat. # X9096), following manufacturer's instructions.

3.5 ChIP-seq Data Processing and Analysis

A brief description of the process follows.

3.5.1 Alignment and Quality Filtering

1. Use fastp with default settings for adapter-trimming and quality-prune of reads. For libraries prepared using commercial Adaptase kit, reads must be trimmed one more time, eliminating the first 10 bases of both R1 and R2 (fastp can also be used for the second trim).
2. Align reads to GRCm38/mm10 using BWA-MEM [13] (Version 0.7.15) with default settings except for option "-M" for Picard compatibility.
3. Use Picard (version 2.21.2, http://broadinstitute.github.io/picard/) and SAMtools (version 1.11) [14] to obtain mapping quality metrics, remove duplicates and filter reads. Keep primary alignment reads that are not duplicated, are properly paired, and have a MAPQ >30. Discard mitochondrial reads. Use MultiQC (version 1.7) to generate metrics reports [15]. Table S1 contains INO80 ChIP-seq and input mapping statistics (all supplementary data can be accessed at GEO, gene expression omnibus, ncbi.nlm.nih.gov/geo/query/acc.cgi?acc=GSE190590).

3.5.2 ChIP-seq Quality Control Metrics

1. Proceed with cross-correlation analysis using phantompeakqualtools (version 1.2, [16, 17]) with default parameters. Results can be found in Fig. 3. Ideally, NSC value should be above 1.05 and RSC value should be above 0.8. Profiles with broad peaks are expected to perform worse on these two statistics when compared to profiles with narrow peaks. In our case, the profile has broad peaks and both replicates have an RSC > 0.8 (ideal) and a NSC < 1.05 (not ideal). We still consider them good quality based on further analysis (reproducibility, FRIP, manual examination of peaks, etc.).

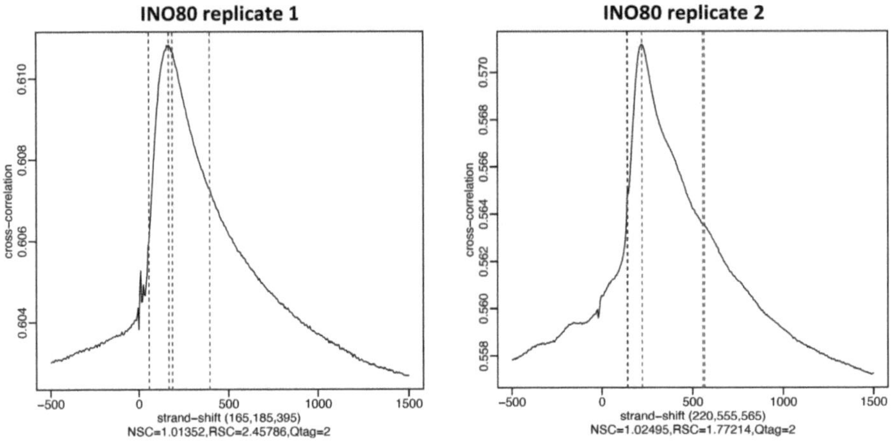

Fig. 3 Cross-correlation analysis corresponding to INO80 ChIP-seq peaks. Graphs correspond to replicate 1 and replicate 2

2. Use samtools view to calculate fraction of reads in peaks (FRIP). The list of peaks we use for each replicate are not black-greylist filtered. An FRIP >1% is recommended by ENCODE.
3. Obtain mapping statistics using samtools stats (number of mapped reads) or samtools flagstat (number of uniquely mapped reads).
4. Greylist regions can be prepared for each input using the GreyListChIP R-package (R package version 1.24.0, https://bioconductor.org/packages/release/bioc/html/GreyListChIP.html) together with the Bsgenome.Mmusculus.UCSC.mm10 package (https://bioconductor.org/packages/release/data/annotation/html/BSgenome.Mmusculus.UCSC.mm10.html) using default settings (greyListBS command). Merge both lists to create a single greylist, which is later used to filter peaks.

3.5.3 Reproducibility Analysis

To analyze the reproducibility of replicates, analyze how many peaks are present in both replicates. Alternatively, Irreproducible Discovery Rate (IDR) can be used as recommended by ENCODE consortia.

1. Use MACS2 (version 2.2.7.1, [18]), using broad mode, default parameters, and corresponding inputs, to call peaks. Filter the lists of peaks for black-greylist regions before intersecting them.
2. Analyze reproducibility using Irreproducible Discovery Rate (IDR) [19]. To do so, create two self-pseudo replicates of each INO80 ChIP-seq real replicate and the input by shuffling

the reads and then splitting them in two equal parts. Then, merge self-pseudo replicate 1 of real replicate 1 with self-pseudo replicate 1 of real replicate 2 to create a pooled-pseudo replicate. Proceed similarly for replicate 2. Call peaks for all the bam files (real replicates, self-pseudo replicates, and pooled replicates) using MACS2, broad mode, and a lenient setting of -p 0.001. Proceed to IDR analysis using IDR software (version 2.0.3, https://www.encodeproject.org/software/idr/) on the lists of peaks already black-greylist filtered and p-value sorted. Use the peaks with an adjusted IDR value >= 540 (IDR =< 0.05) for further analysis.

We detected 5104 and 7757 peaks in INO80 ChIP-seq replicate 1 and 2, respectively. Most peaks detected on replicate 1 are also found in replicate 2 (92%, Fig. 2c). The number of peaks with an IDR <= 0.05 when comparing pooled-pseudo replicates is 1.63 times larger than the one found when comparing true replicates. This is well under the value of 2.0, which is the maximum recommended value. We conclude that Percoll enriched spermatocyte fractions we obtained can be used to obtain robust and reproducible genome wide profiles of chromatin binding proteins.

3.5.4 Peak Annotation, Genome, and Gene Ontology Analysis

1. Use HOMER's annotatePeaks.pl script ([20], version 4.10) to annotate peaks with IDR =< 0.05 (true replicates IDR analysis). Annotated peaks can be found in Table S3.
2. Genome ontology analysis can be done using HOMER with same "annotatePeaks.pl" script plus the "-genomeOntology option". The results can be found in Table S3.

   INO80 chromatin binding profiles can be further analyzed using the list of peaks that had an IDR <= 0.05 (true replicates comparison). Genome ontology analysis revealed that INO80 is found at CpG islands and protein-coding genes (promoters, exons, UTR-5′) at a much higher frequency than that expected by chance. However, on the other hand, INO80 has reduced presence in most non-protein-coding genes (rRNA, miRNA, snRNA, snoRNA, scRNA), intergenic regions, and centromeres (Fig. 2d and Table S3).
3. Use DAVID with default parameters [21, 22] for functional annotation analysis. We generated the list of genes fed to DAVID using INO80 annotated peaks, keeping only genes with an INO80 peak closer than 5000 bp

   Functional analysis revealed that several genes found near an INO80 peak (less than 5000 bp) are genes that code for proteins that regulate transcription (Table S4). Indeed, "Transcription Regulation" is the second most significant Uniprot keyword associated with genes close to INO80 (389 hits and an FDR of 1.95 × 10^{-47}), and “transcription, DNA-templated” is the third most enriched GO-Biological Process term (403 hits and an FDR of 3.95 × 10^{-33}). Further, we searched for genes with recognized meiotic functions among genes near INO80 peaks. We used in our search a manually cured list of candidate genes as well as the genes annotated with the following GO terms: chromatin remodeling, DNA repair, meiosis I, chromosome pairing at meiosis, and synaptonemal complex assembly. We identified genes involved in critical meiotic processes such as telomere-led rapid prophase movements (SUN1 and SUN2) and homologous directed repair of double-strand break (MND1, ERCC4, and MSH2) (complete list of genes is found in Table S5). Overall, our results suggest that INO80 could have a broad effect on transcription in zygotene spermatocytes, both by direct action at gene promoters and by indirect action, by regulating transcription factors.

Our work has also revealed the genome-wide localization of INO80 chromatin remodeler in zygotene spermatocytes (the most advanced spermatocyte seen in INO80 knockout mouse [9]). This is the first step in understanding the meiotic role of this meiotic-indispensable chromatin remodeler.

4. Further processing and plotting were done using R (https://www.r-project.org/, version 4.1.1) and packages within tidyverse (https://joss.theoj.org/papers/10.21105/joss.01686). Functional analysis results can be found in Table S4

3.5.5 Generation of Heatmaps

Bigwig coverage tracks can be generated using bamCoverage tool from deepTools (version 3.4.3, [23]) with a bin size of 1 bp, no smoothing, and RPKM normalization. Plot average aggregate profiles and heatmaps using deepTools (computeMatrix and plotHeatmap).

Heatmaps (Fig. 2a), visualization of specific genomic sites (Fig. 2b), and Venn diagram (Fig. 2c) obtained from two independent replicates show a robust and reproducible signal for INO80 genome wide.

Proceed with statistical tests using R (version 4.1.1) (in our case, reported in the figure legends or main text). INO80 ChIP-seq data (raw and processed) are available at the Gene Expression Omnibus under accession number GSE190590.

In summary, we describe here a methodology for enrichment of synchronized mouse spermatocytes at different stages of prophase I that is fast and unexpensive. This provides valuable material for experiments involving ChIP followed by massive sequencing.

4 Notes

1. We first attempted to separate unfixed spermatocytes obtained from synchronized mice testis by density gradients and centrifugation using a discontinuous Percoll gradient [24]. We obtained an enriched fraction of spermatocytes (zygotene/pachytene) post-Percoll gradient corresponding to $63.7 \pm 11.6\%$ ($n = 3$ independent experiments) (before purification, $49.4 \pm 13.1\%$, $n = 12$). This relatively low spermatocyte enrichment obtained after Percoll purification may be caused by cell aggregation and cell lability, thus, we reasoned that paraformaldehyde fixation of spermatocytes before separating by Percoll gradient may prevent aggregation and stabilize cells, resulting in increased reproducibility and improved purification efficiency ($81.3 \pm 11.8\%$, $n = 55$) (Fig. 1d).
2. We note that the Percoll cell enrichment protocol we describe here uses a fixed amount of time (20 min) regardless the number of mice or cell processed. This is an advantage compared to protocol such as FAC sorting, in which the time of processing is proportional to the number of cells to be purified (in our experience 5–6 h FACS hours to collect approximately 10 million cells).

Acknowledgments

This work was supported by R01-NIGMS R01 GM125803 and R21-NICHD R21 HD103562 to RP.

References

1. Bellve AR et al (1977) Spermatogenic cells of the prepuberal mouse. Isolation and morphological characterization. J Cell Biol 74:68–85
2. Bryant JM, Meyer-Ficca ML, Dang VM, Berger SL, Meyer RG (2013) Separation of spermatogenic cell types using STA-PUT velocity sedimentation. J Vis Exp. https://doi.org/10.3791/50648
3. Barchi M, Geremia R, Magliozzi R, Bianchi E (2009) Isolation and analyses of enriched populations of male mouse germ cells by sedimentation velocity: the centrifugal elutriation. Methods Mol Biol 558:299–321
4. Gaysinskaya V, Soh IY, van der Heijden GW, Bortvin A (2014) Optimized flow cytometry isolation of murine spermatocytes. Cytometry A 85:556–565
5. Romer KA, de Rooij DG, Kojima ML, Page DC (2018) Isolating mitotic and meiotic germ cells from male mice by developmental synchronization, staging, and sorting. Dev Biol 443:19–34
6. Hogarth CA et al (2013) Turning a spermatogenic wave into a tsunami: synchronizing murine spermatogenesis using WIN 18,446. Biol Reprod 88:40
7. Patel L et al (2019) Dynamic reorganization of the genome shapes the recombination landscape in meiotic prophase. Nat Struct Mol Biol 26:164–174
8. Chen Y et al (2020) Refined spatial temporal epigenomic profiling reveals intrinsic connection between PRDM9-mediated H3K4me3

and the fate of double-stranded breaks. Cell Res 30:256–268
9. Serber DW, Runge JS, Menon DU, Magnuson T (2016) The mouse INO80 chromatin-remodeling complex is an essential meiotic factor for spermatogenesis. Biol Reprod 94:8
10. Runge JS, Raab JR, Magnuson T (2018) Identification of two distinct classes of the human INO80 complex genome-wide. G3 (Bethesda) 8:1095–1102
11. de Castro RO, Previato de Almeida L, Carbajal A, Gryniuk I, Pezza RJ (2022) PBAF chromatin remodeler complexes that mediate meiotic transitions in mouse. Development 149
12. Guiraldelli MF et al (2018) SHOC1 is a ERCC4-(HhH)2-like protein, integral to the formation of crossover recombination intermediates during mammalian meiosis. PLoS Genet 14:e1007381
13. Li H (2013) Aligning sequence reads, clone sequences and assembly contigs with BWA-MEM. arXiv:1303.3997
14. Li H et al (2009) The sequence alignment/map format and SAMtools. Bioinformatics 25: 2078–2079
15. Ewels P, Magnusson M, Lundin S, Kaller M (2016) MultiQC: summarize analysis results for multiple tools and samples in a single report. Bioinformatics 32:3047–3048
16. Kharchenko PV, Tolstorukov MY, Park PJ (2008) Design and analysis of ChIP-seq experiments for DNA-binding proteins. Nat Biotechnol 26:1351–1359
17. Landt SG et al (2012) ChIP-seq guidelines and practices of the ENCODE and modENCODE consortia. Genome Res 22:1813–1831
18. Zhang Y et al (2008) Model-based analysis of ChIP-Seq (MACS). Genome Biol 9:R137
19. Li QH, Brown JB, Huang HY, Bickel PJ (2011) Measuring reproducibility of high-throughput experiments. Ann Appl Stat 5: 1752–1779
20. Heinz S et al (2010) Simple combinations of lineage-determining transcription factors prime cis-regulatory elements required for macrophage and B cell identities. Mol Cell 38:576–589
21. Huang DW, Sherman BT, Lempicki RA (2009) Bioinformatics enrichment tools: paths toward the comprehensive functional analysis of large gene lists. Nucleic Acids Res 37:1–13
22. Huang DW, Sherman BT, Lempicki RA (2009) Systematic and integrative analysis of large gene lists using DAVID bioinformatics resources. Nat Protoc 4:44–57
23. Ramirez F et al (2016) deepTools2: a next generation web server for deep-sequencing data analysis. Nucleic Acids Res 44:W160–W165
24. Kogo H et al (2010) Screening of genes involved in chromosome segregation during meiosis I: toward the identification of genes responsible for infertility in humans. J Hum Genet 55:293–299

Chapter 5

A Method to Study the Meiotic Recombination Map in Human Preimplantation Blastocysts

Yuanlin Ma, Jing Wang, and Yanwen Xu

Abstract

Homologous recombination plays pivotal roles in physical attachments and genetic diversity. In the past, it was studied among individuals from different populations. However, only few gametes from individual could generate offspring, which limits its exploration in nature selection. In the last few years, preimplantation blastocysts based on trio SNP-chip data were available in individuals for preimplantation genetic testing (PGT). In this protocol, we demonstrate how to detect meiotic recombination events and construct the genetic map based on trio SNP-chip data, obtained from biopsied blastocysts and their related individuals in PGT cycles, which may allow better understanding of recombination events in nature selection.

Key words Homologous recombination, Blastocyst, SNP-array, Genetic map, Pedigree

1 Introduction

During meiosis, recombination plays a crucial role in establishing physical attachments between homologous chromosomes, ensuring accurate chromosome segregation during meiosis. This process also contributes to genetic diversity by introducing new allele combinations [4]. Owing to its importance in biological and evolutionary processes, it was expected the pattern of recombination was conservative. However, heterogeneity was observed in rates and patterns of recombination among individuals, sexes, and populations [12]. Since the first human genetic map called Marshfield was constructed by linkage disequilibrium (LD) in 1998 [3], several genetic maps have been constructed [1, 5, 11] among individuals of European or African ancestry [14]. To the best of our knowledge, only two studies focused on Asians (Korean and Mongolian pedigrees [2] and Chinese blastocysts [10]) were published.

Authors Yuanlin Ma and Jing Wang have equally contributed to this chapter.

Jesús A. Carballo (ed.), *Meiosis: Methods and Protocols*, Methods in Molecular Biology, vol. 2818,
https://doi.org/10.1007/978-1-0716-3906-1_5,

One of the most commonly used methods to study meiotic recombination is to estimate the crossover locations from LD patterns in population [13] or genomic segments shared among relatives [5, 7]. Nevertheless, it is important to note that only a few gametes per individual generate offspring. Therefore, meiotic recombination obtained from individuals may not provide a complete representation of the recombination pattern. This leads to "missing data," which is crucial for understanding the overall pattern of recombination and nature selection. Theoretically, the most powerful method to study meiotic recombination is observing meiosis processes in gametes directly [6, 9]. Genotyping or sequencing of gametes would provide detailed recombination information [1, 11]; however, these methods were challenging due to scarcity of human oocytes and difficulties in detection at single cell level. Alternatively, it can be done more easily by analyzing the meiotic recombination in preimplantation blastocysts.

In this chapter, we present the utilization of trio SNP-chip data, obtained from preimplantation blastocysts and their related individuals from preimplantation genetic testing (PGT) cycles, to detect meiotic recombination events and construct genetic map.

2 Material

1. Data: A pedigree array data files (.idat) are prepared, containing father, mother, proband, and offspring.
2. Beeline (version 1.0.37, Illumina, USA) for data conversion (https://www.illumina.com.cn/techniques/microarrays/array-data-analysis-experimental-design/beeline.html).
3. Bluefuse Multi (version 4.0, Illumina, USA) for data analysis (https://www.illumina.com.cn/clinical/clinical_informatics/bluefuse.html).
4. R (version 4.2.2) for data analysis (R: The R Project for Statistical Computing (r-project.org)).
5. HEPES, Quinn's Advantage® Medium w/HEPES is purchased from CooperSurgical, Inc., USA.
6. SureMDA kit is purchased from Illumina, USA.

   SureMDA kit contains REPLI-g SC DNA Polymerase, REPLI-g SC Reaction Buffer, Buffer DLB, Stop Solution, PBS SC, DTT, Water SC, and PCR tubes.

   Buffer DLB is prepared by adding Water SC to the lyophilized buffer.

   Buffer D2 is prepared from reconstituted Buffer DLB and DTT.

   Amplification Master Mix is prepared by mixing 9 μL Water SC, 29 μL REPLI-g SC Reaction Buffer, and 2 μL REPLI-g SC DNA Polymerase.
7. Infinium Analysis Kit is purchased from Illumina, USA.

Infinium Analysis Kit contains Multi-Sample Amplification 1 Mix (MA1), Multi-Sample Amplification Master Mix (MA2), Random Primer Mix (MSM), fragmentation solution (FMS), precipitation solution (PM1), PB1, humidifying buffer used during hybridization (PB2), resuspension, hybridization, and wash solution (RA1), XStain BeadChip solution 1 (XC1), XStain BeadChip solution 2 (XC2), XStain BeadChip solution 3 (XC3), XStain BeadChip solution 4 (XC4), Two-Color Extension Master Mix (TEM), Superior Two-Color Master Mix (STM), and Anti-Stain Two-Color Master Mix (ATM).

PB1 is reagent used to prepare BeadChips for hybridization and wash BeadChips after hybridization and in XStain reaction.

8. GeneAmp PCR System 9700 (Applied Biosystems, USA).
9. OctaxShot$_{TM}$ laser system, RI, England.
10. Prepare 0.1 N NaOH.
11. Instruments are purchased from Illumina, USA: 96-well plate, Illumina Hybridization Oven, Hybex incubator, Hyb Chambers, and BeadChip.

3 Methods

This protocol contains three parts: 1, controlled ovarian stimulation, fertilization, blastocyst culture, and blastocyst biopsy (Subheading 3.1); 2, whole genome amplification and hybridization to SNP-chip (Subheading 3.2); 3, construction of genetic map (Subheading 3.3) (Fig. 1). All data analyses are performed under Windows operating system.

3.1 Controlled Ovarian Stimulation, Fertilization, Blastocyst Culture, and Blastocyst Biopsy

1. After routine controlled ovarian stimulation protocols [15], oocytes are obtained by vaginal ultrasound-guided aspiration.
2. Intracytoplasmic sperm injection (ICSI) is used in all cases for PGT, aiming to eliminate the contamination risk from extraneous sperm DNA.
3. Embryos selected by embryologist are cultured into blastocysts, as described in detail by Lu et al. [8].
4. Embryos are observed on the morning and afternoon of the fifth day after oocyte retrieval. Early-stage blastocysts will be kept for watching to the next day. Only fully expanded blastocysts with adequate number of trophectoderm cells meet the criteria for biopsy (the stage of blastocyst ≥4 and trophectoderm cells score A/B).
5. The expanded blastocyst is drilled with a laser to form a hole with a diameter of 18-20 mm on the zona pellucida.

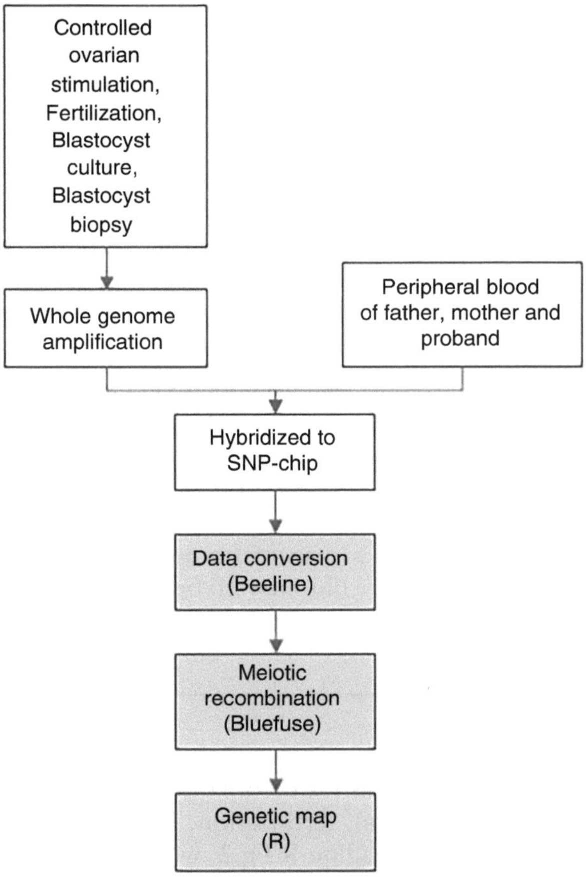

Fig. 1 A flowchart summarizing the analysis steps involved in the construction of genetic map

6. Embryos that hatch with trophectoderm cells outside the zona pellucida are biopsied in the afternoon of day 5 or morning of day 6.
7. A biopsy needle with a diameter of 30 μm is used to aspirate 5–10 trophectoderm cells from each blastocyst and the Octax-Shot$_{TM}$ laser system is used for cell isolation.
8. The biopsied blastocysts are vitrified separately after the biopsy.

3.2 Whole Genome Amplification and Hybridization to SNP-Chip

3.2.1 Embryo Biopsy Sample Collection

1. Aspirate the embryo biopsy sample directly from the biopsy HEPES droplet with a low-adsorption embryonic pipette.
2. Wash embryo biopsy samples three times in a cell culture dish with PBS (containing 10 mg/mL BSA).
3. Put them into 3.5 μL whole genome amplification kit equipped with PBS PCR reaction tube.

3.2.2 The First-Time Whole Genome Amplification

1. Centrifuge PCR reaction tube for several seconds.
2. Add 3 μL Buffer D2 into PCR reaction tube and then centrifuge for several seconds.
3. Tubes were incubated at 65 °C for 10 min on PCR thermocycle instrument (PE9700) and then placed on ice immediately.
4. Add 3.5 μL Stop Solution into PCR reaction tube and then centrifuge for several seconds.
5. Add 40 μL Amplification Master Mix into PCR reaction tube.
6. Tubes are incubated at 30 °C for 8 h on PCR thermocycle instrument (PE9700) and then at 65 °C for 3 min.

3.2.3 The Second Amplification

1. Add 20 μL MA1 to every well of a column in the 96-well plate.
2. Prepare DNA sample from peripheral blood of parents and proband and whole genome amplification embryo biopsy sample.
3. Add 4 μL (50 ng/μL) DNA samples into every well of a column in the 96-well plate.
4. Add 4 μL 0.1 N NaOH into every well of a column in the 96-well plate and seal.
5. Reactions are run with initial denaturation for 1 min at 1600 rpm, followed by a final 1 min at 280 g.
6. Tubes are incubated for 10 min at room temperature.
7. Add 34 μL MA2 and 38 μL MSM into every well of a column in the 96-well plate and seal.
8. Reactions are run for 1 min at 1600 rpm, followed by a final 1 min at 280 g.
9. The 96-well plate is then placed in an Illumina Hybridization Oven and incubated at 37 °C for 13 h.

3.2.4 DNA Fragmentation

1. The 96-well plate is centrifuged at 50 g for 1 min.
2. Add 50 μL FMS into every well of a column in the 96-well plate and seal.
3. Reactions are run with initial denaturation for 1 min at 1650 rpm, followed by a final 1 min at 50 g 22 °C.
4. The plate is then incubated at 37 °C for 45 min in a Hybex incubator.

3.2.5 DNA Precipitation

1. Remove the cover, add 50 μL PM1 into every well of a column in the 96-well plate and seal.
2. Shock the plate for 1 min at 1650 rpm.
3. Incubate the plate at 37 °C for 5 min in a Hybex incubator.
4. Centrifuge the plate at 50 g 22 °C for 1 min.

5. Add 100 μL 100% isopropanol into every well of a column in the 96-well plate and seal.
6. The sample is pipetted up and down 10 times to mix and incubated at 4 °C for 25 min.
7. Centrifuge the plate at 3000 g 4 °C for 20 min.
8. After removing the cover, discard the supernatant.
9. The plate is then left in a rack for 40 min at room temperature.

3.2.6 DNA Suspension

1. Add 23 μL RA1 into every well of a column in the 96-well plate and seal it by heat-sealing metal film.
2. The 96-well plate is then placed in an Illumina Hybridization Oven and incubate at 48 °C for 40 min.
3. Remove the cover, shock the plate for 1 min at 1800 rpm, followed by 1 min at 280 g.

3.2.7 DNA Sample Hybridized to SNP-Chip

1. Incubate the plate at 95 °C for 20 min in a Hybex incubator.
2. The plate is left at room temperature to cool down for 5 min.
3. Pulse centrifuge the MSA3 plate to 280 g.
4. Remove the foil seal from the MSA3 plate, and then dispense 15 μL of each DNA sample onto the appropriate BeadChip section.
5. Load the Hyb Chamber inserts containing BeadChips into the Illumina Hyb Chamber (*see* **Note 1**).
6. Place the Hyb Chamber in the 48 °C Illumina Hybridization Oven.
7. Start the rocker, setting the speed to 5.
8. Incubate at 48 °C for 12 h or overnight but no more than 24 h.

3.2.8 Wash the BeadChip

1. Remove each Hyb Chamber from the Illumina Hybridization Oven.
2. Leave it at room temperature to cool down for 15 min.
3. Remove a BeadChip from the Hyb Chamber and then remove its cover seal.
4. Immediately and carefully slide the BeadChip into the wash rack, making sure that the BeadChip is submerged in the PB1.
5. When all the BeadChips are in the wash rack, move the wash rack up and down for 1 min, breaking the surface of the PB1 with gentle, slow agitation.
6. Move the wash rack to the other wash dish containing clean PB1. Make sure the BeadChips are submerged.
7. Move the wash rack up and down for 1 min, breaking the surface of the PB1 with gentle, slow agitation.

8. Fill the BeadChip Alignment Fixture with 150 mL PB1.
9. Place each BeadChip into the Alignment Fixture. Each BeadChip should be fully immersed in PB1.
10. Chip assembly.
11. Wash the Hyb Chamber reservoirs with DiH_2O.
12. Turn on the water circulator and set it to a temperature that brings the Chamber Rack to 44 °C at equilibrium.

3.2.9 Extend and Stain (XStain) BeadChips

1. Place each flowthrough chamber assembly into the Chamber Rack.
2. Into the reservoir of each flow-through chamber, dispense:

   150 μL RA1, incubate for 30 s, repeat 5 times

   450 μL XC1, incubate for 10 min

   450 μL XC2, incubate for 10 min

   200 μL TEM, incubate for 15 min

   450 μL 95% formamide/1 mM EDTA, incubate for 1 min, repeat one time, incubate 5 min

   450 μL XC3, incubate for 1 min, repeat one time
3. Set the Chamber Rack to 32 °C at equilibrium.
4. Into the reservoir of each flowthrough chamber, dispense:

   250 μL STM, incubate for 10 min

   450 μL XC3, incubate for 1 min, repeat one time, wait 5 min

   250 μL ATM, incubate for 10 min

   450 μL XC3, incubate for 1 min, repeat one time, wait 5 min

   250 μL STM, incubate for 10 min

   450 μL XC3, incubate for 1 min, repeat one time, wait 5 min

   250 μL ATM, incubate for 10 min

   450 μL XC3, incubate for 1 min, repeat one time, wait 5 min

   250 μL STM, incubate for 10 min

   450 μL XC3, incubate for 1 min, repeat one time, wait 5 min
5. Immediately remove the flowthrough chambers from the Chamber Rack.
6. Place the BeadChips in the staining rack while it is submerged in 310 mL PB1.
7. Slowly move the staining rack up and down 10 times.
8. Soak for 5 min.
9. Remove the staining rack from the PB1 dish and place it directly into the wash dish containing 310 mL XC4.

10. Slowly move the staining rack up and down 10 times.
11. Soak for 5 min.
12. Place the tube rack in the vacuum desiccator.
13. Dry under vacuum for 50–55 min.

3.2.10 Extend and Stain (XStain) BeadChips

1. BeadChips are scanned by iScan Reader.
2. BeadChips are stored at room temperature in a vacuum desiccator.

3.3 Construction of Genetic Map

1. A pedigree data is prepared containing father, mother, proband and offspring.
2. Array data files (.idat) are converted to .gtc files using Beeline version 1.0.37.
3. Array data files (.gtc) are imported into BlueFuse Multi version 4.0.
4. Run Bluefuse and generate haplotype.
5. Locations of crossovers are determined from haplotype phase transitions in parent-proband pairs based on informative markers where the parents are heterozygous. Initially, the location is determined by the two closest heterozygous markers to the crossover, providing upper and lower bounds for the location (Fig. 2) (*see* **Note 2**).
6. Construct genetic map

   The genetic map is computed using an expectation-maximization (E-M) algorithm by R Version 4.2.0 (R Foundation for Statistical Computing, Vienna, Austria; https://www.R-project.org/) [5, 10].

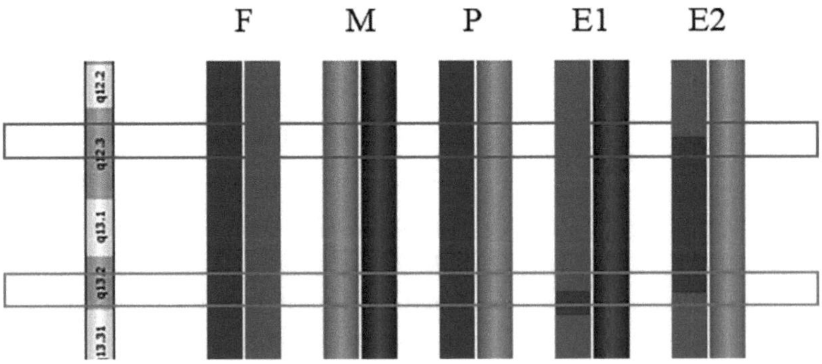

Fig. 2 Schematic diagram shows the location of recombination, which are determined as areas where genotypes are exchanged on the detailed haploblock chart. For example, red box shows a recombination in E2. Green box shows a recombination in both E1 and E2. Since these two embryos present the same location of recombination in this pedigree, as highlighted by the green box, we would consider this recombination actually occurs in the proband. F father, M mother, P proband, E embryo

Read data file as *crossover* and *snp*. The *crossover* file contains case id, chromosome number, up bound of crossover, low bound of crossover. The *snp* file contains all SNP information of SNP-chip, such as chromosome number and chromosome position. For every chromosome, *a0* is the first SNP chromosome position, *an* is the last SNP chromosome position, and *chromlength* is the difference between *an* and *a0*. *dai* and *daj* are the range of crossover and closest SNP. Assume that *mn* and *sn* are the number of homologous crossovers and SNP in chromosome *n*. *k* is the amount of meiosis.

– `dn <- 100*mn/k`

dn is the genetic distance of chromosome *n*.

– `rho0 = dn/chromlength`

Assume the crossover rate (unit: cM/Mb).

– `snp$rjml <- dn/chromlength*snp$daj`

Approximate every marker on each chromosomal has the same constant recombination rate.

– `lidx = list()`
– `uidx = list()`
– `for (i in 1:mn) {lidx[[i]] = which(snp$Begin==crossover$lbnd[i]); uidx[[i]] = which(snp$End==crossover$ubnd[i])}`

Create index structures to point from *crossover* to *snp*.

– `xoidx = list()`
– `for (i in 1:sn) {xoidx[[i]] = which(crossover$lbnd<=snp$Begin[i] & crossover$ubnd>=snp$End[i])}`

Create index structure to point from *snp* to *crossovers*.

– `repeat {`
– `for (i in 1: mn) {crossover$ri[i] <- sum(snp$rjml[lidx[[i]]:uidx[[i]]])}`
– `for (i in 1:sn) {data1 <- crossover[xoidx[[i]],]; data1$p <- snp$rjml[i]/data1$ri; sump <- sum(data1$p); snp$rj[i] <- 100*sump/k}`
– `snp$rhoml <- snp$rjml/snp$daj`
– `snp$rho <- snp$rj/snp$daj`
– `snp$result <- snp$daj * (snp$rhoml-snp$rho)^2`
– `sumresult <- sum(snp$result)/chromlength/rho0**2`
– `if (sumresult < e*chromlength) break`
– `snp$rjml <- snp$rj}`

Perform EM loop. An updated genetic map is computed from these probabilities and crossover rates. According to the updated genetic map, probabilities and crossover rates are calculated. Repeat calculation and update. The iterations are continued until convergence is achieved.

After convergence is achieved, the final genetic map is constructed. The paternal and maternal maps are constructed by projecting the paternal and maternal crossovers onto the combined one (*see* **Note 3**).

The standardized recombination rate is calculated by dividing the regional crossover recombination rate by the genomic average, for each sex separately. The homologous recombination hotspots are defined as any region with the standardized recombination rate ≥10, for each sex separately. Recombination resolution is defined as the minimum SNP distance to discover half of the crossover [5].

4 Notes

1. BeadChips are glass and should be inspected for broken edges and handled with care.
2. At least two informative SNPs are required to determine the true recombination event.
3. A combined map is computed using all the crossovers, *i.e.*, both the paternal and maternal ones, and then the paternal and maternal maps are constructed by projecting the paternal and maternal crossovers onto the combined one.

Acknowledgments

We gratefully acknowledge all the staff in the department of assisted reproduction in Sun Yat-sen University First Affiliated Hospital for their support and cooperation.

References

1. Bell AD, Mello CJ, Nemesh J, Brumbaugh SA, Wysoker A, McCarroll SA (2020) Insights into variation in meiosis from 31,228 human sperm genomes. Nature 583(7815):259–264. https://doi.org/10.1038/s41586-020-2347-0
2. Bleazard T, Ju YS, Sung J, Seo J (2013) Fine-scale mapping of meiotic recombination in Asians. BMC Genet 14(1):19. https://doi.org/10.1186/1471-2156-14-19
3. Broman KW, Murray JC, Sheffield VC, White RL, Weber JL (1998) Comprehensive human genetic maps: individual and sex-specific variation in recombination. Am J Hum Genet 63(3):861–869. https://doi.org/10.1086/302011
4. Chen Y, Lyu R, Rong B, Zheng Y, Lin Z, Dai R, Zhang X, Xie N, Wang S, Tang F, Lan F, Tong M (2020) Refined spatial temporal epigenomic profiling reveals intrinsic

connection between PRDM9-mediated H3K4me3 and the fate of double-stranded breaks. Cell Res 30(3):256–268. https://doi.org/10.1038/s41422-020-0281-1

5. Halldorsson BV, Palsson G, Stefansson OA, Jonsson H, Hardarson MT, Eggertsson HP, Gunnarsson B, Oddsson A, Halldorsson GH, Zink F, Gudjonsson SA, Frigge ML, Thorleifsson G, Sigurdsson A, Stacey SN, Sulem P, Masson G, Helgason A, Gudbjartsson DF, Thorsteinsdottir U, Stefansson K (2019) Characterizing mutagenic effects of recombination through a sequence-level genetic map. Science 363(6425):u1043. https://doi.org/10.1126/science.aau1043
6. Ioannou D, Fortun J, Tempest H (2018) Meiotic nondisjunction and sperm aneuploidy in humans. Reproduction 157(1):R15–R31. https://doi.org/10.1530/REP-18-0318
7. Kong A, Thorleifsson G, Gudbjartsson DF, Masson G, Sigurdsson A, Jonasdottir A, Walters GB, Jonasdottir A, Gylfason A, Kristinsson KT, Gudjonsson SA, Frigge ML, Helgason A, Thorsteinsdottir U, Stefansson K (2010) Fine-scale recombination rate differences between sexes, populations and individuals. Nature 467(7319):1099–1103. https://doi.org/10.1038/nature09525
8. Lu MM, Wen YX, Liu YL, Ding CH, Zhou CQ, Xu YW (2020) Trophectoderm biopsy reduces the level of serum β-human chorionic gonadotropin in early pregnancy. Fertil Steril 114(4):801–808. https://doi.org/10.1016/j.fertnstert.2020.05.015
9. Lynn A, Koehler KE, Judis L, Chan ER, Cherry JP, Schwartz S, Seftel A, Hunt PA, Hassold TJ (2002) Covariation of synaptonemal complex length and mammalian meiotic exchange rates. Science 296(5576):2222–2225. https://doi.org/10.1126/science.1071220
10. Ma Y, Wang J, Li R, Ding C, Xu Y, Zhou C, Xu Y (2023) Mapping of meiotic recombination in human preimplantation blastocysts. G3-Genes Genom Genet 13(4). https://doi.org/10.1093/g3journal/jkad031
11. Ottolini CS, Newnham L, Capalbo A, Natesan SA, Joshi HA, Cimadomo D, Griffin DK, Sage K, Summers MC, Thornhill AR, Housworth E, Herbert AD, Rienzi L, Ubaldi FM, Handyside AH, Hoffmann ER (2015) Genome-wide maps of recombination and chromosome segregation in human oocytes and embryos show selection for maternal recombination rates. Nat Genet 47(7):727–735. https://doi.org/10.1038/ng.3306
12. Stapley J, Feulner PGD, Johnston SE, Santure AW, Smadja CM (2017) Variation in recombination frequency and distribution across eukaryotes: patterns and processes. Philos Trans R Soc B Biol Sci 372(1736):20160455. https://doi.org/10.1098/rstb.2016.0455
13. Vergara-Lope A, Jabalameli MR, Horscroft C, Ennis S, Collins A, Pengelly RJ (2019) Linkage disequilibrium maps for European and African populations constructed from whole genome sequence data. Sci Data 6(1):208. https://doi.org/10.1038/s41597-019-0227-y
14. Weber JL (2002) The Iceland map. Nat Genet 31(3):225–226. https://doi.org/10.1038/ng920
15. Xiong Y, Chen Q, Chen C, Tan J, Wang Z, Gu F, Xu Y (2021) Impact of oral antibiotic treatment for chronic endometritis on pregnancy outcomes in the following frozen-thawed embryo transfer cycles of infertile women: a cohort study of 640 embryo transfer cycles. Fertil Steril 116(2):413–421. https://doi.org/10.1016/j.fertnstert.2021.03.036

Chapter 6

Generation of Meiotic Mouse Models Using CRISPR/Cas9 Technology

Manuel Sánchez-Martín, Fernando Sánchez-Sáez, Elena Llano, and Alberto M. Pendás

Abstract

In recent years, targeted genome editing has emerged as an indispensable tool for creating animal models, facilitating a comprehensive exploration of the molecular mechanisms governing a myriad of biological processes. Within this scientific landscape, the investigation of meiosis in mice has attracted considerable attention across numerous research laboratories. The precision and versatility of the CRISPR/Cas9 genome editing system have revolutionized our ability to generate mice with tailored genetic alterations, including point mutations and null mutations. These genetic modifications have provided invaluable insights into the intricate functionality of various meiotic genes and their associated variants. In this context, we present a detailed state of the art protocol for the creation of novel mouse models, each bearing specific genetic modifications within key meiotic genes, through the application of CRISPR/Cas9 technology. Furthermore, we showcase two distinct genetic modifications, accomplished within our laboratory, that can serve as valuable reference points for researchers seeking to elucidate the molecular intricacies of meiosis in mammals.

Key words Meiosis, Genome editing, CRISPR/Cas9, Mouse gametogenesis

1 Introduction

The process of generating genome-edited murine models is essential for understanding gene functions in vivo. Until a few years ago, these models had been pursued using traditional strategies relying on homologous recombination in embryonic stem (ES) cells [1]. The time required to generate these models by gene targeting, including the cloning of the targeting vector, electroporation of ES cells, screening for identification and expansion of the targeted clone, generation of chimeric mice, breeding and genotyping of chimeric mice, and finally, the generation of homozygous mice, was

Authors Manuel Sánchez-Martín and Fernando Sánchez-Sáez have equally contributed to this chapter.

Jesús A. Carballo (ed.), *Meiosis: Methods and Protocols*, Methods in Molecular Biology, vol. 2818,
https://doi.org/10.1007/978-1-0716-3906-1_6,

at the very least 1 year (more frequently 2 years). Furthermore, the genetic background of the ES cells used was limited to a few strains. Genome editing approaches experienced a major expansion following the description and development of Clustered Regularly Interspaced Short Palindromic Repeats (CRISPR) and their CRISPR-associated proteins (Cas) nearly a decade ago [2]. The CRISPR/Cas9 system, characterized by RNA-mediated adaptive defense mechanisms, have evolved to safeguard organisms from the intrusion of viruses and plasmids [3]. Specifically, the Cas9-type endonuclease functions as molecular scissors, capable of precisely cleaving DNA at locations determined by a short RNA guide molecule that is complementary to the target DNA sequence. The catalytic activity of the Cas9 nuclease relies on the presence of protospacer adjacent motifs (PAM), consisting of three nucleotides (5′-NGG-3′) located at the 3′ end of the target sequence [4]. Recognition of the PAM sequence results in the formation of a stable R-loop configuration and governs DNA cleavage 3 bp upstream of the PAM [5]. Another crucial component of the CRISPR/Cas9 system is the trans-activator RNA (tracrRNA), which partially hybridizes with the guide crRNA (CRISPR RNA) and the Cas9 endonuclease, facilitating the formation of a ribonucleoprotein complex [2]. The mode of action of genome-editing nucleases relies on the induction of double-strand breaks (DSBs), which activate endogenous cellular mechanisms of DNA repair. One of these mechanisms is non-homologous end-joining (NHEJ), in which a set of enzymes mediate the capture of both ends of the broken DNA molecule, the formation of a bridge that reconnects the two DNA ends, and subsequently, the religation of the broken DNA molecule [6]. This pathway does not necessitate the presence of a homologous DNA molecule and possesses the unique feature that it can introduce various insertions, deletions, and even larger chromosomal rearrangements when both DNA ends are religated [7]. For this reason, this strategy is particularly valuable when aiming to generate a null allele, as it can result in frameshift alterations due to the introduced insertions/deletions. The other major mechanism of DNA repair is termed homology-directed repair (HDR). This pathway involves the utilization of a homologous DNA molecule that serves as a template [8]. Homologous repair can occur by gene conversion, where little or no sequence information is lost during the repair process or through single-strand annealing. CRISPR/Cas9 technology harnesses this mechanism, enabling the introduction of an exogenous donor DNA sequence, such as a single-strand oligodeoxynucleotide (ssODN), at the specific site of the double-strand break (DSB). In this manner, HDR allows for the introduction of point mutations into the genome, the replacement of entire genes, the generation of new variants, or the insertion of large DNA sequences, thereby facilitating the creation of knock-in models.

In 2013, CRISPR/Cas9 system demonstrated its efficacy in the genome editing in mammals and human cells, enabling the introduction of mutations (generating insertions and deletions by the NHEJ pathway) and the insertion of donor DNA (utilizing the HDR mechanism) [9]. Since then, this molecular tool has been extensively employed for generating animal models, facilitating the exploration of a wide range of genes and medical conditions. CRISPR/Cas9 reagents can be delivered directly to single-cell mouse zygotes of any strain [10–12]. Among the various techniques for delivering CRISPR/Cas9 reagents, zygote pronuclear (PN) microinjection has emerged as the most commonly utilized approach [13–15], especially when willing to generate knock-in mice.

Our research group is committed to elucidating the intricate molecular mechanisms that underlie meiosis and infertility. While recent years have witnessed remarkable advancements in the in vitro differentiation of Spermatogonial Stem Cells (SSCs) to emulate meiosis and, more broadly, gametogenesis, it is evident that these approaches still fall short of faithfully recapitulating the entire intricacy of spermatogenesis as observed in mouse models [16, 17]. Consequently, there remains a compelling necessity to employ mutant mouse models for comprehensive in vivo investigations of this complex process. To this end, we employ cutting-edge CRISPR/Cas9 technology and adhere to state-of-the-art transgenic methodologies to shed light on the complete spectrum of spermatogenesis in vivo [18–21]. Through these efforts, we have effectively uncovered the functional mechanisms of multiple meiotic genes and their pathogenic variants. Prominent instances encompass the synaptonemal complex protein SIX6OS1, the deubiquitylating enzyme USP26, the spermatoproteasome subunit PSMA8, the transcriptional regulator FOXL2, the recombinase RAD51B, and the meiotic partners HSF2BP-BRME1 [22–27].

As a reference, we provide a schematic outline for the generation of both a point mutation and a null allele in various genes. Specifically, we successfully created a mouse model carrying a point mutation (c.613C>T) in the central element protein of the synaptonemal complex, SYCE1, which was identified in patients affected by premature ovarian failure (POF) [28]. The cellular, molecular, and biochemical analyses of this model enabled us to confirm that the SYCE1-c.613C>T variant causes infertility due to the disruption of one of its binding domains with the protein SIX6OS1, hindering the proper assembly of the synaptonemal complex and impeding meiotic progression. To develop the edited *Syce1* model (Fig. 1), we targeted its intron 10 with a crRNA (5′-TGACTTCT TTCCACACTATC-3′) that was annealed with the tracrRNA to generate the mature sgRNA. We designed a 200 bp-long single-strand oligodeoxynucleotide (ssODN) for genome editing, encompassing the residue to be mutated. This sequence included changes

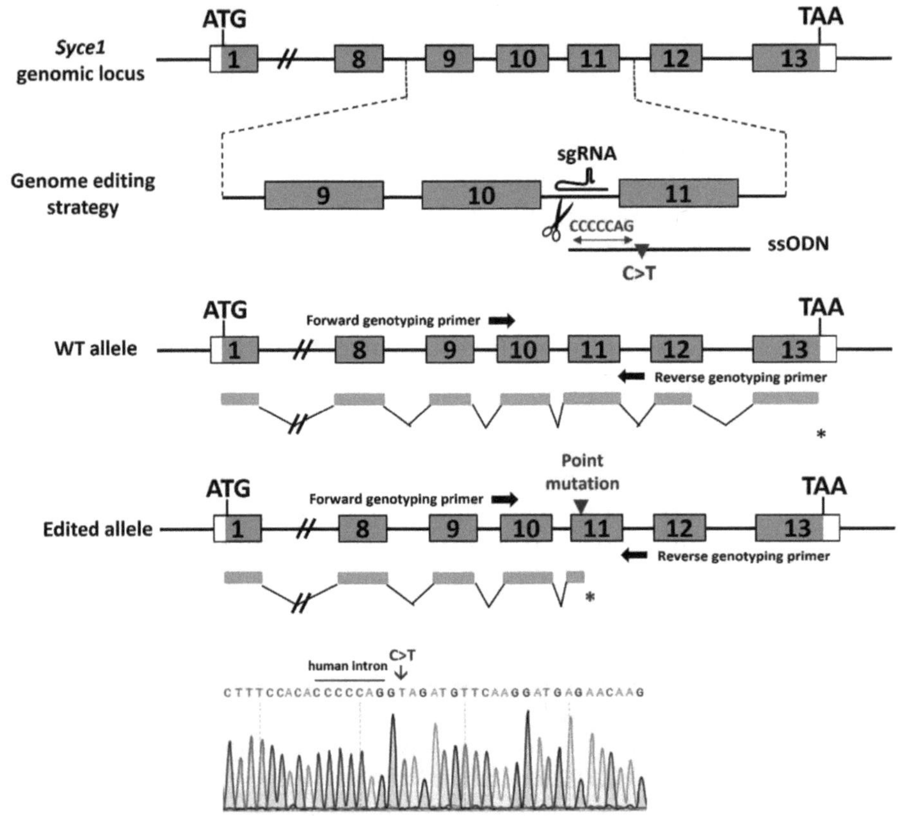

Fig. 1 Generation of *Syce1*-edited mouse model harboring the c.613C>T mutation sequenced in infertile patients. Diagrammatic representation of the murine *Syce1* locus (WT) and the genome editing strategy showing the crRNA located on intron 10 and the ssODN targeting a region between exons 10 and 11 carrying the point mutation. PAM sequence was substituted by the human intron sequence. The corresponding coding exons (blue) and non-coding exons (open boxes) are represented. Thin (non-coding) and green thick (coding sequences) lines under exons represent the expected transcript derived from WT and *Syce1*-edited allele. ATG, initiation codon; TAA and *, stop codon. Primers for PCR genotyping (forward and reverse) are represented by arrows. The nucleotide sequence of the edited region derived from PCR amplification of DNA from *Syce1* homozygous mutant mice is indicated

in the PAM residues, preventing repeated cleavage by the Cas9 enzyme once the template DNA had been integrated into the genome through HDR. Microinjection of a mixture containing the mature sgRNA, the ssODN, and recombinant Cas9 protein into F2 zygotes (see Materials and Methods) enabled us to obtain a founder, which was subsequently bred to establish the edited murine line.

Likewise, within our laboratory, we have recently contributed to elucidate the role of oocyte-specific histone H1FOO in fertility and pluripotency [29]. This is particularly significant, given that several in vitro studies have underscored the indispensable role of mammalian H1FOO in regulating the extrusion of the first polar

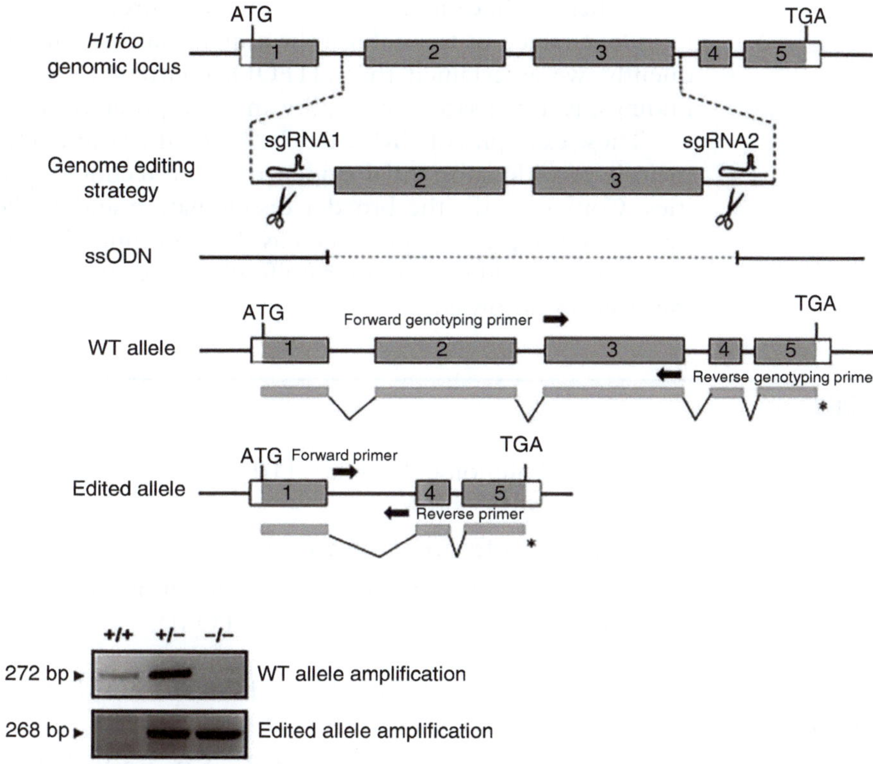

Fig. 2 Generation of *H1foo* KO mice by CRISPR–Cas9 genome editing. Two sgRNAs flanking coding exons 2 and 3, respectively, of the *H1foo* genomic locus are represented. Coding (blue) and non-coding (open box) regions are indicated. Thick lines under the exons represent the expected transcripts derived from WT and *H1foo*-edited alleles and the "*" symbol points translation termination. ATG, start codon; TGA, stop codon. Primers required for the amplification of both WT and edited alleles are also represented by arrows. Under the diagram, PCR analysis of genomic DNA from *H1foo*$^{+/+}$, *H1foo*$^{+/-}$, and *H1foo*$^{-/-}$ pups is shown

body and oocyte maturation [30, 31]. As a result, we embarked on creating the first in vivo murine model lacking H1FOO, with the aim of providing a comprehensive understanding of its involvement in mammalian gametogenesis. Murine H1FOO is encoded by five exons, with its ATG codon located in the initial exon and the STOP codon in the fifth exon. To generate a null mutant, we deleted exons 2 and 3 of *H1foo*, covering over 70% of the coding sequence, by designing two distinct crRNAs (Fig. 2). The first crRNA targeted the 5′ region of exon 2 (5′-GCAGGCCTTGACCGGGG CAG-3′), while the other crRNA targeted the 3′ region after exon 3 (5′-AAGGCTGAAGAGCTGCCAGG-3′). We also included a 200 bp single-strand oligodeoxynucleotide (ssODN) surrounding exons 2 and 3 in the microinjection mixture. This ssODN facilitated the repair of the breakthrough HDR, allowing controlled rejoining of the flanking ends. PAM sequences associated with both guides were altered following cleavage, obviating the need for further modifications to their sequences. Breeding of a selected founder enabled us to establish the mutant murine line, and subsequently, we ascertained that H1FOO, contrary to prior in vitro findings, is dispensable for fertility and pluripotency in mammals.

These examples of CRISPR/Cas9 tool utilization in the field of meiosis provide substantial evidence of the technology's capabilities. Consequently, the broader dissemination and application of genome editing techniques, such as those outlined in this chapter (*see* **Note 1**), hold paramount significance in advancing and enriching scientific knowledge.

2 Materials

2.1 Design of the Editing Strategy

1. Ensembl genome browser 110 (https://www.ensembl.org/index.html).
2. Snapgene software (Dotmatics).
3. IDT gRNA design tool (https://eu.idtdna.com/site/order/designtool/index/CRISPR_CUSTOM).

2.2 Preparation of the Microinjection Solution

1. IDT buffer (1 mM Tris–HCl, pH 7.5; 0.1 mM EDTA prepared in nuclease-free water). Store at 4 °C.
2. Recombinant, high-purity Alt-R® S.p. Cas9 nuclease (IDT, stock solution 10 μg/μL, working solution 300 ng/μL, resuspended in IDT buffer). Store at −80 °C.
3. Alt-R® CRISPR tracrRNA (stock solution 2.5 μg/μL, working solution 250 ng/μL, resuspended in IDT buffer). Store at −80 °C.
4. Alt-R® CRISPR-Cas9 crRNAs (working solution 250 ng/μL, resuspended in IDT buffer). Store at −80 °C.
5. ssODN (Ultramer™ DNA oligo 200 pb, working solution 200 ng/μL, resuspended in IDT buffer) if necessary. Longer ssODNs may be designed if the editing strategy requires it. Store at −20 °C.
6. Block heater (VWR).
7. Glass beaker.
8. Thermometer.

2.3 Microinjection Platform

To visualize and introduce the CRISPR-Cas9 reagents into the pronuclear (PN) cell compartment, a microinjection station is indispensable (Fig. 3). The station comprises an inverted microscope equipped with phase contrast (such as Olympus YX70, YX71 or a similar model) with appropriate magnification, two manipulator arms (e.g., Narishige, Eppendorf Transferman NK-2, or a similar system) equipped with both a holding pipette and a

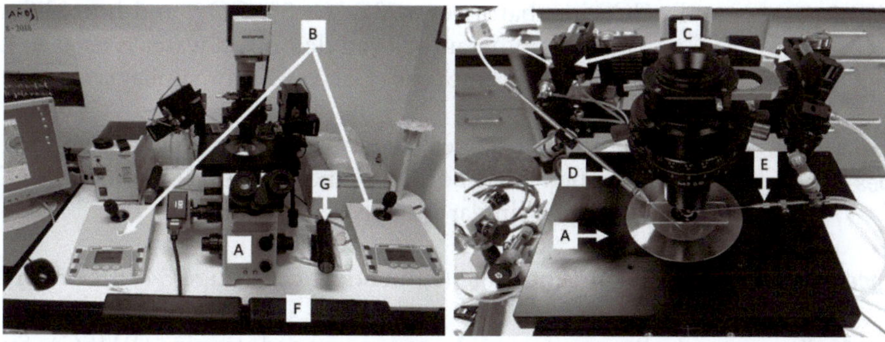

Fig. 3 Microinjection station. Figure legend: Inverted microscope (A) with micromanipulators (B) connected to both arms (C) with holding (D) and microinjection (E) pipette systems on an anti-vibration air-table (F). CellTram device (G)

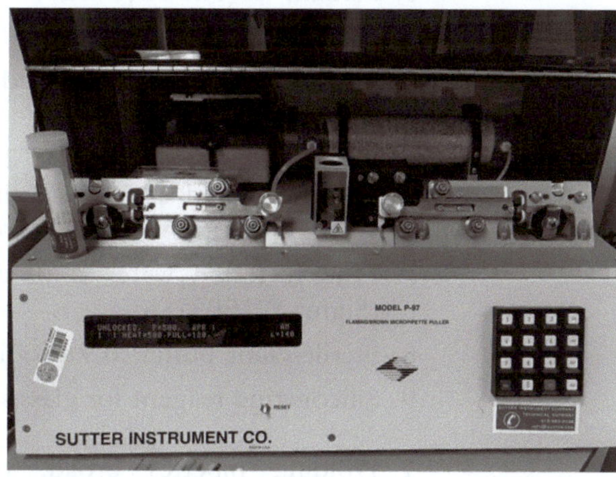

Fig. 4 Needle puller

microinjection pipette system. The holding pipette is connected through either oil or air-filled tubing to a device used for gripping and releasing the zygotes by adjusting pressure to create either under- or overpressure conditions, respectively (e.g., Eppendorf CellTram Air or a similar system). The injection capillary is ultimately connected to an electronic device (e.g., Eppendorf Femtoject or a similar device) or a mechanical device (a 50 mL glass syringe) designed to inject the solution contained within the capillary into the pronucleus (PN) of the zygote. To prevent vibrations from the surrounding environment that might disrupt the microinjection process, a vibration isolation table is also essential (e.g., Newport antivibration table or a similar model).

2.4 Needle Puller

1. Needle puller (Fig. 4) with FB255B 2.5 mm square box filament (Sutter Instruments Model P-97).

2.5 CO_2 Incubator

1. CO_2 incubator set at 37 °C, 5% CO_2, and humidified with a pan containing autoclaved deionized water is used to keep and grow mouse embryos (Galaxi 48R or similar).

2.6 Stereo-microscopes

1. Stereomicroscopes for embryo collection and embryo transfer equipped with both top and bottom light sources (Olympus SZX12 or similar models).

2.7 Microcentrifuge

1. Microcentrifuge (Eppendorf centrifuge 5415R or similar).

2.8 Drugs, Media, and Chemicals

1. Pregnant mare's serum gonadotropin (PMSG) Sincropart PMSG, CEVA, 6000 I.E.
2. Human chorion gonadotropin (hCG) (CG5-1VL hCG, Sigma) (*see* **Note 2**).
3. Paraffin oil (NidOil, Nidacon).
4. M2 medium (Sigma M7167-100ML).
5. KSOM embryo culture medium with essential AA (Merk, MR-106-D).
6. Hyaluronidase (Sigma H4272) prepared as a 20 mg/mL stock solution stored at −20 °C in aliquots of 50 μL per tube. Working solution at 300 μg/mL in M2 medium.
7. Embryo-tested water is used to prepare the M2 medium (Sigma W1503).
8. Sterile saline buffer (0.9% sodium chloride).
9. Siliconizing reagent for glass slides (Sigma, SL2-25ML).

2.9 Lab Consumables

1. Holding pipettes broad, bent, pulled, and broken at 125–130 μm. Opening diameter: 20–25 μm. Bent angle ~35°, bent length ~600 μm (BioMedical Instruments).
2. Capillaries for embryo handling and transfer, retransfer capillaries, Blaubrand, 2 mm, fire polished, ID = 108–114 μm (BioMedical Instruments).
3. Thin wall glass borosilicate capillaries with inner filament (WPI, ref. TW100F-4) are used to generate microinjection pipettes.
4. 3.5-cm petri dishes (Falcon, ref. 351008).
5. Microloader, Eppendorf No. 5242 956.003.
6. Mouth aspirator tube assembly (Sigma A5177 5EA).
7. Excavated slides (75 mm x 25 mm slide blank with round excavation in the center).
8. Syringes ICO plus3, 1 mL, W/O needle ref. N15663.
9. 30G sterican needles (Braun, ref. 46,563,000).
10. Sterile disposable pipette tips: 10 μL, 200 μL, and 1000 μL.

11. Sterile plastic centrifuge tubes for aliquots of media and solutions: 15 mL, 50 mL.
12. Plastic tubing for transfer pipette assembly.
13. Standard pipettes: P20, P200, and P1000.

2.10 Surgical Instruments and Supplies for Embryo Transfer

1. Isoflurane and a tank of O_2 for anesthesia.
2. O_2 regulator such as the E-Z 7000 Classic System (manufactured by E-Z Systems).
3. Surgical gloves.
4. Sterile, mounted absorbent swabs (Sugi eyespear pointed tip ref.30901).
5. Mouth-controlled pipette.
6. Two pairs of tweezers Dumont #55 Biology (ref.14099).
7. One curved tissue forceps (WPI ref.501216).
8. Straight-bladed Mayo scissors for opening the skin.
9. Spring (microdissection) scissors with sharp points for opening the body wall.
10. Reflexclip applier 7 mm (WPI ref.500343).
11. Wound reflexclips, 7 mm (WPI ref.500344).
12. Bulldog Serrefine (Clamp) Straight 28 mm.
13. Ear punch.
14. Stainless steel instrument sterilization trays for storage of sterilized surgery tools.
15. Heating stage and warming plate or pad (*see* **Note 3**).

2.11 Miscellaneous Items

1. Combi 4 °C refrigerator and −20 °C freezer.
2. Ultra-freezer (−80 °C).
3. Precision balances for weighing.
4. Water bath.
5. Magnetic stir plate.
6. Bunsen burner or microburner.
7. Diamond pencil and rubber bulbs.
8. Glassware for media preparation, i.e., beakers and cylinders.
9. Magnetic stir bars.
10. Spoons and weighing papers (boats) for media preparation.
11. pH strips.
12. Sterile filter units with 0.22 μm filters and 150 mL, 500 mL, and 1000 mL volume capacities.
13. Sterile disposable plastic transfer pipettes.

14. Ice and ice bucket.
15. Paper towels.
16. Sharps disposal containers.

2.12 Mice

1. Donor female mice: 20–40 juvenile females, 3–5 weeks old, from inbred or hybrid strains.
2. Stud males: 20–30 stud males of the same strain as the donors, 2–10 months of age.
3. Vasectomized males: 10–20 vasectomized outbred males such as ICR (CD-1) or Swiss Webster are used to induce pseudopregnancy.
4. Recipients (pseudo-pregnant females): 50–100 outbred females from strains such as ICR (CD-1) or Swiss Webster, older than 6 weeks and weighing 30–35 g.

2.13 Genotyping of Genetically Modified Mice

1. Milli-Q® water.
2. Scissors and ear punch.
3. 50 mM NaOH.
4. Block heater (VWR).
5. 1 M Tris–HCl, pH 8.
6. Vortex.
7. PCR oligonucleotides (IDT).
8. 10X Expand Long Template Buffer 2.
9. dNTPs NZYSet (NZYTech).
10. NZYTaq II DNA polymerase (NZYTech).
11. Veriti 96-well thermal cycler (Applied Biosystems).
12. Agarose (NZYTech).
13. Weighing machine (GRAM, AHZ Series).
14. Erlenmeyer flask.
15. 1X TAE buffer.
16. Ethidium bromide (Merck Millipore).
17. Electrophoresis gel cassette and 20-well combs.
18. Wide Mini-Sub Cell GT system (Bio-Rad).
19. PowerPac Basic electrophoresis power supply (Bio-Rad).
20. pBR322 (Takara), HaeIII enzyme (Takara), 10X M buffer (Takara) and a 37 °C incubator for marker V preparation.
21. 10X loading buffer (62.5 mL of 80% glycerol, 20 mL of 50X TAE, 17.5 mL of water, and bromophenol blue until a consistent blue coloration is reached).
22. Gel Doc 2000 gel transilluminator (Bio-Rad).

2.14 Animal Welfare

According to European Union regulations, mice were housed in pathogen-free and temperature-controlled facilities using individually ventilated cages, standard diet, and a 12-hour light-dark cycle with lights on at 8 a.m. and off at 8 p.m., at the Animal Experimentation Service (University of Salamanca). Mice protocols were approved by the Ethics Committee for Animal Experimentation of the University of Salamanca, and we made every effort to minimize suffering and to improve animal welfare.

3 Methods

3.1 Design of the Editing Strategy

Effective planning of a genomic editing design significantly enhances the efficiency of the process. To achieve this, the genomic nucleotide sequence of the target gene should be retrieved from the Ensemble Genome Browser (https://www.ensembl.org/index.html). Subsequently, open the .dna file in Snapgene software and proceed to select and identify the specific exon or region to be targeted. Copy the flanking sequence, up to 1000 base pairs, in FASTA format and input it into the IDT gRNA design tool (https://eu.idtdna.com/site/order/designtool/index/CRISPR_CUSTOM), specifying the species of interest. Among the various crRNAs provided by the software, prioritize those with higher on-target and off-target scores as they are the most suitable options. Select as many crRNAs as required by your editing strategy. In the event that your genome editing strategy necessitates the use of a single-strand oligodeoxynucleotide (ssODN), design it using the Snapgene software. Pay particular attention to modifying the Protospacer Adjacent Motif (PAM) sequence if it is not naturally altered due to Cas9 cleavage. This modification is essential to prevent new cleavage events initiated by the endonuclease when integrating the exogenous DNA into the genome.

3.2 Preparation of the Microinjection Solution

1. Upon receipt, reconstitute tracrRNA, crRNA, and ssODNs in IDT buffer at both stock and working dilutions (refer to Materials, Subheading 2 and **Note 4**), and store them in an ice bucket.
2. In an individual Eppendorf tube for each crRNA, combine 5 μL of 250 ng/μL crRNA with 10 μL of 250 ng/μL tracrRNA to facilitate annealing and produce mature sgRNAs.
3. Incubate the mixtures at 95 °C for 5 min using a block heater. In the meantime, bring a glass beaker filled with boiling water.
4. Swiftly transfer the tube(s) from the block heater to the boiling water in the glass beaker and allow them to gradually equilibrate to room temperature (this process will take up to 2 h).

5. For each sgRNA, take 1 μL of the annealed sgRNAs and combine it with 1.2 μL of 300 ng/μL Cas9 and 9.2 μL of IDT buffer. Prepare separate mixtures for each sgRNA; do not combine them at this stage.
6. Incubate the tubes at room temperature for 15 min to facilitate the formation of ribonucleic complexes.
7. In case you are using more than one sgRNA, consolidate the mixtures into a single tube and supplement it with the ssODN (200 ng/μL), resulting in a final concentration of 10 ng/μL (0.6 μL for one sgRNA, 1.2 μL for two sgRNAs). The final concentrations for Cas9 and sgRNAs at this point will be 30 ng/μL and 20 ng/μL, respectively.
8. Utilize this mixture directly for pronuclear microinjection (refer to Methods, Subheading 3.7).

3.3 Mice Superovulation

Superovulation is a reproductive technique commonly employed to maximize the release of oocytes while minimizing the use of animals. The oocyte yield can vary depending on factors such as the strain and age of donor females, hormone dosage, and light cycle. Typically, C57BL6J or B6J/CBA mice are used prior to their first natural estrus, at an age of 4–5 weeks. All hormone injections are administered intraperitoneally according to the following schedule:

1. Day 1: Inject 5 I.U. of PMSG in 0.1 mL sterile saline buffer intraperitoneally between 12 p.m. and 2 p.m.
2. Day 3: After 46–48 h, administer 5 I.U. of hCG in 0.1 mL saline buffer using the same intraperitoneal route. Subsequently, mate the females with stud males (*see* **Note 5**).

3.4 Harvesting of Zygotes

At day 4, zygotes are harvested at 10 a.m. as follows:

1. Set up media dishes in the morning at least 1 h before embryo collection. Prepare dishes of KSOM culture medium under a laminar hood. Plate 3 mL into a 35 mm Petri dish. Plate 8 drops (30 μL) into a 35 mm Petri and cover the drops with paraffin oil (Fig. 5). Keep plates in a 5% CO_2 incubator at 37 °C to equilibrate medium.
2. Euthanize the donor females by cervical dislocation (maximum of six mice at a time) and open their abdomens with sharp scissors. Remove oviducts and put all of them into a 35 mm Petri dish containing 3 mL M2.
3. Transfer groups of two oviducts into single drops of 50 μL M2, and under a stereomicroscope, open ampullas with fine forceps (Fig. 6).

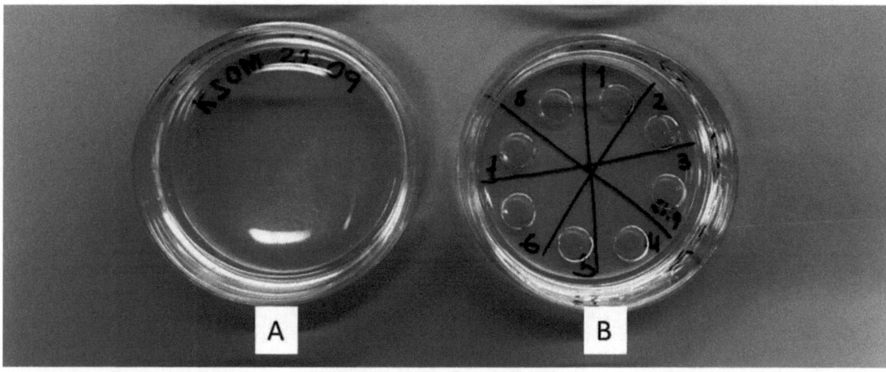

Fig. 5 KSOM embryo culture Petri dishes. (**a**) 3 mL of KSOM in a 35 mm Petri dish. (**b**) KSOM drops covered with paraffin oil

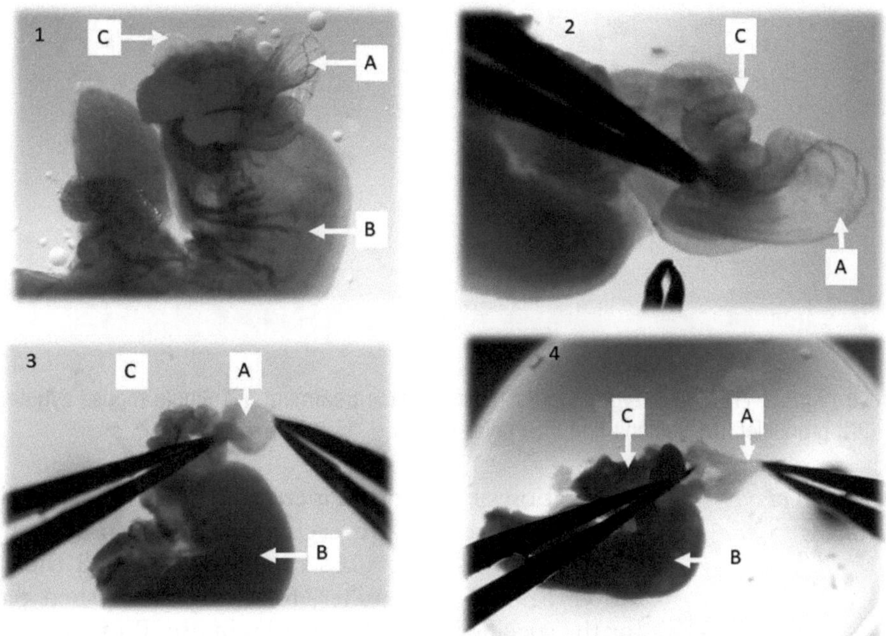

Fig. 6 Ampullas and cumulus collection process (1–4). (A) Ampullas, (B) uterus, and (C) oviduct

4. Add one thawed aliquot of hyaluronidase to each drop to disperse cumulus (Fig. 7) and keep the drops at RT 5 min. Pick up each single zygote with a mouth-controlled pipette (see glass tools below) and wash all of them in a new Petri dish containing 3 mL M2 (Fig. 8).
5. Transfer and wash embryos into a 35 mm Petri dish containing 3 mL equilibrated KSOM. Pick up them and transfer into 30 μL drops of KSOM under mineral oil (30 embryos/drop) and keep the Petri dish into a cell culture incubator.

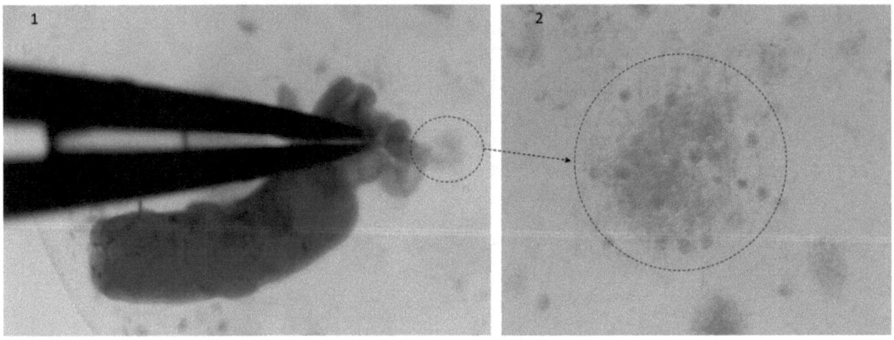

Fig. 7 Cumulus. Cumulus at different magnification (1–2) can be observed surrounded by discontinued line

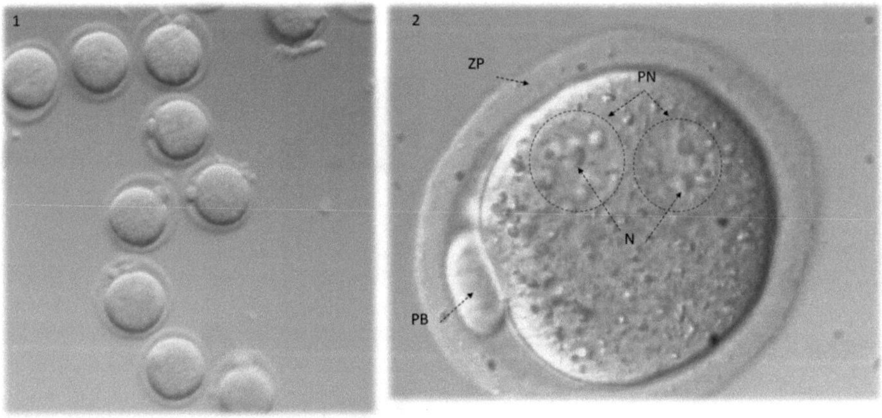

Fig. 8 Mouse zygotes. (1) Mouse zygotes after hyaluronidase treatment. (2) Single mouse zygote. (ZP) zone pellucida, (PN) pronuclei, (N) nucleoli and (PB) polar body

3.5 Preparation of Glass Microtools

1. Preparation of Injection Pipettes: Injection pipettes should be freshly prepared before each microinjection session. Utilize thin-wall borosilicate glass capillaries and pull them using a Sutter P-97 pipette puller, following established parameters for the specific batch of capillaries in use. Presently, the recommended parameters are as follows: Heat 520, Pull 90, Velocity 140, Time 100 (*see* **Note 6**).
2. Preparation of Mouth-Controlled Pipettes for Embryo Handling and Transfer: These pipettes can be easily fabricated from drawn-out Pasteur pipettes (see Fig. 9). Hold both ends of the pipette with your fingers and, using a flame, gently heat the shoulder area of the pipette just above the tip until the glass becomes soft. Carefully pull the glass in opposite directions and break it at the thinnest point using a diamond pencil. Depending on the internal diameter obtained (150 μM is the optimal size), the larger piece can be utilized as a handling/embryo-transfer pipette (*see* **Note 7**).

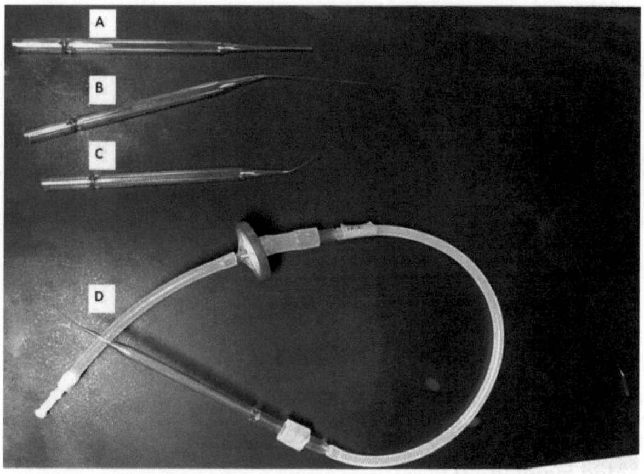

Fig. 9 Pipettes for embryo handling. Glass Pasteur pipettes: (**a**) normal, (**b**) pulled, (**c**) broken, and (**d**) complete mouth-controlled pipette

3. Assembly of the Handling Pipette: Attach the handling pipette to one end of a piece of plastic tubing, while affixing a mouthpiece to the other end and inserting a syringe filter somewhere along the tubing (see Fig. 9).

3.6 Preparation of the Injection Chamber and Injection Setup

1. A thin film of silicone reagent is applied to the bottom side of the excavated glass slide to make the injection chamber. The excavated area is filled with 30 μl M2 medium that is then overlaid with mineral oil. The injection chamber is mounted on the microinjection station.
2. The holding capillary is assembled and immersed into M2 drop. Aspirate M2 medium from the drop (up to 1 cm).
3. Charge the injection capillaries with 2 μL of the CRISPR solution using micro loader pipettes and keep it 5 min at RT in a vertical position until the CRISPR solution reaches the tip. After that, assemble the injection capillary on the microinjection pipette system.

3.7 Pronuclear Microinjection

1. Using mouth-pipetting, transfer the zygotes from a KSOM drop into the prepared injection chamber. Typically, 30–40 zygotes are transferred simultaneously, but the number may vary based on the operator's expertise. Organize the embryos into three distinct regions within the injection chamber: (1) above for discarded zygotes, (2) at the center for uninjected zygotes, and (3) below for injected zygotes (see Fig. 10).
2. Fix one zygote in place by gently rotating it with the holding capillary until two separate pronuclei are visible (refer to Fig. 8). If only one pronucleus is observed, or if there are three pronuclei, discard the zygote (position 1). Choose the optimal position where the larger pronucleus is closest to the injection needle, and focus on it.

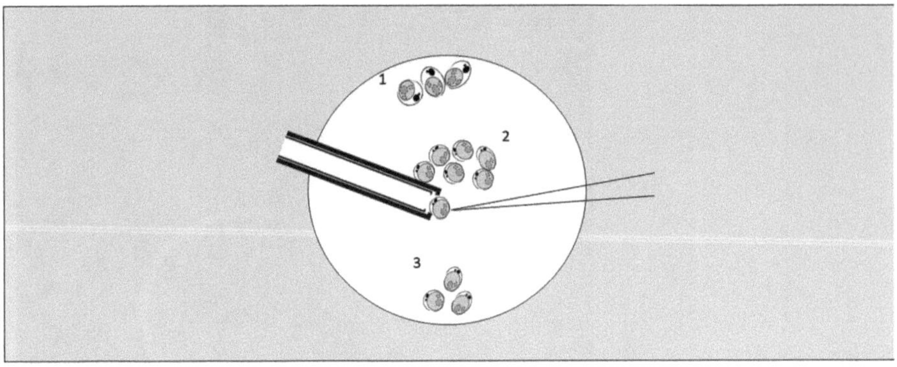

Fig. 10 Disposition of mouse zygotes inside the injection chamber. Discarded zygotes (1), uninjected zygotes (2), and injected zygotes (3)

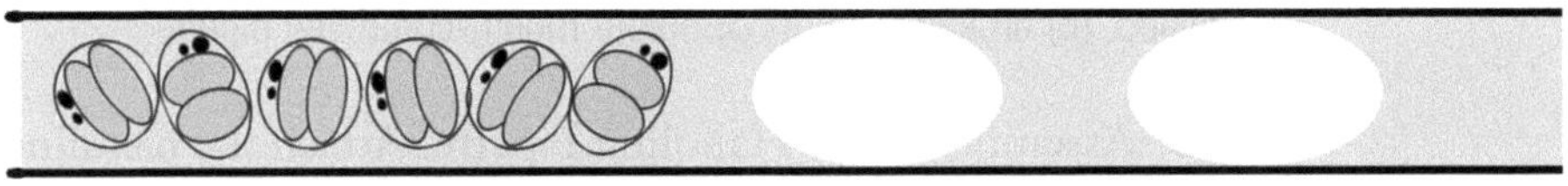

Fig. 11 Embryo transfer pipette with two embryos inside

3. Carefully insert the needle through the zona pellucida, the cell wall, and the nuclear membrane into one of the pronuclei. Then, apply gentle pressure until a noticeable swelling of the pronucleus becomes evident. Following a successful injection, place the zygote at location 3 and repeat the process with uninjected oocytes situated at position 2.
4. Collect all injected zygotes and transfer them into a 35 mm KSOM Petri dish. Subsequently, place them into a 30 μL KSOM drop under mineral oil and return them to the incubator. Incubate the zygotes overnight to assess the proportion that develops into the two-cell stage. Concurrently, culture uninjected embryos under identical conditions for control purposes.

3.8 Embryo Transfer

At day 5, unilateral embryo transfer is used to place two-cell embryos into the reproductive tracts of foster females. Two-cell embryos are transferred to recipient females as follows (*see* **Note 8**):

1. Females showing a vaginal plug are identified and separated early on the morning (at 8–9 a.m.).
2. The recipient female is anesthetized by isoflurane/O_2 inhalation and placed on a warming blanket under the stereoscope.
3. Wash 25 two-cell embryos in a dish of M2 medium and load into the embryo transfer pipette (see Fig. 11) as follows: 2–3 mm of medium followed by a 2-mm air bubble, 2–3 mm M2 medium followed by a second 2-mm air bubble,

and finally 25 two-cell embryos in as little M2 medium as possible.

4. Wash the skin with 70% ethanol, and with a pair of dissecting scissors, make a single incision (~1.0 cm) through the skin of the left flank above the kidney zone and slightly lateral to the spine. Dissect the musculature exposed on the peritoneum and open the peritoneal cavity. Lift the body wall and pull out the ovary and oviduct with a pair of forceps, and keep the ovary and oviduct outside the body with the bulldog serrefine clamp attached to the fat pad.
5. Using the stereomicroscope, localize the infundibulum and tear the bursa with the pairs of tweezers Dumont #55. Remove the blooding liquid with an absorbent swab.
6. Take the embryo transfer pipette with two-cell embryos with the right hand, and with the left hand hold the infundibulum with a tweezers Dumont #55. Place the tip of the pipette into the infundibulum and blowing the with mouthpiece expel gently the embryos until air bubbles can be seen in the oviduct.
7. Detach the bulldog serrefine clamp from the fat pad and introduce the internal tissues carefully back into the peritoneal cavity. Suture the musculature hole and close the skin with two sterile stainless steel wound clips. Apply any analgesic (i.e., buprenorphine 0.05–0.1 mg/kg or similar) for pain relief.
8. Repeat **steps 2–7** until all injected embryos have been transferred into recipients.
9. Place the foster female in a clean cage and keep the cage on a warming plate until she has recovered from the anesthesia.

3.9 Genotyping of Genetically Modified Mice

Mice genotyping is accomplished through PCR amplification. To extract genomic DNA from genetically modified mice, perform small tail biopsies. Place individual biopsies in separate Eppendorf tubes and add 600 μL of 50 mM NaOH. Incubate for 30 min at 95 °C using a block heater. Subsequently, vortex the tubes to break up the tails and neutralize the solution by adding 50 μL of Tris–HCl (pH 8.8). Use 1.5 μL of this DNA as the template for PCR amplification.

Design 20–22 bp oligonucleotides flanking the genomic region of interest. Prepare a PCR master mix, including 0.4 μM of each oligonucleotide, 200 μM dNTPs, 1X polymerase buffer, and 0.75 U of NZYTaq DNA II Polymerase in a final volume of 25 μL. Perform the following PCR cycle using a Veriti 96-well thermal cycler:

Initial denaturation for 2 min at 94 °C.

35 cycles consisting of:

(a) 20 seconds denaturation at 94 °C
(b) 20 seconds annealing at 55–68 °C (based on the melting temperature of the oligonucleotides)
(c) Elongation at 72 °C for 1 min per kb to amplify

Following PCR amplification, analyze the PCR product using standard 2% agarose gel electrophoresis with ethidium bromide as the gel stain. For molecular weight determination, we recommend preparing Marker V as a molecular ladder. To do this, digest 120 μL of pBR322 (1 μg/μL) with 15 μL of HaeIII enzyme and M buffer in a final volume of 200 μL overnight. Run 5 μL of the digested Marker V alongside the PCR products (mixed with 10X loading buffer to ensure proper migration) to verify their molecular weight. Visualize the gel using a Gel Doc 2000 gel transilluminator.

4 Notes

1. The equipment described for this procedure is those we have found to work best in our laboratory. Several supplies are constantly changing, and new and better ones are being released frequently.
2. Pregnant mare's serum gonadotropin (PMSG) and human chorionic gonadotropin (hCG) are supplied as lyophilized powders. Dissolve each in sterile saline (0.9% NaCl) to prepare working solutions of 50 IU/mL. Make 1 mL aliquots and store at −80 °C. Final dose: 5 IU/female.
3. A heating stage is used for keeping the embryo handling medium warm, while the warming plate or pad is used to prevent heat loss.
4. When preparing the microinjection solution, working in RNase-free conditions is extremely suggested: use disposable material and filter pipette tips.
5. Alternatively, to further increase the yield of oocytes, a new ultra-superovulation technique has also been developed using inhibin antiserum (IAS) and eCG (IASe) to promote follicle growth [32].
6. Run a Ramp Test for each new batch of glass capillaries on the Sutter P-97 following the manufacturer's instruction.
7. These pipettes can be either prepared as described above or purchased from commercial companies.
8. Foster mice are previously mated to vasectomized on the day before (day 4) the actual microinjection. CD-1 outbred strain is used in our laboratory, but other outbred mice are widely used.

Acknowledgments

A.M.-P.'s laboratory is funded by MINECO, grant number PID2020-120326RB-I00, and by *Junta de Castilla y León*, grant numbers CSI239P18 and CSI148P20. F.S.-S. received funding from *Junta de Castilla y León* fellowship (EDU/556/2019). CIC-IBMCC is supported by the *Programa de Apoyo a Planes Estratégicos de Investigación de Estructuras de Investigación de Excelencia* cofunded by the *Castilla y León* autonomous government and the European Regional Development Fund (CLC–2017–01).

References

1. Tröder SE, Zevnik B (2022) History of genome editing: from meganucleases to CRISPR. Lab Anim 56:60–68. https://doi.org/10.1177/0023677221994613
2. Jinek M, Chylinski K, Fonfara I, Hauer M, Doudna JA, Charpentier E (2012) A programmable dual-RNA-guided DNA endonuclease in adaptive bacterial immunity. Science 337: 816–821. https://doi.org/10.1126/SCIENCE.1225829
3. Fineran PC, Charpentier E (2012) Memory of viral infections by CRISPR-Cas adaptive immune systems: acquisition of new information. Virology 434(2):202–209. https://doi.org/10.1016/j.virol.2012.10.003
4. Gleditzsch D, Pausch P, Müller-Esparza H, Özcan A, Guo X, Bange G, Randau L (2019) PAM identification by CRISPR-Cas effector complexes: diversified mechanisms and structures. RNA Biol 16:504–517. https://doi.org/10.1080/15476286.2018.1504546
5. Sternberg SH, Lafrance B, Kaplan M, Doudna JA (2015) Conformational control of DNA target cleavage by CRISPR–Cas9. Nature 527:110–113. https://doi.org/10.1038/NATURE15544
6. Weterings E, Chen DJ (2008) The endless tale of non-homologous end-joining. Cell Res 18(1):114–124. https://doi.org/10.1038/cr.2008.3
7. Heidenreich E, Novotny R, Kneidinger B, Holzmann V, Wintersberger U (2003) Non-homologous end joining as an important mutagenic process in cell cycle-arrested cells. EMBO J 22:2274–2283. https://doi.org/10.1093/EMBOJ/CDG203
8. Liang F, Han M, Romanienko PJ, Jasin M (1998) Homology-directed repair is a major double-strand break repair pathway in mammalian cells. Proc Natl Acad Sci USA 95: 5172–5177. https://doi.org/10.1073/PNAS.95.9.5172/ASSET/9A6AA555-9F43-4132-9332-620C2298F906/ASSETS/GRAPHIC/PQ0884324007.JPEG
9. Cong L, Ran FA, Cox D, Lin S, Barretto R, Habib N, Hsu PD, Wu X, Jiang W, Marraffini LA, Zhang F (2013) Multiplex genome engineering using CRISPR/Cas systems. Science 339:819–823. https://doi.org/10.1126/SCIENCE.1231143
10. Shen B, Zhang J, Wu H, Wang J, Ma K, Li Z, Zhang X, Zhang P, Huang X (2013) Generation of gene-modified mice via Cas9/RNA-mediated gene targeting. Cell Res 23:720–723. https://doi.org/10.1038/CR.2013.46
11. Wang H, Yang H, Shivalila CS, Dawlaty MM, Cheng AW, Zhang F, Jaenisch R (2013) One-step generation of mice carrying mutations in multiple genes by CRISPR/Cas-mediated genome engineering. Cell 153:910–918. https://doi.org/10.1016/J.CELL.2013.04.025
12. Yang H, Wang H, Shivalila CS, Cheng AW, Shi L, Jaenisch R (2013) One-step generation of mice carrying reporter and conditional alleles by CRISPR/Cas-mediated genome engineering. Cell 154:1370–1379. https://doi.org/10.1016/J.CELL.2013.08.022
13. Gordon JW, Scangos GA, Plotkin DJ, Barbosa JA, Ruddle FH (1980) Genetic transformation of mouse embryos by microinjection of purified DNA. Proc Natl Acad Sci USA 77:7380–7384. https://doi.org/10.1073/PNAS.77.12.7380
14. Brinster RL, Chen HY, Trumbauer M, Senear AW, Warren R, Palmiter RD (1981) Somatic expression of herpes thymidine kinase in mice following injection of a fusion gene into eggs. Cell 27:223–231. https://doi.org/10.1016/0092-8674(81)90376-7
15. Costantini F, Lacy E (1981) Introduction of a rabbit beta-globin gene into the mouse germ line. Nature 294:92–94. https://doi.org/10.1038/294092A0

16. Yoshimatsu S, Kisu I, Qian E, Noce T (2022) A new horizon in reproductive research with pluripotent stem cells: successful in vitro gametogenesis in rodents, its application to large animals, and future in vitro reconstitution of reproductive organs such as "uteroid" and "oviductoid". Biology (Basel) 11:987. https://doi.org/10.3390/BIOLOGY11070987/S1
17. Sato T, Katagiri K, Kubota Y, Ogawa T (2013) In vitro sperm production from mouse spermatogonial stem cell lines using an organ culture method. Nat Protoc 8:2098–2104. https://doi.org/10.1038/NPROT.2013.138
18. Nagy A, Gertsenstein M, Vintersten K, Behringer R (2003) Manipulating the mouse embryo: a laboratory manual. Cold Spring Harbor Laboratory Press, Cold Spring Harbor
19. Liu C, Du Y (2019) Microinjection: methods and protocols. Springer methods in molecular biology. ISSN 1064-3745. Springer
20. Pease S, Saunders TL (2011) Advanced protocols for animal transgenesis. Springer, Heidelberg. https://doi.org/10.1007/978-3-642-20792-1
21. Saunders TL (2023) Transgenesis: methods and protocols. Springer methods in molecular biology. ISSN 1064-3745. Springer
22. Gómez-H L, Felipe-Medina N, Sánchez-Martín M, Davies OR, Ramos I, García-Tuñón I, De Rooij DG, Dereli I, Tóth A, Barbero JL, Benavente R, Llano E, Pendas AM (2016) C14ORF39/SIX6OS1 is a constituent of the synaptonemal complex and is essential for mouse fertility. Nat Commun 7(1):1–16. https://doi.org/10.1038/ncomms13298
23. Llano E, Todeschini AL, Felipe-Medina N, Corte-Torres MD, Condezo YB, Sanchez-Martin M, López-Tamargo S, Astudillo A, Puente XS, Pendas AM, Veitia RA (2023) The oncogenic FOXL2 C134W mutation is a key driver of granulosa cell tumors. Cancer Res 83: 239–250. https://doi.org/10.1158/0008-5472.CAN-22-1880
24. Franca MM, Condezo YB, Elzaiat M, Felipe-Medina N, Sánchez-Sáez F, Muñoz S, Sainz-Urruela R, Martín-Hervás MR, García-Valiente R, Sánchez-Martín MA, Astudillo A, Mendez J, Llano E, Veitia RA, Mendonca BB, Pendás AM (2022) A truncating variant of RAD51B associated with primary ovarian insufficiency provides insights into its meiotic and somatic functions. Cell Death Differ 29: 2347–2361. https://doi.org/10.1038/S41418-022-01021-Z
25. Felipe-Medina N, Caburet S, Sánchez-Sáez F, Condezo YB, de Rooij D, Gómez-H L, García-Valiente R, Todeschini AL, Duque P, Sánchez-Martín M, Shalev SA, Llano E, Veitia RA, Pendás AM (2020) A missense in HSF2BP causing primary ovarian insufficiency affects meiotic recombination by its novel interactor C19ORF57/BRME1. elife 9:1–93. https://doi.org/10.7554/ELIFE.56996
26. Gomez-H L, Felipe-Medina N, Condezo YB, Garcia-Valiente R, Ramos I, Suja JA, Barbero JL, Roig I, Sanchez-Martin M, De Rooij DG, Llano E, PendasI AM (2019) The PSMA8 subunit of the spermatoproteasome is essential for proper meiotic exit and mouse fertility. PLoS Genet 15:e1008316. https://doi.org/10.1371/JOURNAL.PGEN.1008316
27. Felipe-Medina N, Gómez-H L, Condezo YB, Sanchez-Martín M, Barbero JL, Ramos I, Llano E, Pendás AM (2019) Ubiquitin-specific protease 26 (USP26) is not essential for mouse gametogenesis and fertility. Chromosoma 128: 237–247. https://doi.org/10.1007/S00412-019-00697-6
28. Sánchez-Sáez F, Gómez-H L, Dunne OM, Gallego-Páramo C, Felipe-Medina N, Sánchez-Martín M, Llano E, Pendas AM, Davies OR (2020) Meiotic chromosome synapsis depends on multivalent SYCE1-SIX6OS1 interactions that are disrupted in cases of human infertility. Sci Adv 6. https://doi.org/10.1126/SCIADV.ABB1660
29. Sánchez-Sáez F, Sainz-Urruela R, Felipe-Medina N, Condezo YB, Sánchez-Martín M, Llano E, Pendás AM (2022) The oocyte-specific linker histone H1FOO is not essential for mouse oogenesis and fertility. Cells 11. https://doi.org/10.3390/CELLS11223706
30. Furuya M, Tanaka M, Teranishi T, Matsumoto K, Hosoi Y, Saeki K, Ishimoto H, Minegishi K, Iritani A, Yoshimura Y (2007) H1foo is indispensable for meiotic maturation of the mouse oocyte. J Reprod Dev 53:895–902. https://doi.org/10.1262/JRD.19008
31. Yun Y, An P, Ning J, Zhao GM, Yang WL, Lei AM (2015) H1foo is essential for in vitro meiotic maturation of bovine oocytes. Zygote 23: 416–425. https://doi.org/10.1017/S0967199414000021
32. Takeo T, Nakagata N (2015) Superovulation using the combined administration of inhibin antiserum and equine chorionic gonadotropin increases the number of ovulated oocytes in C57BL/6 female mice. PLoS One 10: e0128330. https://doi.org/10.1371/JOURNAL.PONE.0128330

Part II

Cultures, Cell Isolation, Microscopy & Live Imaging

Chapter 7

Isolation of Homogeneous Sub-populations of Spermatocytes from Mouse Testis

Yao Chen, Zhen Lin, Xi Zhang, Meixia Wu, and Ming-Han Tong

Abstract

Mammalian meiosis is a highly specialized cell division process, resulting in the production of genetically unique haploid cells. However, the molecular mechanisms governing meiosis remain largely unknown, primarily due to the difficulty in isolating pure sub-populations of spermatocytes. Definitive molecular, biochemical, and functional investigations of the meiosis process require the isolation of these individual homogeneous sub-populations of spermatocytes. Here, we present an approach that enables the purification of homogeneous spermatocytes from mouse testis at desired sub-stages. This approach consists of two strategic steps. The first is to synchronize spermatogenesis, aiming to minimize the diversity and complexity of testicular germ cells. The second involves utilizing mouse models with germ cell-specific fluorescent markers to differentiate the desired subtype from other cells in the testis. By employing fluorescence-activated cell sorting (FACS), this approach yields highly pure populations of spermatocytes at each sub-stage. When combined with other massively parallel sequencing techniques and in vitro cell culture methods, this approach will enhance our comprehension of the molecular mechanisms underlying mammalian meiosis and promote in vitro gametogenesis.

Key words Mammalian meiosis, Isolation of spermatocytes, Fluorescent-labeled mouse line, Spermatogenesis synchronization, FACS

1 Introduction

Mammalian meiosis, the critical step in gametogenesis, is a complex and highly organized developmental process during which recombination occurs between paternal and maternal chromosomes to produce genetically unique haploid cells [1]. Taking mouse spermatogenesis as an example, undifferentiated spermatogonia undergo differentiation followed by a series of mitotic divisions that generate A1, A2, A3, A4, intermediate (In), and type B spermatogonia [2–4]. Type B spermatogonia then give rise to preleptotene to enter the meiotic phase, which leads to the production of haploid spermatids by two rounds of cell divisions, meiosis I and meiosis II, preceded by a single premeiotic round of DNA

Jesús A. Carballo (ed.), *Meiosis: Methods and Protocols*, Methods in Molecular Biology, vol. 2818,
https://doi.org/10.1007/978-1-0716-3906-1_7,

replication. In this phase, primary spermatocytes undergo a prolonged meiotic prophase I to complete complex meiosis-specific events, such as pairing, synapsis, and recombination of homologous chromosomes, which ensure the precise segregation of homologues and genetic diversity [5, 6]. According to cytological patterns, meiotic prophase I can be divided into the leptotene, zygotene, pachytene, and diplotene sub-stages [5, 6]. At diakinesis, primary spermatocytes are reduced to diploid secondary spermatocytes, which rapidly undergo meiosis II to generate haploid round spermatids [5, 6].

Spermatogenesis is a continuous process and each step concomitantly takes place in the seminiferous tubules of the testis [7]. Because the transition from Aal to A1 spermatogonia (known as spermatogonial differentiation) is asynchronous, mammalian spermatogenesis is asynchronous, which is essential for the continual production of sperm in large numbers [1]. As a result, the testis contains spermatogenic cells at all stages of spermatogenesis, meaning the spermatogenic cell population is highly heterogeneous. This heterogeneity has made it difficult to isolate pure populations of spermatogenic cells, such as different sub-stages of meiotic spermatocytes, thereby hindering the understanding of the molecular mechanisms underlying male meiosis. During the last half century, considerable efforts have been devoted to isolating male germ cell populations from adult testis to studying the molecular mechanisms of spermatogenesis. At least three methods have been developed to enrich and isolate specific spermatogenic cell populations: velocity sedimentation at unit gravity (known as STA-PUT), centrifugal elutriation, and flow cytometry (FACS) [8–11]. Both STA-PUT and centrifugal elutriation rely on cell size and density as criteria for cell separation, and indeed, a heterogeneous population of spermatogonia, pachytene spermatocytes, and round spermatids can be successfully separated by these methods [8, 10, 12]. Based on DNA content and light scattering properties, FACS can sort a mixed population of preleptot'ene, meiotic prophase I spermatocytes, and haploid spermatids [9, 11]. However, these approaches still failed to obtain a homogeneous population of spermatogenic cells. For instance, while both FACS and STA-PUT methods can yield high purity of pachytene sub-stages from adult testis [10, 11], the transcriptome and epigenetic modifications of early pachytene spermatocytes are significantly different from those of late pachytene spermatocytes [13, 14]. For this reason, it has been difficult to determine the nature of these cells at specific stages of development. Furthermore, these methods cannot isolate rare spermatogenic cell types (such as G1 phase of preleptotene spermatocytes and secondary spermatocytes) of interest from testis as these cell types remain for only a few hours before beginning the next process. Therefore, it is necessary to develop a new approach for isolating high-purity homogeneous populations of male germ cells

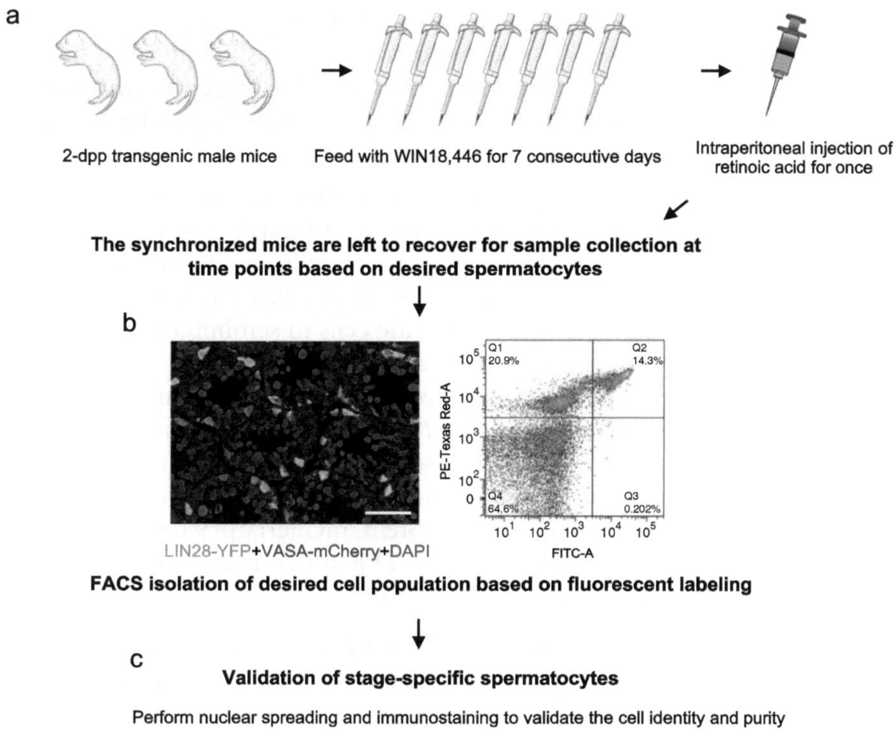

Fig. 1 Overview of the workflow for isolation of homogeneous populations of stage-specific murine spermatocytes. (**a**) Synchronization in the neonatal mouse testes of transgenic mouse models with WIN 18,446 and RA. Then, these synchronized mice are left to recover for sample collection at the time point based on the desired spermatogenic cell types. (**b**) FACS isolation of the desired spermatogenic cell population at the specific time point. Scale bar, 50μm. (**c**) After collection, experiments should be performed to further validate the identity and purity of the sorted cells

at any desired stage of spermatogenesis. Here we provided details of an approach to purify all types of homogeneous spermatocytes from mice to allow the study of the omics at defined sub-stages. We have combined this approach with RNA-seq, NOMe-seq (nucleosome occupancy and methylome sequencing), and ChIP-seq (chromatin immunoprecipitation sequencing) to establish a comprehensive transcriptome atlas of spermatogenesis and generate epigenomic profiles of meiotic prophase I [13, 14].

In order to purify homogeneous stage-specific meiotic cells, we devised an approach consisting of two strategic steps:

First, we reduced the heterogeneity and complexity of male germ cells within the testis by synchronization of spermatogenesis [15, 16] (Fig. 1a). We applied a potent inhibitor of the enzyme responsible for retinoic acid (RA) synthesis, WIN 18,446, to 2-day-old male mice for 7 consecutive days to block spermatogonial differentiation. Thus, a single injection of exogenous RA into these mice will drive only one cohort of undifferentiated spermatogonia to differentiate into A1 spermatogonia [14, 17]. Because of synchronous initiation of spermatogonial differentiation, spermatogenesis will be synchronous, resulting in a testis that is enriched for given stages of spermatogenic cells at a certain time point [7, 18].

Second, we generated mouse models carrying germ cell-specific fluorescent markers to distinguish a single defined germ cell subtype from other germ cells (predominant undifferentiated spermatogonia) and somatic cells. There are still several types of germ cells and somatic cells in seminiferous tubules after spermatogenesis has been synchronized. To overcome these challenges, we engineered two knock-in mouse models: *Vasa*-mCherry and *Lin28*-YFP knock-in mouse lines. To distinguish germ cells from somatic cells, we utilized a *Vasa*-mCherry knock-in mouse line in which the spermatogenic cells express a red fluorescent protein, mCherry [19]. Undifferentiated spermatogonia in *Lin28*-YFP knock-in mouse line express a yellow fluorescent protein, YFP.

In general, for the isolation of homogeneous spermatocyte populations, we generated male mice carrying both *Vasa*-mCherry and *Lin28*-YFP alleles and synchronized germ cell development in these animals to allow testes containing predominantly defined germ cell sub-stages. As shown in Fig. 1b, according to fluorescent labeling, testicular cells in these animals can be divided into three types: (i) mCherry and YFP double-positive cells are the undifferentiated spermatogonia; (ii) mCherry and YFP double-negative cells are the somatic cells; and (iii) mCherry single positive cells are the advanced germ cells we wanted. Thus, we then used FACS to isolate the mCherry^{+}YFP^{-} advanced germ cells based on mCherry, YFP, DNA content, and cell size. The purified synchronous and homogeneous germ cell subtype populations were applied to next step experiments such as RNA-seq, ChIP-seq, ATAC (assay for targeting accessible chromatin), Hi-C (high-throughput chromosome conformation capture), metabolomics, and proteomics. After collection, we performed experiments to validate the identity and purity of the sorted cells (Fig. 1c).

The workflow for isolation of homogeneous sub-populations of murine spermatocytes is summarized in Fig. 1.

2 Materials

All solutions, including stock solutions, should be prepared under sterile conditions.

2.1 Genotyping

1. *Vasa*-mCherry knock-in mouse line (*see* **Note 1**).
2. *Lin28*-YFP knock-in mouse line (*see* **Note 1**).
3. PCR Mix. Store at −20 °C.
4. Primers for PCR genotyping of *Vasa*-mCherry and *Lin28*-YFP mice: *Vasa*-mCherry _F: AGA CTT CAC AGG ATT ACA TTG; *Vasa*- mCherry_wtR: AAA GCA CAT CAC ATC CTA TTG; *Vasa*-mCherry _kiR: CCC TTG CTC ACC ATA GGA CCA. *Lin28*-YFP_F: CCC CAG TTC TCA GGG AAA GCC; *Lin28*-YFP_wtR: CCA CCC TTA CCC CCA CTT TCT; *Lin28*-YFP_kiR: TGA ACT TGT GGC CGT TTA CGT. Store at −20 °C.
5. 0.2 mL PCR tubes (e.g., Axygen, cat. no. PCR-02-C).
6. PCR instrument (e.g., Eppendorf, model no. nexus SXI).

2.2 Synchronization

1. *1% (m/vol) gum tragacanth* Dissolve 0.5 g of gum tragacanth powder in 500 mL of sterile ddH_2O and store at 4 °C.
2. *WIN 18,446 (50 mg/mL)* Place up to 75 mg WIN 18,446 powder into 1.5 mL microtube under sterile condition, and add 1% (m/vol) gum tragacanth to make the final concentration of WIN 18,446 50 mg/mL. Use a contactless ultrasonic instrument to mix well and store at 4 °C.
3. *Retinoic acid (RA) solution (20 mg/mL)* Dissolve 50 mg of RA powder in 2.5 mL of dimethyl sulfoxide (DMSO). The RA solution can be stored at −20 °C for at least 2 months. Avoid freeze–thaw cycles; cover to protect from light.
4. 1.5 mL microtubes.
5. Analytical balance.
6. Contactless ultrasonic instrument.
7. P20 pipette (e.g., Eppendorf, model no. 3120000291).
8. Insulin syringe.

2.3 FACS

1. *Collagenase I solution* Prepare a 120 U/mL solution in 1× PBS, 3 mL is enough for one mouse, and prepare this fresh each time.
2. 0.25% Trypsin–EDTA. Store at 4 °C.
3. *DNase I stock solution (10 mg/mL)* Dissolve 50 mg of DNase I in 5 mL of sterile ddH_2O. The DNase stock solution can be stored at −20 °C for at least 2 months. Avoid freeze–thaw cycles that could diminish the enzyme activity; make 150μL aliquots, and thaw only the necessary amount for each experiment.

4. *Hoechst 33342 solution (10 mg/mL)* Dissolve 10 mg of Hoechst 33342 in 1 mL of sterile ddH_2O, make 100μL aliquots, stored at −20 °C for at least 2 months. Store thawed Hoechst 33342 at 4 °C. Cover to protect from light.
5. *10% (vol/vol) FBS in DMEM* Add 5 mL FBS in 45 mL DMEM, mix well, and place on ice. 5 mL of 10% (vol/vol) FBS in DMEM is needed per sample. Prepare this fresh each time under sterile condition.
6. *FACS collection tubes* Add 1 mL of 10% (vol/vol) FBS in DMEM per FACS collection tube, and place on ice. Prepare this fresh each time.
7. 75% (vol/vol) ethanol.
8. 3.5 cm dish.
9. 15 mL conical centrifuge tubes.
10. 40μm and 70μm cell strainer.
11. FACS collection tubes, 5 mL polypropylene round-bottom tubes.
12. Surgical straight standard scissors.
13. Surgical straight standard tweezers.
14. Refrigerated centrifuge.
15. Hybridization oven with rotisserie (e.g., UVP, model no. HL-2000).
16. FACS sorter (e.g., Becton Dickinson, model no. FACSAria SORP).

2.4 Nuclear Spreading and Immunostaining

1. *10 mM sodium acetate* Dissolve 0.041 g sodium acetate in 50 mL ddH_2O. Store at room temperature.
2. *100 mM PMSF* Dissolve 0.1742 g PMSF in 10 mL isopropanol. The PMSF stock solution can be stored at −20 °C for at least 2 months. Avoid freeze–thaw cycles; make 100μL aliquots, and thaw only the necessary amount for each experiment.
3. *170 mM sodium citrate* Add 2.50 g sodium citrate dihydrate to 47.5 mL ddH_2O. Adjust to 50 mL with ddH_2O. Store at room temperature.
4. *50 mM boric acid* Dissolve 0.1546 g boric acid in 50 mL ddH_2O. Store at room temperature.
5. *500 mM EDTA, pH 8.2* Add 7.31 g EDTA to 47.5 mL ddH_2O. Adjust pH to 8.2. Adjust to 50 mL with ddH_2O. Store at room temperature.
6. *500 mM DTT* Dissolve 0.7712 g DTT in 10 mL 10 mM sodium acetate. The DTT stock solution can be stored at −20 °C for at least 2 months. Avoid freeze–thaw cycles; make 100μL aliquots, and thaw only the necessary amount for each experiment.

7. *500 mM sucrose* Add 8.56 g sucrose to 47.5 mL ddH$_2$O. Adjust to 50 mL with ddH$_2$O. Store at 4 °C.
8. *100 mM sucrose* Add 1 mL 500 mM sucrose to 4 mL ddH$_2$O. Store at 4 °C.
9. *600 mM Tris–HCl, pH 8.2* Add 3.63 g Tris–HCl base to 47.5 mL ddH$_2$O. Adjust pH to 8.2. Adjust to 50 mL with ddH$_2$O. Store at room temperature.
10. *10 mL Hypo-Extraction Buffer* Add 500μL 600 mM Tris–HCl, pH 8.2, 1 mL 500 mM sucrose, 1 mL 170 mM sodium citrate, 100μL 500 mM EDTA, pH 8.2 to 5 mL ddH$_2$O. Adjust pH to 8.2–8.4 using 50 mM boric acid. Adjust to 10 mL with ddH$_2$O. Add 50μL 500 mM DTT and 50μL 100 mM PMSF just prior to using solution (*see* **Note 2**).
11. *0.1% PBST* Dissolve 1 mL Triton X-100 in 1000 mL PBS. Store at room temperature.
12. *1% PFA* Add 0.25 g PFA to 22.5 mL ddH$_2$O. Add one drop of 1 M NaOH and incubate at 60 °C for at least 20 min to dissolve PFA. Cool to room temperature. Adjust pH to 9.2–9.4 using 50 mM boric acid. Dissolve 50μL of TritonX-100 in the PFA solution. Adjust to 25 mL with ddH$_2$O. Prepare this fresh each time.
13. *0.2% Photo-Flo* Add 2 mL Photo-Flo 200 to 1000 mL PBS. Store at room temperature.
14. *10 x PBS* Dissolve 80 g NaCl, 2 g KCl, 14.4 g Na_2HPO_4, 2.4 g KH_2PO_4 in 800 mL of ddH$_2$O. Adjust pH to 7.4 using HCl. Adjust to 1 L with ddH$_2$O. Sterilize the solution (e.g., autoclaving). Store at room temperature.
15. *4% PFA* Add 40 g of paraformaldehyde powder to 800 mL of ddH$_2$O, stir the mixture at 60 °C in ventilation hood, add 5 drops of 1 M NaOH to help dissolve the PFA powder, cool the solution to room temperature, add 100 mL 10 x PBS, and adjust the volume to 1 L using ddH$_2$O. Check the pH to 7.4 and adjust with HCl and NaOH if required. Stored at −20 °C for at least 2 months. In 5–10 mL aliquots, thaw only the necessary amount for each experiment.
16. *Blocking buffer* Add 1 mL normal donkey serum (e.g., Jackson ImmunoResearch, cat. no. 017–000-121) to 9 mL 0.1% PBST. The blocking buffer can be stored at –20°C for at least one month. Avoid freeze-thaw cycles, make 500 μL aliquots, and thaw only the necessary amount for each experiment.
17. Alexa Fluor 488 AffiniPure Donkey Anti-Mouse IgG (e.g., Jackson ImmunoResearch, cat. no. 715-545-151).
18. Alexa Fluor 594 AffiniPure Donkey Anti- Rabbit IgG (e.g., Jackson ImmunoResearch, cat. no. 711-585-152).
19. Rabbit anti-SYCP3 (Abcam, cat. no. ab15093).
20. Mouse anti- γH2AX (Millipore, cat. no. 05-636).

21. Mounting medium with DAPI-aqueous, fluoroshield (e.g., Abcam, cat. no. ab104139).
22. Microscope slides.
23. Humidifier chamber.

2.5 Fluorescence Microscopy Imaging and Bright Field Imaging

1. Fluorescence microscope (e.g., Zeiss, model no. AXIO scope A1).
2. Inverted phase contrast microscope (e.g.. Nikon, model no. ECLIPSE Ts2).

3 Methods

3.1 Preparation of Mouse Models

1. Set up the breeding cages. Male mice carrying both *Vasa*-mCherry and *Lin28*-YFP alleles can be used to obtain desired sub-stages of meiotic cell populations (*see* **Note 3**).
2. Check daily when the mice are about to give birth. And be sure to obtain an accurate birth time (*see* **Note 4**).

3.2 Synchronization in the Neonatal Mouse Testes

1. Prepare the working solutions (WIN 18,446, and RA solution) as described in the Materials.
2. 2-dpp mice are pipette-fed 100μg/g body weight WIN 18,446, once per day for 7 consecutive days (*see* **Note 5**).
3. 9-dpp mice are given an intraperitoneal injection of RA (33μg/g body weight), record the exact injection time, and then these mice are left to recover for sample collection at the given time points (*see* **Note 6**).
4. Genotyping for *Vasa*-mCherry and *Lin28*-YFP knock-in mice is performed via PCR using genomic DNA extracted from mouse tail before collection (*see* **Note 7**).
5. Choose the collection time point based on the desired meiotic cell types. Collect specific stages of meiotic cell populations after RA treatment on the C57BL/6 genetic background mice at the following time-points:

Cell subtypes	G1	ePL	mPL	lPL	L	Z	eP	mP	lP	D
Time after RA (hrs)	144	162	174	186	194	216	264	312	348	384

G1: G1 phase of preleptotene spermatocyte, ePL: early S phase of preleptotene spermatocyte, mPL: middle S phase of preleptotene spermatocyte, lPL: late S phase of preleptotene spermatocyte, L: leptotene spermatocyte, Z: zygotene spermatocyte, eP: early pachytene spermatocyte, mP: middle pachytene spermatocyte, lP: late pachytene spermatocyte, D: diplotene spermatocyte.

6. When establishing the timing of meiotic cell development after RA treatment, a series of experiments should be performed: (i) fix one testis in Bouin's solution and the other in 4% (m/vol) PFA, embed both of them in paraffin, section, and stain with hemotoxylin-eosin (H&E) to assess synchronization efficiency; (ii) when the advanced spermatogenic cells are at desired meiotic sub-stages, perform nuclear spreading and immunostaining for specific markers, such as SYCP3 and γH_2AX to validate the cell identity [20–22] (*see* **Note 8**).

3.3 FACS Isolation of Stage-Specific Spermatogenic Cells

1. Prepare the working solutions (DNase I solution, collagenase I solution, 10% (vol/vol) FBS in DMEM, and FACS collection tubes) as described in the Materials. Put DNase I solution on ice, and warm collagenase I solution, 0.25% Trypsin–EDTA, and DMEM to 32 °C.
2. Clean all dissection tools with 75% (vol/vol) ethanol and wipe them dry.
3. Add 3 mL of cold 1× PBS into 3.5-cm dish and place on ice. Prepare the same number of dishes as the mice.
4. Euthanize mice according to institutional guidelines, such as by CO_2 euthanasia (*see* **Note 9**).
5. Harvest the testes, perform body and testis weight measurements, and then place the testes in the 3.5 cm dish filled with cold 1× PBS on ice (*see* **Note 10**).
6. Cut a small piece of one testis, place the tissue with cold 1× PBS on ice, then do the following experiments while or after sorting: (i) fix the tissue in Bouin's solution for 4 h, embed it in paraffin, section, and stain with hemotoxylin-eosin (H&E) and immunostaining for STRA8 when the synchronous advanced spermatogenic cells are at preleptotene spermatocyte stage; (ii) perform nuclear spreading and immunostaining for SYCP3 and γH_2AX when the synchronous advanced spermatogenic cells are at prophase I meiotic stages [21] (*see* **Note 11**).
7. Remove the tunica albuginea of each testis, then transfer a pair of testes from one mouse into a 15 mL conical tube containing 3 mL pre-warmed collagenase I solution, incubate for 5 min at 32 °C in a hybridization oven with constant rotation (*see* **Note 12**).
8. Add 5 mL pre-warmed 0.25% Trypsin–EDTA with 10μg/mL DNase I to the conical tube and gently pipette up and down 5 times to mix well.
9. Incubate the conical tube for 5 min at 32 °C in the hybridization oven with constant rotation. Halfway into the 5 min digestion period, gently pipette the tubules up and down 10 times. At the end of the incubation time, gently pipette 10 times again (*see* **Note 13**).

10. Add 0.5 mL FBS to inactivate trypsin, mix thoroughly by gently pipetting.
11. If the cell suspension is viscous, add an additional 10μg/mL DNase I and gently pipette up and down 10 times (*see* **Note 14**).
12. Pre-wet a sterile 70μm nylon cell strainer and use this to filter the cell suspension into a new 15 mL conical tube.
13. Rinse the digestion tube once with 2 mL of 10% (vol/vol) FBS in DMEM to collect the remaining cells and add them to the conical tube. Gently pipette up and down 5 times and leave a 100μL aliquot for cell counting and viability estimation. Pellet the cells by centrifugation at 500 g at 4 °C for 5 min (*see* **Note 15**).
14. While cells are being centrifuged, count cell number and prepare the Hoechst 33342 staining solution as described in the Materials. Add 1μL of Hoechst 33342 stock solution (10 mg/mL) to 5 mL pre-warmed DMEM with 10μg/mL DNase I for each sample.
15. After the centrifugation step is complete, discard the supernatant by decanting and resuspend the pellet in Hoechst 33342 staining solution (10 mg/mL) to a concentration of 1 million cells/mL.
16. Incubate for 10 min at 32 °C in the hybridization oven with slow constant rotation.
17. Add an additional 0.4μL Hoechst 33342 stock solution (10 mg/mL) per million cells for the final 6μg Hoechst 33342/million cells [11], gently pipette up and down 5 times, and incubate for another 10 min at 32 °C in the hybridization oven with slow constant rotation (*see* **Note 16**).
18. After the Hoechst 33342 staining is complete, pellet the cells by centrifugation at 500 g at 4 °C for 5 min.
19. After the centrifugation step is complete, discard the supernatant by carefully pipetting and resuspend the cells in 10% (vol/vol) FBS in DMEM with 10μg/mL DNase I to a concentration of 1×10^7 cells /mL.
20. Pre-wet a sterile 40μm nylon cell strainer and use this to filter the resuspended cells into a 5 mL FACS polypropylene round-bottom tube.
21. Rinse the conical tube once with 100–200μL of 10% (vol/vol) FBS in DMEM to collect the remaining cells and filter them into the FACS tube.
22. Place samples and FACS collection tubes on ice for transport to a FACS facility. Keep cells on ice and protect from light (*see* **Note 17**).

23. Gating strategies are different for isolating different types of spermatogenic cells or different downstream applications. Isolate G1 phase of preleptotene spermatocyte or cells without Hoechst 33342 as described in Step A. And isolate prophase I meiotic cells with Hoechst 33342 as described in Step B (*see* **Note 18**).
    (A) Isolation of preleptotene spermatocyte without Hoechst 33342
        (i) Use the forward-scatter-A (FSC-A) and the side-scatter-A (SSC-A) channels to exclude debris (Fig. 2, top left).
        (ii) Use pulse-processing of forward-scatter-W (FSC-W) versusfForward-scatter-H (FSC-H) and side-scatter-W (SSC-W) versus side-scatter-H (SSC-H) to exclude cell doublets and aggregates (Fig. 2, top right and bottom left).
        (iii) Collect preleptotene spermatocyte (*Vasa*-mCherry^{+}, *Lin28*-YFP^{-}) (Fig. 2, bottom right) in 5 mL FACS collection tubes (*see* **Note 19**).
    (B) Isolation of meiotic cells with Hoechst 33342
        (iv) Use the forward-scatter-A (FSC-A) and the side-scatter-A (SSC-A) channels to exclude debris (Fig. 3, top left).
        (v) Use pulse-processing of forward-scatter-W (FSC-W) versus forward-scatter-H (FSC-H) and side-scatter-W (SSC-W) versus side-scatter-H (SSC-H) to exclude cell doublets and aggregates (Fig. 3, top right and bottom left).
        (vi) Gate 4N cells using DAPI-A and Indo-1(blue)-A channels (Fig. 3, bottom middle).
        (vii) Collect target cells (*Vasa*-mCherry^{+}, *Lin28*-YFP^{-}) (Fig. 3, bottom right) in 5 mL FACS collection tubes (*see* **Note 20**).
24. Keep sorted cells on ice after FACS. We highly recommend that cells be processed for next applications as soon as possible, as further delays may compromise viability.

3.4 Post-FACS Processing

1. After FACS, to validate the identity and purity of the sorted cells, leave a 100μL aliquot on ice for validation. Spin down the remaining sorted cells at 500 *g* at 4 °C for 5 min. Resuspend the cell pellet in an appropriate volume for downstream applications. For single-cell profiling, resuspend cells at the desired concentration in a solution appropriate for platform to be used. For ChIP, crosslink the sorted cells with 1% formaldehyde for 10 min at room temperature and stop by adding 125 mM glycine. Alternatively, cell pellets may be used immediately for RNA isolation.

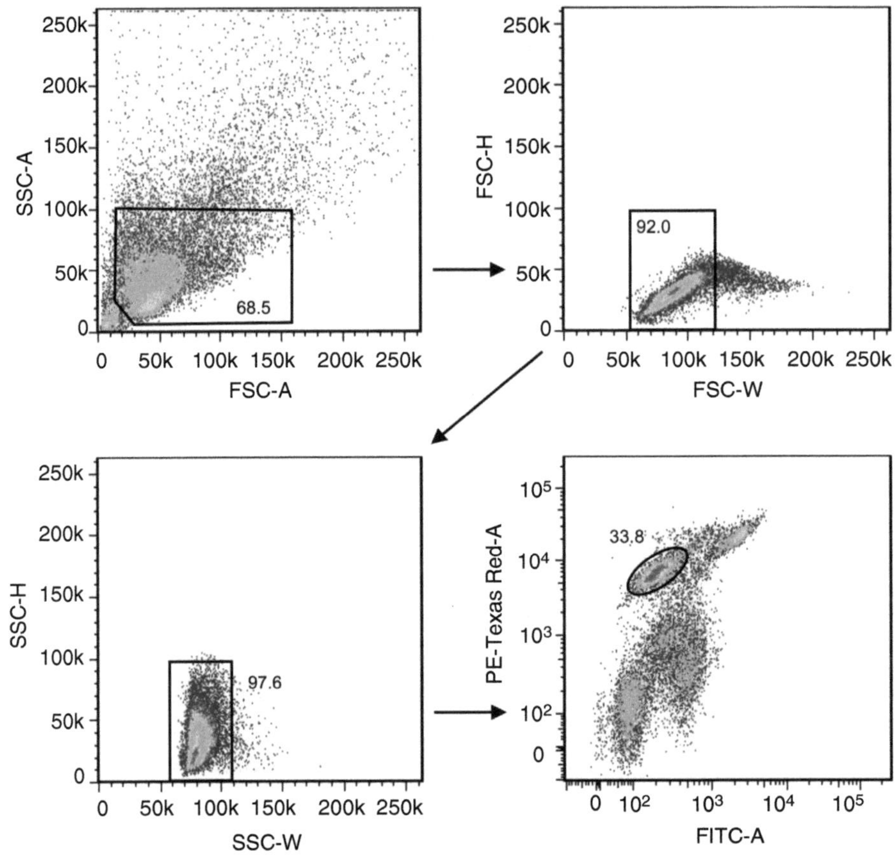

Fig. 2 Gating for preleptotene spermatocyte. Representative FACS plots showing the gating strategy for the purification of preleptotene spermatocyte from synchronized mouse testes without Hoechst 33342 staining. After excluding debris (top left) and doublets (top right and bottom left), PE-Texas Red$^+$ FITC$^-$ cells are selected and collected (bottom right). Bottom right image was displayed using auto biexponential display. Percentages refer to the proportion of cells in the previous parent gate. Plots show 30,000 events. FSC-A, forward-scatter area; FSC-W, forward-scatter width; FSC-H, forward-scatter height; SSC-A, side-scatter area; SSC-W, side-scatter width; SSC-H, side-scatter height. PE-Texas Red channel was used to detect *Vasa*-mCherry fluorescence, and FITC-A channel was used to detect *Lin28*-YFP fluorescence

3.5 Validation of the Obtained Cells

1. Use fluorescence microscope to observe the fluorescence intensity of a 25μL aliquot of the sorted cells, and use inverted phase contrast microscope to observe the cell size of a 25μL aliquot of the sorted cells.
2. Prepare the working solutions (Hypo-Extraction Buffer, 100 mM sucrose, 1% PFA, 0.2% Photo-Flo, and 0.1% PBST) as described in the Materials.
3. Mix 50μL Hypo-Extraction Buffer with another 50μL aliquot of sorted cells and incubate at room temperature for 20 min.
4. After the incubation is complete, pellet the cells by centrifugation at 500 g at room temperature for 5 min.

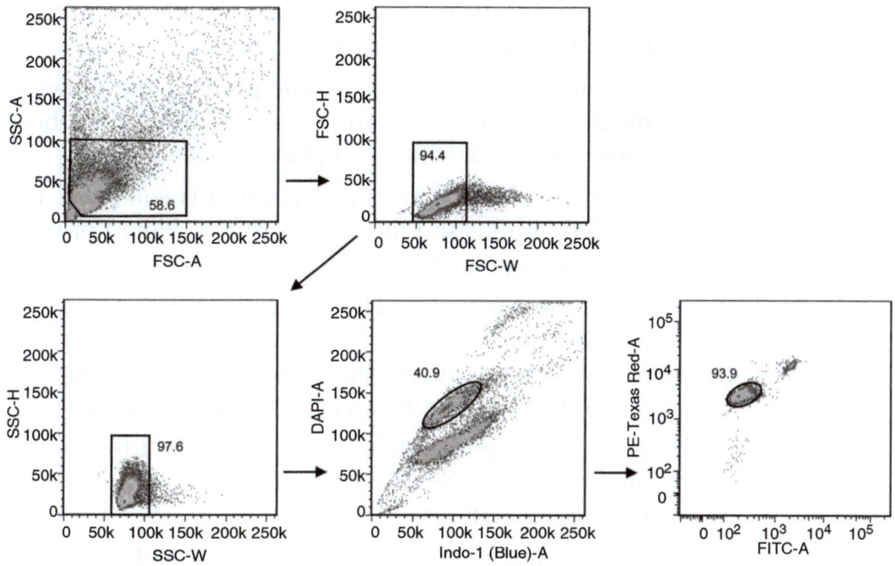

Fig. 3 Gating for zytotene spermatocyte. Representative FACS plots showing the gating strategy for the purification of meiotic cells from synchronized mouse testes with Hoechst 33342 staining. After excluding debris (top left) and doublets (top right and bottom left), gate 4 N cells (bottom middle) and then PE-Texas Red$^+$ FITC$^-$ meiotic cells are collected (bottom right). Bottom right image was displayed using auto biexponential display. Percentages refer to the proportion of cells in the previous parent gate. Plots show 30,000 events. FSC-A, forward-scatter area; FSC-W, forward-scatter width; FSC-H, forward-scatter height; SSC-A, side-scatter area; SSC-W, side-scatter width; SSC-H, side-scatter height. PE-Texas Red channel was used to detect *Vasa*-mCherry fluorescence, and FITC-A channel was used to detect *Lin28*-YFP fluorescence

5. After the centrifugation step is complete, carefully remove 95μL supernatant and resuspend cells with 10μL of 100 mM sucrose. Avoid air bubbles during resuspension by setting the volume of the pipette to less than the total volume.
6. Dip a glass slide in 1% PFA, and apply~15μL cell suspension to the slide.
7. Dry the cell nuclei overnight in a slightly opened humidified chamber (*see* **Note 21**).
8. Wash the slides for 10 min with 0.2% Photo-Flo, dry at room temperature, and store the slides at −20 °C if they will be used within a few days or place at −80 °C for long-term storage. For immunostaining, proceed to the next step.
9. Thaw the frozen slides at room temperature for 10 min. In the meantime, prepare blocking buffer as described in the Materials and place on ice.
10. Rinse the slides in PBS for 5 min and permeabilize with 0.1% PBST for 10 min.
11. Rinse the slides in PBS for 5 min and draw a grease ring around the spread nuclei.

12. Add 50–100μL of blocking buffer inside the grease ring and block for 1 h at room temperature.
13. During incubation, prepare the primary antibodies diluted in blocking buffer (e.g., anti-SYCP3 1:500, anti-phospho-Histone γH2A.X 1:1000) and place on ice.
14. After the blocking buffer incubation is complete, remove the blocking buffer and incubate with the diluted primary antibodies overnight at 4 °C.
15. Wash the slides 3 × 10 min with PBST. In the meantime, prepare the secondary antibody diluted in blocking buffer (e.g., donkey anti-rabbit 594 1:1000, donkey anti-mouse 488 1:1000), place on ice, and protect from light.
16. After washing the slides, rinse them in PBS for 5 min and incubate with the diluted secondary antibodies for 1 hr. at room temperature.
17. After incubation is complete, wash the slides 3 x 10 min with PBST and mount with mounting medium with DAPI-aqueous.
18. Image the samples with a fluorescence microscope. To validate the cell identity and its purity, 100 spread cells from sorted cell population should be scored and classified after each sort.

4 Notes

1. All mice were maintained on the C57BL/6 J (B6) background. These mouse lines were described previously [14]. All animal experiments should be performed according to relevant guidelines and regulations. In this protocol, all animal experiments were conducted in accordance with the guidelines in the Animal Care and Use Committee at Center for Excellence in Molecular Cell Science, Chinese Academy of Science.
2. Hypo-Extraction buffer is good for 2 h. Prepare this fresh each time.
3. To increase the reproducibility of developmental timing, we strongly recommend using genetically uniform mice. All mice used in our experiments were on the C57BL/6 genetic background. And the time-course of development after synchronization should be reestablished if any other genetic background mice are used.
4. It is important to know the exact time of birth. Differentiating spermatogonia first appears in 3-dpp (days postpartum) mouse testis [16, 23], so WIN 18,446 should be used before 2-dpp to prevent spermatogonial differentiation. We strongly

recommend keeping male pups only to ensure they have sufficient milk intake. Find a foster mother if the birth mother does not feed well. Developmental timing after synchronization is less consistent if the pups are in poor health.

5. Shake the WIN 18,446 suspension well before feeding the pups. WIN 18,446 should be given at the same time every day during 7 consecutive days' treatment.
6. Body weight is an important indicator of health status. If the 9-dpp mice weight is over 4.5 g, they are ready for an injection. If the pups have severe growth retardation (less than 4 g), it is better to euthanize them. If the weight is 4–4.5 g, an extra day of WIN 18,446 can be administered to allow for further weight gain before RA injection. If the overall mortality rate among the RA-injected pup mice is higher than 50%, a lower dose of RA would be better. However, the effects of RA proved to be dose dependent [17], so try to keep a high RA dosage.
7. Under the premise of complying with relevant guidelines and regulations, genotyping should be performed as early as possible to reduce the litter size and help the remaining pups recover from RA treatment.
8. Carefully monitor body weight and testis weight of the synchronized mice at collection time points; the predictable timing of spermatogenic cell development depends on whether the mice are healthy and of normal weight.
9. Follow appropriate institutional guidelines with regard to animal husbandry and experimental procedures. Our experiments were performed with the approval of the Animal Care and Use Committee at Center for Excellence in Molecular Cell Science, Chinese Academy of Sciences.
10. After establishing the timetable for collecting different stages of spermatogenic cells in a new mouse strain, body weight and testis weight serve as key indicators for obtaining consistent spermatogenic cell types at specific collection time points.
11. This step will help you to validate the degree and the exact seminiferous epithelial stage of synchronous spermatogenesis and determine the cell identity. When the synchronous advanced spermatogenic cells are at meiotic stages, choose Subheading 3.3, **step 6** or Subheading 3.5 of this Methods, both of which could validate the cell identity.
12. After the collagenase digestion step is complete, seminiferous tubules should become dispersed.
13. Add a small volume of digested sample, ~10–20μL, to a glass slide and visually inspect the sample under the microscope. If intact tubular fragments are present, increase the digestion time. If the sample has only single cells, the sample may have been overdigested.

14. After this step is complete, the seminiferous tubules should be reduced to a single-cell suspension.
15. Subheading 3.3, **steps 6–12** describe a general testis digestion protocol. For isolation of G1 phase of preleptotene spermatocyte or cells without Hoechst 33342, proceed to Subheading 3.3, **step 18**. For isolation of meiotic cells with Hoechst 33342 staining, proceed to Subheading 3.3, **step 13**.
16. There are only several spermatogenic cell populations in the synchronized testis, each type of cell populations can be easily separated by DNA staining with *Vasa*-mCherrry *Lin28*-YFP; too much Hoechst 33342 incubation time will compromise cell viability, resulting in lower cell yield.
17. FACS should be performed as soon as possible.
18. Choose an appropriate FACS sorter. Hoechst 33342 is effectively excited by a 355 nm laser, and the dye's wide emission spectrum detected in two distinct channels: the "DAPI" (450/50 nm filter) and the "Indo-1(blue)" (530/30 nm filter).
19. YFP could be detected in FITC channel. The population that is *Vasa*-mCherry$^+$ *Lin28*-YFP$^+$ at this step is undifferentiated spermatogonia without/with early stage of differentiating spermatogonia. A1–A4 spermatogonia sub-stages in *Lin28*-YFP knock-in mouse line still show weak signal of YFP.
20. The population that is *Vasa*-mCherry$^+$ *Lin28*-YFP$^+$ at this step is early stage of differentiating spermatogonia. In meiotic stages, combining DNA staining with the synchronized transgenic mice could further improve the purity of cell populations. However, DNA staining procedures can be skipped to obtain cells with higher viability for downstream applications such as in vitro cell culture. The gating strategy for meiotic cells such as zygotene spermatocytes without Hoechst 33342 staining is shown in Fig. 4. All cells must be collected at low pressure to optimize viability. Use a 100-μm nozzle aperture while collecting late pachytene spermatocytes and diplotene spermatocytes if possible. Collect cells in prefilled FACS tubes to cushion them. It is important to note that the volume of liquid in the collection tube will increase as cells are collected.
21. Optimize the humidity and temperature; they are two important factors of the nuclear chromatin spreading during the drying-down step. Rapid or slow drying caused by a variable and/or inappropriate humidity will have effects on the intensity of the signal [21].

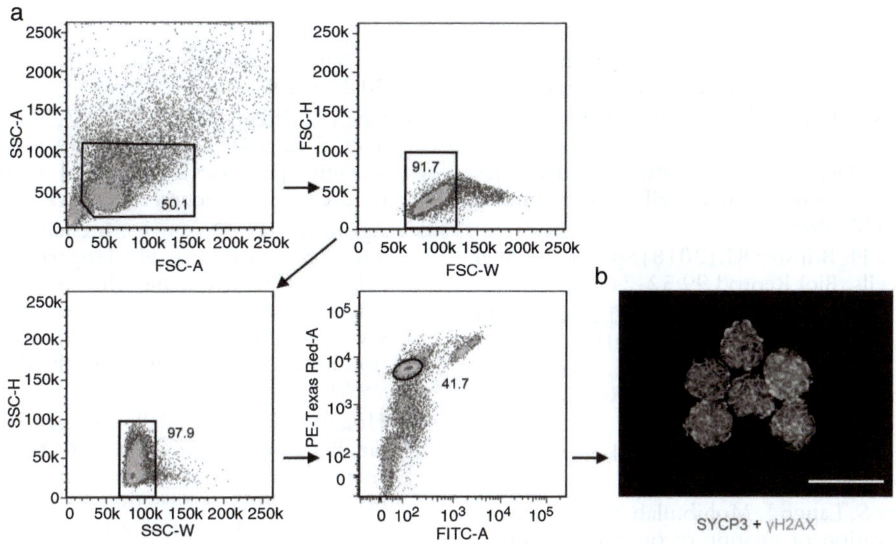

Fig. 4 Gating for zygotene spermatocytes without Hoechst 33342 staining. (**a**) Representative FACS plots showing the gating strategy for the purification of meiotic cells from synchronized mouse testes without Hoechst 33342 staining. After excluding debris (top left) and doublets (top right and bottom left), PE-Texas Red$^+$ FITC$^-$ cells are selected and collected (bottom right). Bottom right image was displayed using auto biexponential display. Percentages refer to the proportion of cells in the previous parent gate. Plots show 30,000 events. FSC-A, forward-scatter area; FSC-W, forward-scatter width; FSC-H, forward-scatter height; SSC-A, side-scatter area; SSC-W, side-scatter width; SSC-H, side-scatter height. PE-Texas Red channel was used to detect *Vasa*-mCherry fluorescence, and FITC-A channel was used to detect *Lin28*-YFP fluorescence. (**b**) Surface-spread spermatocyte nuclei immunofluorescence staining for SYCP3 and γH2AX of the sorted cells. Scale bar, 50μm

Acknowledgments

Y.C. was supported by the Youth Program of National Key Research and Development Program of China Grant No.2022YFC2702800, the Science and Technology Commission of Shanghai Municipality Grant No.23ZR1469700, the National Natural Science Foundation of China Grant No.32170861. M.-H.T. was supported by the National Key Research and Development Program of China (No.2021YFC2700200 and No.2022YFC2702602), the National Natural Science Foundation of China Grant No.31930034. Z.L. was supported by the Youth Program of National Natural Science Foundation of China Grant No.32000585. We thank the Histology, Flow Cytometry, and Vivarium services at CEMCS.

Author Contributions M.H.T. conceived the project, Y.C. designed the project, Y.C. and M.-H.T. wrote the manuscript with contributions from all authors, Y.C., Z.L., X.Z., and M.W. performed the experiments, and Y.C. and Z.L. performed the data analysis.

Competing Interests The authors declare no competing interests.

References

1. Clermont Y (1972) Kinetics of spermatogenesis in mammals: seminiferous epithelium cycle and spermatogonial renewal. Physiol Rev 52: 198–236
2. de Rooij DG (2017) The nature and dynamics of spermatogonial stem cells. Development 144:3022–3030
3. Kubota H, Brinster RL (2018) Spermatogonial stem cells. Biol Reprod 99:52–74
4. Lin Z, Hsu PJ, Xing X, Fang J, Lu Z, Zou Q et al (2017) Mettl3-/Mettl14-mediated mRNA N(6)-methyladenosine modulates murine spermatogenesis. Cell Res 27:1216–1230
5. Hunter N (2003) Synaptonemal complexities and commonalities. Mol Cell 12:533–535
6. Keeney S, Lange J, Mohibullah N (2014) Self-organization of meiotic recombination initiation: general principles and molecular pathways. Annu Rev Genet 48:187–214
7. Oakberg EF (1956) Duration of spermatogenesis in the mouse and timing of stages of the cycle of the seminiferous epithelium. Am J Anat 99:507–516
8. Barchi M, Geremia R, Magliozzi R, Bianchi E (2009) Isolation and analyses of enriched populations of male mouse germ cells by sedimentation velocity: the centrifugal elutriation. Methods Mol Biol 558:299–321
9. Bastos H, Lassalle B, Chicheportiche A, Riou L, Testart J, Allemand I et al (2005) Flow cytometric characterization of viable meiotic and postmeiotic cells by Hoechst 33342 in mouse spermatogenesis. Cytometry A 65:40–49
10. Bryant JM, Meyer-Ficca ML, Dang VM, Berger SL, Meyer RG (2013) Separation of spermatogenic cell types using STA-PUT velocity sedimentation. J Vis Exp. https://doi.org/10.3791/50648
11. Gaysinskaya V, Soh IY, van der Heijden GW, Bortvin A (2014) Optimized flow cytometry isolation of murine spermatocytes. Cytometry A 85:556–565
12. Meistrich ML, Bruce WR, Clermont Y (1973) Cellular composition of fractions of mouse testis cells following velocity sedimentation separation. Exp Cell Res 79:213–227
13. Chen Y, Lyu R, Rong B, Zheng Y, Lin Z, Dai R et al (2020) Refined spatial temporal epigenomic profiling reveals intrinsic connection between PRDM9-mediated H3K4me3 and the fate of double-stranded breaks. Cell Res 30:256–268
14. Chen Y, Zheng Y, Gao Y, Lin Z, Yang S, Wang T et al (2018) Single-cell RNA-seq uncovers dynamic processes and critical regulators in mouse spermatogenesis. Cell Res 28:879–896
15. Agrimson KS, Onken J, Mitchell D, Topping TB, Chiarini-Garcia H, Hogarth CA et al (2016) Characterizing the Spermatogonial response to retinoic acid during the onset of spermatogenesis and following synchronization in the neonatal mouse testis. Biol Reprod 95:81
16. Hogarth CA, Evanoff R, Mitchell D, Kent T, Small C, Amory JK et al (2013) Turning a spermatogenic wave into a tsunami: synchronizing murine spermatogenesis using WIN 18,446. Biol Reprod 88:40
17. Gaemers IC, van Pelt AM, van der Saag PT, de Rooij DG (1996) All-trans-4-oxo-retinoic acid: a potent inducer of in vivo proliferation of growth-arrested A spermatogonia in the vitamin A-deficient mouse testis. Endocrinology 137:479–485
18. Russell LD, Ettlin RA, Hikim APS, Clegg ED. Histological and histopathological evaluation of the testis. (1990)
19. Toyooka Y, Tsunekawa N, Takahashi Y, Matsui Y, Satoh M, Noce T (2000) Expression and intracellular localization of mouse Vasa-homologue protein during germ cell development. Mech Dev 93:139–149
20. Hamer G, Roepers-Gajadien HL, van Duyn-Goedhart A, Gademan IS, Kal HB, van Buul PP et al (2003) DNA double-strand breaks and gamma-H2AX signaling in the testis. Biol Reprod 68:628–634
21. Peters AH, Plug AW, van Vugt MJ, de Boer P (1997) A drying-down technique for the spreading of mammalian meiocytes from the male and female germline. Chromosome Res 5:66–68
22. Yuan L, Liu JG, Zhao J, Brundell E, Daneholt B, Hoog C (2000) The murine SCP3 gene is required for synaptonemal complex assembly, chromosome synapsis, and male fertility. Mol Cell 5:73–83
23. Snyder EM, Small C, Griswold MD (2010) Retinoic acid availability drives the asynchronous initiation of spermatogonial differentiation in the mouse. Biol Reprod 83:783–790

Chapter 8

Analysis of Meiotic Progression by Ex Vivo Culture of Mouse Embryonic Ovaries

Odei Barreñada, Silvia González-Sanz, Alba López-Palacios, Jesús A. Carballo, Jesús del Mazo, and Miguel A. Brieño-Enríquez

Abstract

Oogenesis is the central process required to produce viable oocytes in female mammals. It is initiated during embryonic development, and it involves the specification of primordial germ cells (PGCs) and progresses through the activation of the meiotic program, reaching a crucial phase in prophase I before pausing at diplotene around the time of birth. The significance of meiosis, particularly the prophase I stage, cannot be overstated, as it plays a pivotal role in ensuring the formation of healthy gametes, a prerequisite for successful reproduction. While research has explored meiosis across various organisms, understanding how environmental factors, including radiation, drugs, endocrine disruptors, reproductive age, or diet, influence this complex developmental process remains incomplete. In this chapter, we describe an ex vivo culture method to investigate meiotic prophase I and beyond and the disruption of oogenesis by external factors. Using this methodology, it is possible to evaluate the effects of individual xenobiotics by administering chemicals at specific points during oogenesis. This culture technique was optimized to study the effects of two selected endocrine disruptors (vinclozolin and MEHP), demonstrating that vinclozolin exposure delayed meiotic differentiation and MEHP exposure reduced follicle size. This approach also opens avenues for future applications, involving the exploration of established or novel pharmaceutical substances and their influence on essential events during prophase I, such as homologous recombination and chromosome segregation. These processes collectively dictate the ultimate fitness of oocytes, with potential implications for factors relevant to the reproductive age and fertility.

Key words Oocyte development, Meiosis, Meiotic recombination, Folliculogenesis, Ex vivo culture 3D, Reproductive aging, 3Rs, Reproductive toxicology

1 Introduction

Gametogenesis starts during embryogenesis and is initiated by the specification of primordial germ cells (PGCs) [1, 2]. In the female mouse embryo, PGCs initiate the meiotic program and enter in prophase I of meiosis at 13.5 days post-coitum (dpc) and arrest at diplotene around the time of birth [3]. Meiosis is a specialized cell division characterized by a single round of DNA replication

Jesús A. Carballo (ed.), *Meiosis: Methods and Protocols*, Methods in Molecular Biology, vol. 2818,
https://doi.org/10.1007/978-1-0716-3906-1_8,

followed by two rounds of chromosome segregation, resulting in the formation of haploid gametes. During the first meiotic division (MI), homologous chromosomes first segregate to opposite spindle poles, and this is ensued by a mitosis-like second division (MII) where sister chromatids separate into haploid gametes [4, 5]. During meiotic prophase, the homologous chromosomes must pair, synapse, and complete meiotic recombination to generate crossover events. Pairing of homologous chromosomes occurs through the formation of the synaptonemal complex, while crossover recombination occurs at the DNA level between two non-sister chromatids [6–8]. The correct completion of meiosis is fundamental for sexual reproduction and has been studied in several organisms. However, our understanding of how environmental factors such as radiation, drugs, endocrine disruptors, reproductive aging, or diet impact on fertility remains limited.

Experimental evidence clearly indicates that germ cell fate is strongly influenced by the environment in which these cells differentiate during early embryonic development. Exposure to environmental factors such as environmental pollutants including endocrine disruptor compounds contributes to induction of defects in germ cell development, disruption in oocyte meiotic progression, abnormal folliculogenesis, premature ovarian insufficiency, and epigenetic modifications [9, 10]. Epigenetic modifications like DNA methylation can produce deleterious phenotypic effects that are inherited paternally [11–13], although to a lesser extent, they are also known to be inherited maternally [14]. A great limitation in the study of mammalian female prophase I is based on the fact that the key processes such as pairing synapses and recombination occur during the embryonic development. To shed light on how environmental factors can disrupt meiotic prophase and folliculogenesis, we describe a methodology for ex vivo culture of oocytes. Building upon the initial protocols developed by Morohaku [15] and González-Sanz [16], we further refined and applied these experimental approach to investigate the potential adverse effects of different disruptors on oogenesis including meiotic prophase progression and folliculogenesis. These methods allow to culture ovaries collected at 13 dpc and maintain the culture up to 30 days during which it is possible to get secondary follicles.

In this chapter, we present a comprehensive compilation of procedures, summarized in Fig. 1, encompassing ovary explant culture and visualization techniques that enable us to investigate meiosis and oogenesis stages of cultured explants under normal conditions and upon exposure to various xenobiotic compounds. By administering a range of molecules, including reprotoxic agents, inhibitors, or regulatory compounds at specific time points during oogenesis, we can effectively assess the impact of individual compounds on distinct phases of mammalian meiosis, folliculogenesis, and oocyte development. The techniques described in this protocol

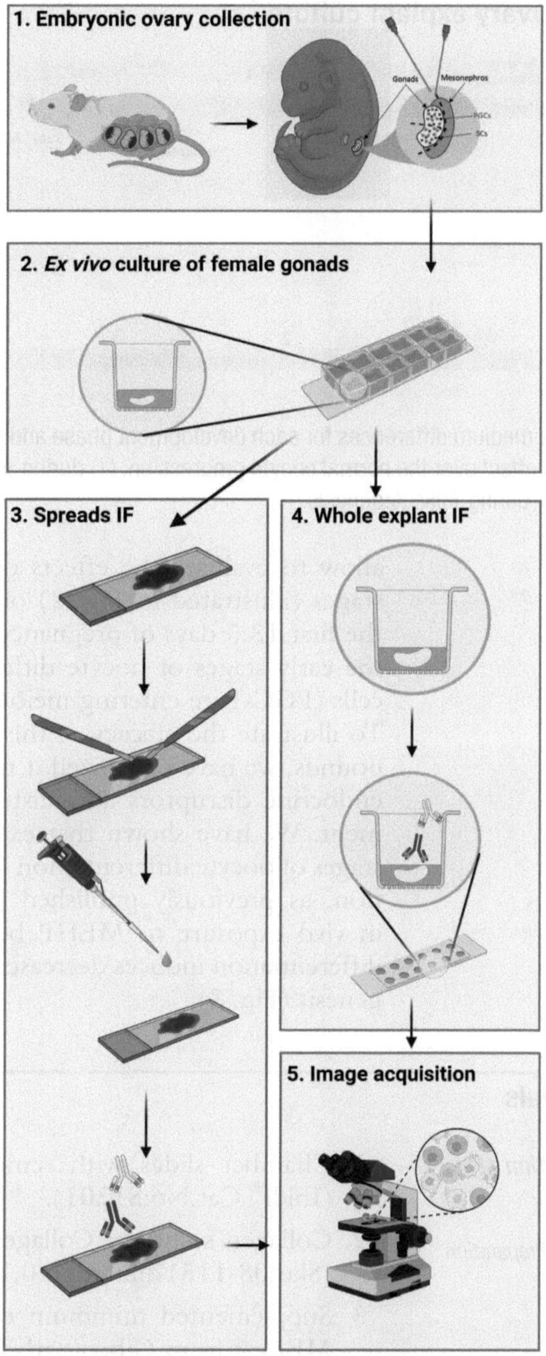

Fig. 1 Schematic diagram of proposed protocol steps including (1) embryonic ovary collection, (2) ex vivo culture of gonads, (3) optional cell disaggregation and spread, and (4) whole explant immunohistochemistry

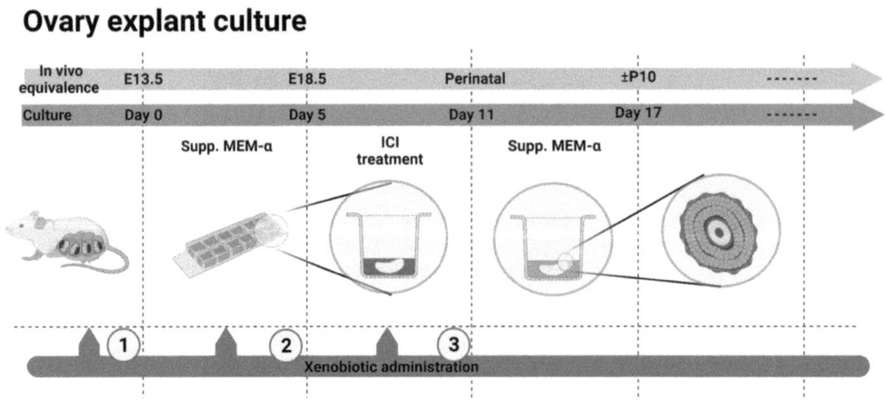

Fig. 2 Culture medium differences for each development phase and exposure to different potential compounds to assess the effect over the normal oocyte progression: (1) during pregnancy or in vivo exposure, (2) first days of culture, (3) during folliculogenesis

allow to evaluate the effects of different treatments at different stages (illustrated in Fig. 2) of ovarian development: (1) during the first 13.5 days of pregnancy (by in vivo exposure), (2) during the early stages of oocyte differentiation, when primordial germ cells (PGCs) are entering meiosis, and (3) during folliculogenesis. To illustrate the efficacy of this method for evaluating such compounds, we have optimized it to study the effects of two different endocrine disruptors administered at various stages of development. We have shown that exposure to vinclozolin during early stages of oocyte differentiation leads to delayed meiotic differentiation, as previously published [16]. Here, we demonstrate that in vivo exposure to MEHP before primordial germ cell (PGC) differentiation induces decreased follicle size during early folliculogenesis (Fig. 3).

2 Materials

2.1 Collection of Ovaries

2.1.1 Plate Preparation

1. Chamber slides with removable 12-well silicone chambers (Ibidi® Cat.No:81201).
2. Collagen solution: Collagen Type I, Rat Tail from MERCK (Sku 08-115) diluted in 0.1 M $NaHCO_3$.
3. Supplemented minimum essential medium alpha (MEM-α): MEM-α from Gibco with 0.25 g of 2-O-alpha-D-glucopyranosyl-L-ascorbic acid, 0.1% penicillin/streptomycin, and 10% fetal bovine serum (FBS), all of them from Sigma-Aldrich.

2.1.2 Animal Dissection

4. Ethanol 70%.
5. Dissecting scissors.
6. Dumont Tweezers #5.

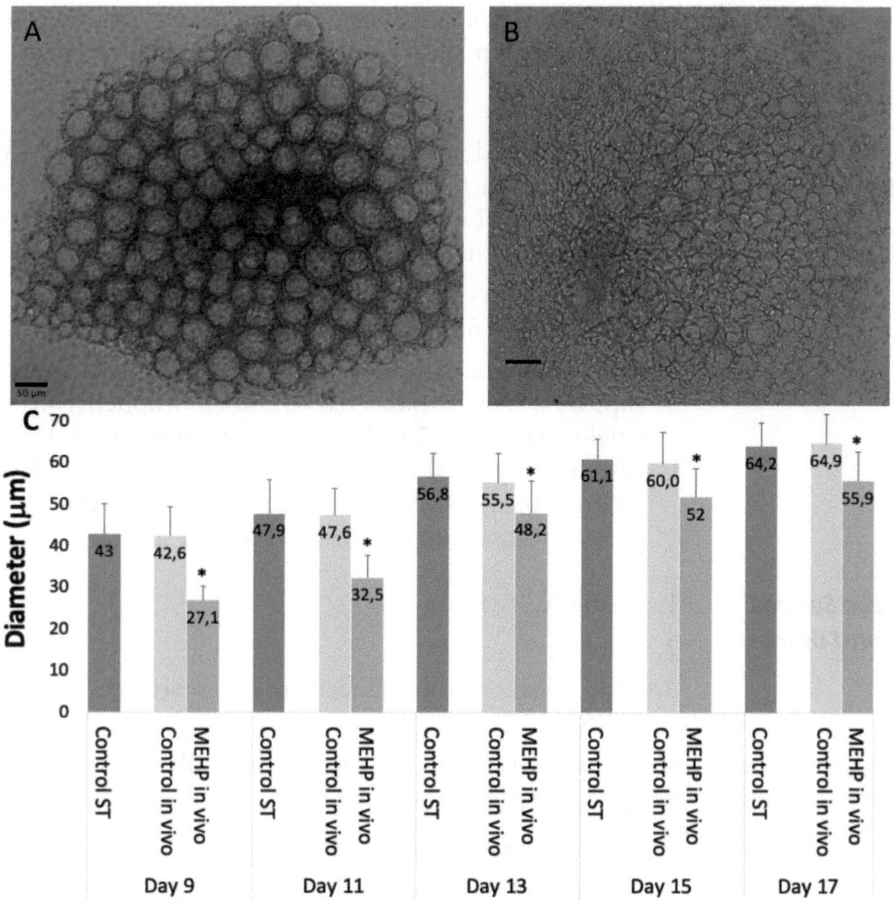

Fig. 3 Representative images of exposing animals in vivo with MEHP or control (with or without DMSO) during the first days of pregnancy (from 0.5 dpc to 13.5 dpc) in the drinking water at a concentration of 35 mg/kg/day with 1.1 mg/kg/day of DMSO. (**a**) Control explants and (**b**) explants exposed to MEHP. (**c**) Analysis of follicular growth of explants exposed to MEHP in vivo. The graph shows that this treatment negatively impacts follicular development, leading to significantly smaller follicle sizes compared to the control across all examined periods. This fact can be observed from early stages of culture, day 9, until day 17, when the experiment finished. Scale bars = 50 µm

7. Supplemented L-15 Leibovitz medium: L-15 Leibovitz media (Sigma-Aldrich, SKU L5520) with 5% FBS (Sigma-Aldrich) and 10 units/mL penicillin–streptomycin (Gibco, Cat. No 15140122).
8. Corning 100 mm × 20 mm polystyrene dish (Ref. 430167).
9. Falcon 35 × 10 mm polystyrene dish (Ref. 351008).
10. Stereomicroscope.
11. Vannas scissors.

2.1.3 Ovary Preparation

12. Tungsten needles (Fine Scientific Tools, cat. no. 10130–05) or could be made from 0.5 mm tungsten filament as indicated in [17].

2.2 Ex Vivo Culture of Female Gonads

1. Supplemented minimum essential medium alpha (MEM-α): MEM-α from Gibco with 0.25 g of 2-O-alpha-D-glucopyranosyl-L-ascorbic acid, 0.1% penicillin/streptomycin, and 10% FBS all of them from Sigma.
2. Estrogen receptor antagonist ICI 182,780 (Sigma Aldrich, Sku: 5310420001) at 5 μM in DMSO.
3. Xenobiotic compound to analyze. We used the endocrine disruptors Vinclozolin (3-(3,5-dichlorophenyl)-5-methyl-5-vinyl-oxazolidine-2,4-dione) and a phthalate MEHP (mono (2-ethylhexyl)phthalate) that were administered into the medium using DMSO as vehicle due to the hydrophobic properties of these compounds (*see* **Notes 1** and **2**).

2.3 Meiotic Spreads from Cultured Ovaries

1. Absolute ethanol.
2. Coplin jar.
3. Slide with cavities (Electron Microscopy Sciences, SKU: 71861–122).
4. Trypsin–EDTA 0.25%, phenol red (Gibco, Cn: 25200056).
5. Dumont Tweezers #5.
6. Sodium citrate solution 6% in deionized water.
7. Sucrose solution: Sucrose 100 mM in deionized water.
8. PFA-Triton solution: Paraformaldehyde (PFA) 4% + 0.2% Triton X-100 in PBS.
9. Hydrophobic Barrier Pap Pen (Thermofisher Cat. No. R3777).
10. PFA 4% in deionized water.
11. Photo-Flo solution: Kodak Photo-Flo 200 (Electronic Microscopy Science) 0.4% in phosphate buffered saline (PBS).
12. PBS solution: 137 mM NaCl, 2.7 mM KCl, 8 mM Na_2HPO_4, and 2 mM KH_2PO_4 in deionized water pH 7.4.
13. Blocking solution: PBS 1x + 5% bovine serum albumin (BSA).
14. Antibodies: Follow manufacturer's instructions for dilution range (*see* **Note 3**) (Table 1).
15. Cover glasses 24 × 24 mm.

Table 1
Antibodies: follow manufacturer instructions for dilution range

Antibody	Manufacturer	Used dilution
Anti-SYCP3 Mouse IgG1	Santa Cruz Biotech	1:500
Anti-SYCP3 Rabbit	Abcam	1:200
Anti-MLH1 Mouse	BD Pharmigen	1:50
Alexa 568 Goat anti-rabbit	Molecular probes	1:700
Alexa 488 Goat anti-mouse	Molecular probes	1:700

2.4 Explants Immunofluorescence

1. Washing solution or PBS-T: PBS with 0.1% Triton X-100 (Sigma-Aldrich, Cat. No. X100).
2. Fixation solutions:
   - Paraformaldehyde 1% in PBS with 0.15% Triton X-100.
   - Paraformaldehyde 2% in PBS with 0.15% Triton X-100.
   - Paraformaldehyde 4% in PBS with 0.15% Triton X-100.
3. Blocking solution: PBS 1x + 5% bovine serum albumin (BSA).
4. Antibodies: Follow manufacturer's instructions for dilution range (*see* **Note 3**) (Table 1).
5. Antifade Mounting Medium VECTASHIELD® with DAPI (Vector Laboratories).
6. Cover glasses 24 × 24 mm.

3 Methods

3.1 Embryonic Ovary Collection

All procedures to use and stabled mice were adhered to federal and institutional guidelines and were approved by the respective animal care committees. Opposite sex CD1 mice pairs were housed in a single cage under a 14:10 light/dark cycle at 28–30 °C and fed ad libitum. The morning in which the vaginal plug is observed is considered the day 0.5 of development.

3.1.1 Plate Preparation

1. The day before the culture, treat the Ibidi® plates with collagen solution. For this purpose, add 200 μL of collagen solution at 34 μg/mL concentration diluted in 0.1 M $NaHCO_3$ in each well and incubate at 37 °C with 5% CO_2 overnight.
2. Remove collagen solution and wash the plate with supplemented Minimum Essential Medium α (α-MEM) twice.
3. Prepare the plate with 50 μL of α-MEM just before putting the gonads.

3.1.2 Animal Dissection

4. Euthanize 13.5 dpc pregnant mice (*see* **Note 4**).
5. Spray with 70% ethanol the abdomen and make a midline incision in the skin to access to the pregnant uterus.
6. Move the uterus to a Corning plate with warmed L-15 medium.

3.1.3 Ovary Preparation

7. Open each embryo sac and put again in warmed supplemented L-15 medium.
8. To access the E13.5 ovaries of embryos, work just as adult mice, but with extra carefulness due that the tissues are more fragile than in adults. The ovaries are in the posterior wall of the abdomen and are morphologically differentiable from the male embryonic testis.
9. Use mesonephros to manipulate the ovary without damaging the tissue. Just before passing into the Ibidi® plates, use tungsten needles to remove the mesonephros, leaving the gonads clean for culture.

3.2 Ex Vivo Culture of Female Gonads

10. Cut the gonads in half (*see* **Note 5**) and put into the collagen-treated Ibidi® plates (*see* **Note 6**).
11. Cover the plates and incubate at 37 °C in a humidified atmosphere with 5% CO_2 for 30–45 minutes.
12. After this time, add 20 μL of medium and return to the incubator in same conditions (*see* **Notes 7** and **8**).
13. Every other day, remove 50% percent (35 μL) of culture media and replace with fresh medium (*see* **Note 9**).

If one wants to maintain the culture for more than 5 days (equivalent to 18.5 dpc in vivo), follow next steps:

14. Replace the full medium with new medium containing the estrogen receptor antagonist ICI 182,780 at 5 μM to improve follicle assembly in vitro [17]. In addition, add the xenobiotic compounds of interest to analyze the effect on the oocytes progression (*see* **Note 10**).
15. The day 11 of culture (equivalent to perinatal period in vivo) replaces again all volume with alpha-MEM media without ICI treatment.
16. Keep changing half of the media every 2–3 days. We were able to maintain the gonads up to 30 days (Fig. 4).

3.3 Meiotic Spreads from Cultured Ovaries

For optimal observation of individual cells, such as during meiotic phase exploration, it is recommended to spread the cells (*see* **Note 11**) as show in Fig. 5.

1. The day before, immerse the slides in absolute ethanol overnight in a Coplin jar.

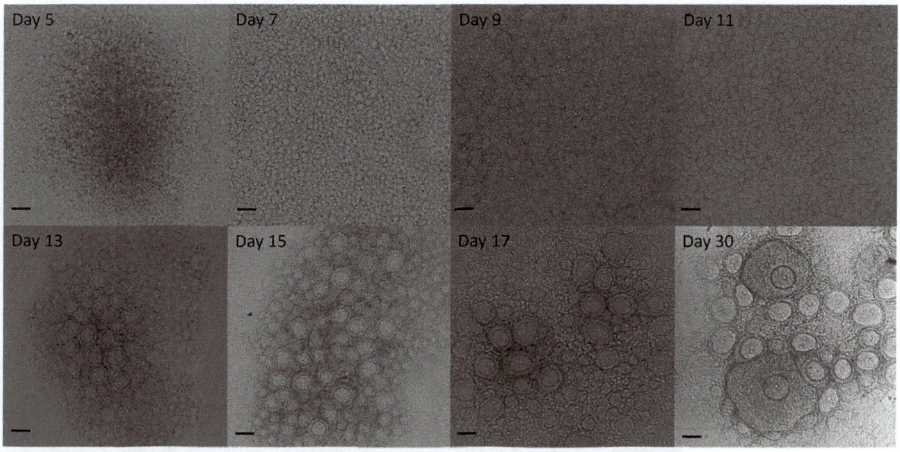

Fig. 4 Development of the ovary explants up to 30th day. Scale bars = 50 μm

2. Remove the medium and incubate cultured explant in 50 μL of Trypsin–EDTA solution at room temperature for 5 min in glass depression slides.
3. To help the disaggregation, use Dumont Tweezers #5 to complete mechanical tissue breakdown.
4. Add 10 μL of 6% sodium citrate and incubate for 30 minutes at room temperature.
5. Slowly replace 60 μL of the solution with a pipette so that cells and tissue debris are retained at the bottom of the recipient for 50 μL of sucrose solution (*see* **Note 12**).
6. During this period, prepare slides with two drops of 50 μL PFA-Triton solution. It is also possible to use a hydrophobic pen at this point surrounding the tissue to avoid reagents dispersion.
7. Dispense 25 μL of cell suspension onto each PFA-Triton solution drop, allowing it to fall from a height of 2 cm (*see* **Note 13**).
8. Incubate the slide overnight in humid chamber at room temperature.
9. Allow the slides to air-dry at room temperature.
10. Wash for 10 min in a Coplin jar with Photo-Flo solution.
11. Air-dry the slides for 10 min at room temperature before washing again with PBS.
12. Incubate with blocking solution for 1 h.
13. Follow protocol from Subheading 3.4, **step 6** for spread immunostaining.

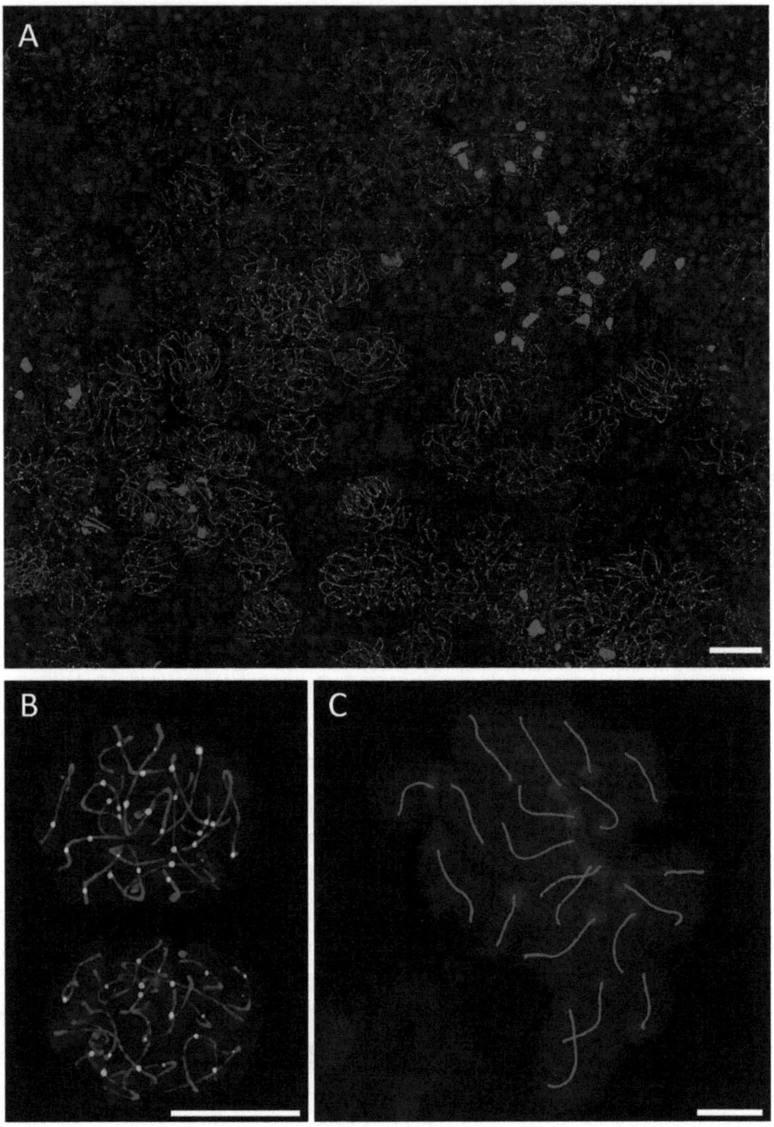

Fig. 5 Differences between whole explant and spread immunostaining. Whole ovary explant immunofluorescence (**a**) with a detailed view of individual cells (**b**) and the result of spreading for meiotic study (**c**). In all cases, DNA was stained using DAPI (blue) and the axial/lateral elements were detected using anti-SYCP3 (red). In A and B, anti-MLH1 (green) was used as a crossover marker. Scale bars = 10 μm

3.4 Explant Immunofluorescence

Upon reaching the desired stage of development, the explants are ready for immunofluorescence staining; follow next steps to proceed:

1. Remove all the medium in the samples and add 200 μL PBS-T.
2. Rinse 2 times in the same solution.

3. Fix ovaries by sequential incubations in 1%, 2%, and 4% fixation solutions for 20 minutes each at room temperature.
4. Rinse for 3 more times in PBS-T.
5. For blocking, leave the ovaries in blocking solution for 30–45 min.
6. Incubate overnight in a humidity chamber at 4 °C with the desired antibodies diluted according to manufacturer's instructions in blocking solution overnight (*see* **Note 14**).
7. Wash in PBS-T three times.
8. Add secondary antibody and incubate at 37 °C for 1 h in humidity chamber.
9. Remove secondary antibody, quit removable 12-well silicone, and wash one last time in PBS-T for 15 min in a coplin jar.
10. Apply 4 μL of VECTASHIELD mounting media per 24 × 24 mm coverslip.

3.5 Image Acquisition

1. Follicle development images were taken using in situ microscopy phase contrast, for example, with a Nikon Eclipse TE300 microscope equipped with a Nikon DXM1200 camera and the ACT-1 for DXM1200 and DXM1200F programs.
2. Whole-mount preparation images were collected using fluorescence microscopy (e.g., Leica) and confocal laser microscopy (Leica SP5).
3. Spread preparation images were taken using fluorescence microscopy, such as BX41 (Olympus, Japan), and laser scanning confocal microscopy SP8 STED 3X (Leica, Germany).

4 Notes

1. Even when using DMSO, it may be difficult to mix properly; in that case, white particles may form in the medium. To ensure proper mixing, try shaking, sonicating, or increasing DMSO concentration.
2. Be careful, a lot of endocrine disruptors such as Vinclozolin are light sensitive. To avoid decay in compound activity, cover the vials with aluminum foil and keep in dark.
3. As a standard we started with 1:100 dilution and kept going to a lower dilution. The dilutions used for these images are indicate in Table 1.
4. At this stage, the number of PGCs has just increased in number before in the female line arrest at meiotic prophase I.
5. The two halves of the embryonic ovary were spaced apart to prevent cell clumping.

6. To introduce the gonads into the Ibidi plates, use glass Pasteur pipettes.
7. It is recommended to use the first column and the last column in Ibidi plate as moisturizer, fill with PBS or filtered water to prevent culture media from evaporating.
8. To avoid the detachment of the gonads, carefully introduce the media by pipetting in the corner of the Ibidi well.
9. When carrying out the gonad culture, replacing half the culture medium retains a portion of the growth factors secreted by the cells [18].
10. The timing of compound addition during the culture period should be determined based on the specific stage of oogenesis that needs to be analyzed or the target of the xenobiotic compound.
11. This disaggregation disrupts the original tissue structure, so this step could be skipped when it is desired to observe the structural development.
12. This step is recommended to be carried out in a stereo microscope.
13. Alternatively, the solution can be gently spread using a pipette tip to facilitate even distribution. To prevent damage to the fragile tissue, careful handling is essential.
14. In our experience, 70 μL per Ibidi chamber is enough.

Acknowledgments

Authors contribution: OB, SG-S, and AL-P have contributed equally to this study. JAC, JdM, and MAB-E share corresponding authorship. Study design and data analysis were performed by JAC, JdM, and MAB-E. Experimental work and data analysis were performed by OB, SG-S, and AL-P. OB, JAC, JdM, and MAB-E wrote and revised the manuscript. We would like to thank the Animal House and the Confocal Microscopy facility at CIB Margarita Salas for their invaluable assistance. JAC was funded by the Ministry of Science and Innovation of Spain, grant number: RYC-2013-13950 and BFU2015-64361-P and JdM by the Ministry of Science and Innovation of Spain (BFU2017-87095-R). Figures 1 and 2 were created with BioRender.com.

References

1. Saitou M, Yamaji M (2012) Primordial germ cells in mice. Cold Spring Harb Perspect Biol 4: a008375
2. Sybirna A, Wong FCK, Surani MA (2019) Genetic basis for primordial germ cells specification in mouse and human: conserved and divergent roles of PRDM and SOX transcription factors. In: Current topics in developmental biology. Elsevier, pp 35–89
3. Feng C-W, Bowles J, Koopman P (2014) Control of mammalian germ cell entry into meiosis. Mol Cell Endocrinol 382:488–497

4. Bolcun-Filas E, Handel MA (2018) Meiosis: the chromosomal foundation of reproduction. Biol Reprod 99:112–126
5. Zickler D, Kleckner N (2015) Recombination, pairing, and synapsis of homologs during meiosis. Cold Spring Harb Perspect Biol 7: a016626
6. Cohen PE, Holloway JK (2015) Mammalian meiosis. In: Knobil and Neill's physiology of reproduction. Elsevier, pp 5–57
7. Svetlanov A (2004) Mismatch repair proteins, meiosis, and mice: understanding the complexities of mammalian meiosis. Exp Cell Res 296: 71–79
8. Gray S, Cohen PE (2016) Control of meiotic crossovers: from double-strand break formation to designation. Annu Rev Genet 50:175–210
9. Ewen KA, Koopman P (2010) Mouse germ cell development: from specification to sex determination. Mol Cell Endocrinol 323:76–93
10. Kocer A, Reichmann J, Best D et al (2009) Germ cell sex determination in mammals. Mol Hum Reprod 15:205–213
11. Brieño-Enríquez MA, García-López J, Cárdenas DB et al (2015) Exposure to endocrine disruptor induces transgenerational epigenetic deregulation of MicroRNAs in primordial germ cells. PLoS One 10:e0124296
12. Chen Q, Yan W, Duan E (2016) Epigenetic inheritance of acquired traits through sperm RNAs and sperm RNA modifications. Nat Rev Genet 17:733–743
13. Gapp K, Bohacek J (2018) Epigenetic germline inheritance in mammals: looking to the past to understand the future. Genes Brain Behav 17: e12407
14. Ge Z-J, Sun Q-Y (2019) Chapter 5 – Maternal epigenetic inheritance. In: Tollefsbol TO (ed) Transgenerational epigenetics, 2nd edn. Academic, pp 75–105
15. Morohaku K, Tanimoto R, Sasaki K et al (2016) Complete in vitro generation of fertile oocytes from mouse primordial germ cells. Proc Natl Acad Sci USA 113:9021–9026
16. González-Sanz S, Barreñada O, Rial E et al (2020) The antiandrogenic vinclozolin induces differentiation delay of germ cells and changes in energy metabolism in 3D cultures of fetal ovaries. Sci Rep 10:18036
17. Morohaku K, Hirao Y, Obata Y (2017) Development of fertile mouse oocytes from mitotic germ cells in vitro. Nat Protoc 12:1817–1829
18. Masters JR, Stacey GN (2007) Changing medium and passaging cell lines. Nat Protoc 2:2276–2284

Chapter 9

The Organotypic Culture of Mouse Seminiferous Tubules as a Reliable Methodology for the Study of Meiosis In Vitro

Pablo López-Jiménez, Inés Berenguer, Irene Pérez-Moreno, Julia González de Aledo, María Teresa Parra, Jesús Page, and Rocío Gómez

Abstract

Male mouse meiosis has been traditionally studied using descriptive methods like histological sections and spreading or squashing techniques, which allow the observation of fixed meiocytes in either wildtype or genetically modified mice. For these studies, the sacrifice of the males and the extraction of the testicles are required to obtain the material of study. Other functional in vivo studies include the administration of intravenous or intraperitoneal drugs, or the exposure to mutagenic agents or generators of DNA damage, in order to study their impact on meiosis progression. However, in these studies, the exposure times or drug concentration are important limitations to consider when acknowledging animal welfare. Recently, several approaches have been proposed to offer alternative methodologies that allow the in vitro study of spermatocytes with a considerable reduction in the use of animals. Here we revisit and validate an optimal technique of organotypic culture of fragments of seminiferous tubules for meiotic studies. This technique is a trustable methodology to develop functional studies that preserve the histological configuration of the seminiferous tubule, aim homogeneity of the procedures (the use of the same animal for different study conditions), and allow procedures that would compromise the animal welfare. Therefore, this methodology is highly recommendable for the study of meiosis and spermatogenesis, while it supports the principle of 3R's for animal research.

Key words Mouse, Organotypic testicular tissue culture, Spermatogenesis, Meiosis

1 Introduction

Animal experimentation with mammals has been used in basic and applied research [1]. In particular, the house mouse, *Mus musculus,* is extensively used as mammalian model for experimental research, contributing to finding solutions to biological and biomedical questions related to human health. However, stemming from the proposal by Russell and Burch for the replacement, reduction, and refinement of studies involving live animals, the 3R's rule [2], many

Jesús A. Carballo (ed.), *Meiosis: Methods and Protocols*, Methods in Molecular Biology, vol. 2818,
https://doi.org/10.1007/978-1-0716-3906-1_9,

laboratories, researchers and scientific agencies have encouraged in the past years the development of alternative methods. This is recommended not only by ethical reasons, avoiding suffering, pain, stress, and/or death of animals, but also by the aim of achieving optimized in vitro methodologies that improve the quantity and quality of the research data.

The production of male gametes is called spermatogenesis, which is the transformation of spermatogonial diploid cells into haploid spermatozoa, a process that occurs over an extended period of time inside the testis [3, 4]. Spermatogenesis comprises three events: i) stem germ cells (spermatogonia) proliferation, ii) meiosis of spermatocytes, and iii) maturation of spermatids into spermatozoa (spermiogenesis) [5]. Meiosis itself implies two rounds of chromosome segregation with a single premeiotic event of DNA replication [6], giving rise to the production of haploid spermatids. These later undergo dramatic morphological changes resulting in the maturation of fully competent motile spermatozoa [7].

At the histological level, spermatogenesis takes place in the seminiferous epithelium of the testis, a highly specialized cell environment. Growth factors and hormones tightly regulate many of these crucial steps leading to the formation of spermatozoa [5]. The organization of the seminiferous epithelium is unique since it is composed of a monolayer of the highly specialized Sertoli cells, which facilitate the progression of germ cells to spermatozoa via direct contact with them and by influencing both the development of spermatogonial germ cells and meiocytes [8, 9]. Therefore, the histological conformation of the seminiferous epithelium is a key requirement of gamete production in mammals [5].

Male mouse meiosis has been traditionally studied using descriptive methods like cell spreading [10] or squashing [11] and histological sections or electron microscopy. In virtually all cases, seminiferous tubules are freshly taken from sacrificed individuals and immediately processed. Nevertheless, several studies have pursued the development of in vitro systems. The first dated attempt was carried out in 1920 [12]. Later, many studies attempted to develop spermatogenesis in vitro by culturing dissociated testis cells. However, those studies reported that the entry of premeiotic cells into meiosis was blocked under in vitro conditions by erroneous or ineffective contact between Sertoli and germ cells [13]. Recent studies have progressed into this exciting context through the use of spermatogonial stem cells for in vitro spermatogenesis and potential in vivo applications for the restoration of fertility [14, 15]. Moreover, key advances in biomedical research have permitted the application of in vitro mouse and human spermatogonial stem cells proliferation to achieve their transplantation into the seminiferous tubules of infertile mouse or men to restore fertility [16–18], yet the isolation of spermatogonial stem cells implies several complex procedures. For this reason, nowadays, the most successful results have been obtained from the organotypic culture of seminiferous tubules. In 2011, a hybrid gas-liquid interphase method was established by Sato and colleagues, where germ stem cells were capable of producing fertile spermatids and sperm in vitro [19]. In that publication, neonatal mouse testes containing primitive spermatogonia could produce spermatids and reproductively competent sperm in vitro with a serum-free culture media over an agarose gel. Those authors also combined the organ culture method with a spermatogonial stem cell transplantation technique [19, 20]. This organotypic culture method set the basis for the use of in vitro controlled manipulation of the micro-environmental conditions required for proper murine spermatogenesis. Further studies succeeded in producing functional sperm from primitive spermatogonia in explanted neonatal mouse testis tissues [21]. Since then, several researchers have used this methodology to display functional approaches to the study of mouse spermatogenesis [22–30]. Latest advances using modified methodologies built upon the foundation of original protocols have proposed that culturing isolated spermatogonia tubules, as opposed to tissue masses, offer distinct advantages in precisely evaluating the diverse environmental factors influencing the progression of spermatogenesis [31]. Alternatively, the polydimethylsiloxane-chip ceiling method is suggested to ensure a uniform supply of nutrients and oxygen throughout the testicular tissue [32]. Moreover, the integration of single polydimethylsiloxane-chip platforms, combined with real-time cellular paracrine support from allogeneic bone marrow mesenchymal stem cells, holds promise in providing a steady and stable microfluidic flow for the advancement of spermatogonial cultures [33]. Nonetheless, accomplishing a complete in vitro spermatogenesis capable of yielding a quantity of spermatozoa comparable to age-matched in vivo controls remains a significant technical challenge, especially in the context of human reproduction.

Besides the exciting advances for the biomedical applications of spermatogonia stem cells research as an approach to deal with human infertility, basic science also can benefit from the use of in vitro methodologies [34]. While the main aim of previous studies has been setting the conditions for successful production of spermatozoa, this can be also applied to more basic research, as for instance the study of meiosis. This special kind of cell division is crucial to increase the variability in the offspring. Furthermore, the proper occurrence of genetic and chromosomal events during meiosis is critical for the integrity of the genome that is transmitted from one generation to the next one. Indeed, many of the inherited chromosome aberrations are produced as defects in chromosome pairing, recombination, or segregation during meiosis.

Contrary to other biological models in which meiotic cells manipulation in culture is feasible, study of meiosis in mammals has been partially obstructed by the technical difficulties to perform in vitro experiments. However, the ethical concerns in the application of physical or chemical treatment that can cause animal stress of pain, as well as the large number of animals needed to carry out some experiments, are pushing toward the development of alternative methods or the use of alternative models. Therefore, the implementation of in vitro techniques for the study of mammalian male meiosis is an interesting and necessary tool in meiosis research. In the past years, we have successfully used some of these methods, the organotypic culture of seminiferous tubules being the most successful and reliable technique. We have particularly used the protocol described by Sato *et al*, with some specifications and modifications. Here we present a detailed description of the improved protocol that we have previously used [35–38], including specific tips and validation data, and also indicate some limitations of use.

2 Materials

2.1 Sample Extraction

1. Testes from prepuberal or adult mice are extracted after euthanasia of the individuals, following local and regional ethics committees' guidelines. Testes are detunicated with the help of precision tweezers in a glass petri dish with 1 mL of PBS 1×. Detunicated seminiferous tubules should remain in PBS 1× at 34 °C until culture.
2. Seminiferous tubules are slightly disentangled with two forceps and cut transversally in small fragments with a sterile razor-blade. These fragments are then be cultured in 12-well plates and maintained at 34 °C in an atmosphere with 5% CO for the duration of the study following the protocol explained below.

2.2 Agarose Gel and Media Preparation

1. 20 mL of a 1.5% agarose gel are prepared by dissolving agarose in distilled water in a microwave. The agarose gel is then let to solidify in a 9 cm diameter glass petri dish.
2. Medium: MEMα culture medium supplemented with 10% KnockOut Serum Replacement (KRS) and 1% antibiotics (penicillin/streptomycin) is prepared and kept at 34 °C until use.
3. For inhibition studies, the selected drug should be diluted in the culture medium at the correspondent concentration.

2.3 Slides Preparations

1. Slides for the squashing technique: the slides are washed in a 1:1 ethanol-acetone solution and covered with poly-L-lysine (Sigma) to promote cell adhesion.

2. Slides preparation for the spreading technique: the slides are washed in a 1:1 ethanol-acetone solution and kept dry until use.

3 Methods

3.1 Organotypic Cultures of Control Seminiferous Tubules

1. 15 × 15mm sections of 3.2 mm thick agarose squares are placed in each well of the culture plate (Fig. 1a) (*see* **Notes 1** and **2**).
2. 2 mL of medium is placed in each well of a sterile 12-well culture plate (Fig. 1a).
3. Culture plates are kept overnight at 4 °C to allow the interchange of water and MEMα between the liquid and agarose. After 24 h, the medium of the wells is refreshened. For control

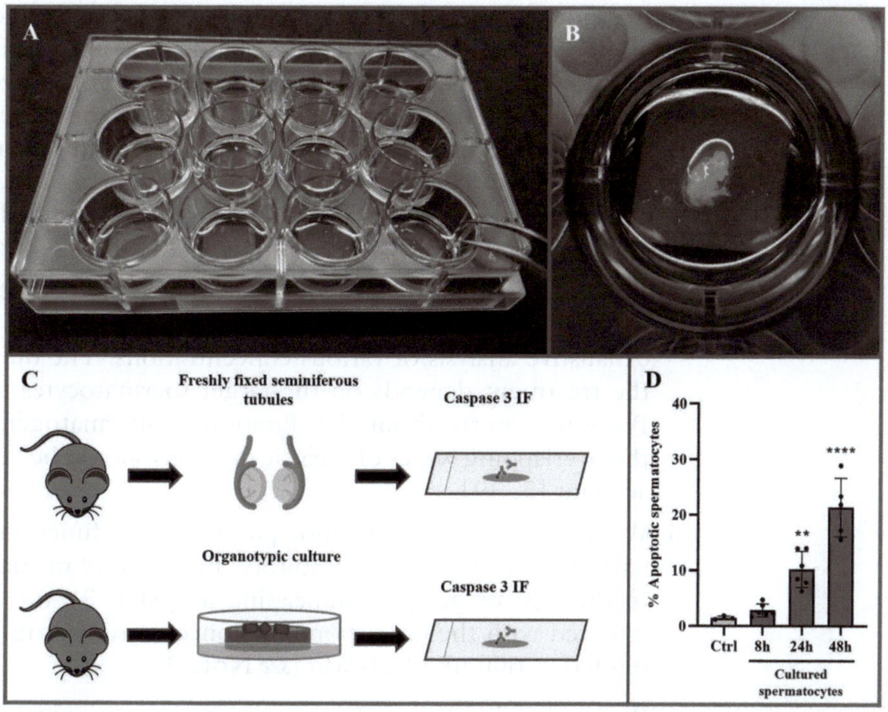

Fig. 1 Organotypic cultures of control seminiferous tubules. (**a**) Photograph of a representative 12-well culture plate containing medium and agarose gels. (**b**) Organotypic culture of seminiferous tubules in a selected well. The image shows the media, the agarose gel, and the fragments of seminiferous tubules. (**c**) Experimental approach for the comparison of the level of apoptosis between freshly fixed seminiferous tubules extracted from dissected testis and the cultured seminiferous tubules with fixation after culture. In both cases, immunolabeling of cleaved Caspase 3 was used to assess the extent of apoptosis. (**d**) Quantitative analysis of the percentage of apoptotic spermatocytes in freshly fixed seminiferous tubules and the cultured seminiferous tubules (at least three biological replicates for each condition, minimum of 500 spermatocytes per condition and individual). Data represents mean ± SD, $p < 0.05$ (**), $p. < 0.005$ (****), one-way ANOVA

studies, 600 μL of fresh medium is added to each well. This should not cover the upper surface of the agarose gel.

4. The seminiferous tubules of freshly detunicated testis are cut into small pieces of approximately 5 mm length. The fragments of seminiferous tubules are placed over the half-soaked agarose gel and are kept in an air-liquid interphase (Fig. 1b) (*see* **Note 3**).
5. Well plates are kept at 34 °C in an atmosphere with 5% CO for the duration of the study.
6. After appropriate time, fragments of seminiferous tubules are moved to a Petri dish containing PBS 1× until used for squashing or spreading techniques.

3.2 Inhibition Studies with the Organotypic Culture Method

1. To inhibit proteins implicated in meiosis progression in cultures of seminiferous tubules, the required concentration of the drug is usually between 10× and 100× higher than the concentration used for comparable studies with somatic cell cultures (*see* **Notes 5** and **6**). This is due to the characteristics of the organotypic culture *per se*, as the drug diluted in the medium is absorbed through the agarose gel and across the seminiferous tubules before reaching the spermatocytes. This procedure is drastically different to adding drugs on top of somatic adherent cells cultures. Examples of these functional studies comparing the inhibition of selected proteins in vitro using organotypic cultures to the equivalent in vivo genetically modified models enhance the reliability of this method [35–38].
2. When designing the experimental approach, the optimized concentration of the drug needs to be identified after an exhaustive analysis of various concentrations. The duration of the treatment depends on the target spermatocytes to study. Previous reports about the duration of spermatogenesis and the overlapping waves of meiotic divisions should be taken into account [5, 39].
3. When developing the organotypic culture for functional assays with the use of drugs or inhibitors, after 24 h of media transfer to the agarose gel (*see* Subheading 3.1, **step 3**), the media is replaced with the corresponding concentration drug of study diluted in 600 μL of MEMα (*see* **Note 3**).

Both control and treated spermatocytes of organotypic cultures can be processed by either the squashing or the spreading methodologies. The steps are provided below in Subheadings 3.3 and 3.4.

3.3 Squashing of Cultured Seminiferous Tubules

1. The cultured fragments of seminiferous tubules are moved to a new glass chamber containing a 2% PFA (BDH AnalaR) + 0.05% Triton X-100 (Sigma) fixation solution for 15 (*see* **Notes 1** and **2**).

2. Preparation of squashed seminiferous tubules is performed following conventional previous protocols [11]. In brief, two or three 2 cm length fragments of the fixed post-cultured seminiferous tubules are deposited in a slide that contains poly-L-lysine (*see* **Note 4**). The fragments of seminiferous tubules are carefully disintegrated with the aid of curved toothed forceps. Next, an additional drop of formaldehyde-Triton fixative was added, and a coverslip is placed on top of the material. Holding the coverslip from one of the corners, the fragments of the seminiferous tubules are softly squeezed by passing a pen over the coverslip. Then, the cells are squashed by applying pressure with the thumb on the coverslip to obtain a monolayer of cells.
3. The samples are then submerged in liquid nitrogen and stored at −80 °C until future use. In case they are to be used immediately, the coverslip is removed with the help of a razor, and slides are rehydrated in PBS. The rest of the preparations can be stored at −80 °C for long periods of time. When needed, they should be immersed again in liquid nitrogen for a few seconds. Then, the coverslip is removed with the help of a razor and the slides are rehydrated in PBS. The following steps are equal to the fresh preparations produced on the day of the organotypic culture procedure.
4. Immunolabeling methods of the squashed preparations can be done following standard protocols. In brief, the slides with the squashed seminiferous tubules are rinsed three times for 5 min in PBS and incubated for 1 h at 37 °C, or overnight at 4 °C, with primary antibodies diluted in PBS. Then, the slides are rinsed three times for 5 min in PBS and incubated for 1 h at room temperature with the secondary antibodies. Finally, the slides are counterstained with 10 μg/mL DAPI for 3 min, rinsed in PBS, and mounted with Vectashield (Vector Laboratories).

3.4 Spreading of Spermatocytes Extracted from Cultures Seminiferous Tubules

1. Preparations of spreaded spermatocytes are performed following conventional previous protocols [10] (*see* **Notes 1** and **2**). In summary, the post-cultured fragments seminiferous tubules are placed in a glass petri dish with 300 μL of PBS, minced with a scalpel, and disaggregated with two curved forceps. 3 mL of PBS is then added, and the solution is poured into a centrifuge tube. The content is resuspended with a Pasteur pipette and let to settle for 5 min to precipitate the larger fragments. The supernatant is then collected and transferred to a new tube that is centrifuged for 8 min at 1200 r.p.m. After that, most of the supernatant is discarded, leaving a small volume for resuspension with the rest of the material. The tube is then gently tapped to homogenize the mixture. Another 4 mL of PBS is added for a second centrifugation of 8 min at 1200

r.p.m. Supernatant is discarded, and 300 μL of 100 mM sucrose was added slowly to the pellet, resuspending every 3 drops. This cell suspension is then spread on slides that had been previously immersed in 1% paraformaldehyde in distilled water (pH = 9) containing 0.15% of Triton X-100. 15 μL of the suspension is added per slide and placed horizontally in a humid chamber. Preparations are kept in humid and dark conditions for 2 h. Afterward, the chamber cover is partially opened, and the samples were let to completely dry. After that, they are washed with 0.4% Photoflo (Kodak) in distilled water twice. Once dried, they are stored at −80 °C, except those that are to be used immediately, which were rehydrated in PBS.

2. Immunolabeling methods of the squashed preparations can be done following standard protocols (explained in the previous section for the squashing technique).

3.5 Validation and Reliability of the of Organotypic Culture Method

In vitro studies are never fully equivalent to the in vivo condition as they lack histological and endocrine regulation; therefore, it is important to carry out validation tests. In order to verify if the organotypic culture method is trustworthy for meiotic studies, we have carried out three experiments that identify the level of suitability and reliability of this methodology.

1. First, we conducted a study to determine the level of spermatocyte apoptosis in organotypic cultured seminiferous tubules at different times. For this, we immunolabeled cleaved Caspase 3 and quantified spermatocytes with positive labeling in freshly fixed seminiferous tubules directly extracted from the testis (in vivo condition) and in squashed preparations of cultures spermatocytes at 8, 24, and 48 h (in vitro condition) (Fig. 1c). In all time points, spermatocytes are still alive, but increasing levels of apoptosis are observed with culture time (Fig. 1d). Although the level of apoptosis after 48 h is around 20% average, the spermatocytes that are not undergoing apoptosis show a normal pattern of distribution of the meiotic usual markers (discussed below). We conclude that the organotypic culture of seminiferous tubules is a trustable methodology for up to 48 h. After that period, researchers should take into consideration the potential side effects of the in vitro methodology.
2. Second, we conducted an experiment aiming to assess if organotypic cultured spermatocytes accumulate DNA damage as a consequence of the culture. To do this, we analyzed by immunolabeling the localization patterns of Synaptonemal Complex Protein 3 (SYCP3), a marker of synapsis progression [40], and γH2AX, a known marker of meiotic DNA double-strand breaks (DSBs) [41]. Results showed that DNA endogenous damage

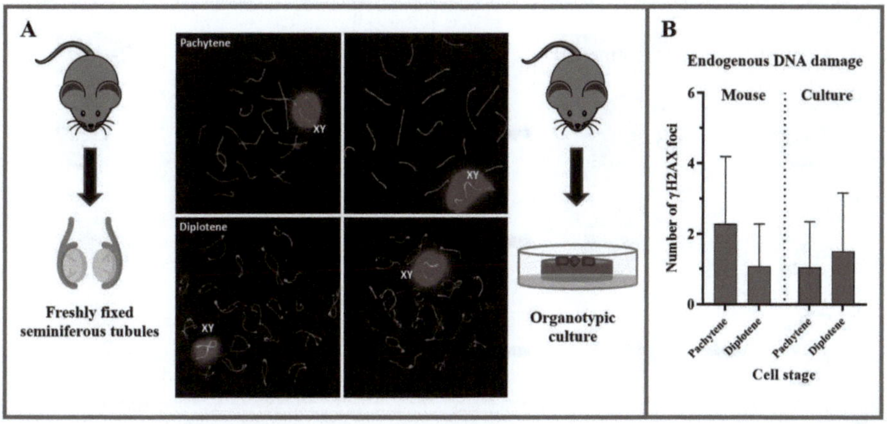

Fig. 2 Validation of the pattern of distribution of meiotic markers in organotypic cultures of control seminiferous tubules. (**a**) Double immunolabeling of SYCP3 (green) and γH2AX (red) of squashed spermatocytes at pachytene and diplotene stages from freshly processed (left) or organotypically cultured seminiferous tubules (right). (**b**) Quantitative analysis of the level of endogenous DNA damage detected by γH2AX in all the experimental conditions ($n = 20$ spermatocytes for each condition in three biological replicates)

patterns are equivalent in spermatocytes directly extracted from the testis and in spermatocytes obtained from seminiferous tubules cultured for 24 h (Fig. 2a). We concluded that there is no significant difference between the pattern of distribution of the meiosis markers between the in vivo and the in vitro conditions, at least during the stages of prophase I, and corroborated that culture do not induce new DSBs (Fig. 2b).

3. A final experiment was conducted to address the suitability of the organotypic culture method for functional studies on DNA repair. To do this, we studied the effect of gamma radiation. We compared three different conditions (Fig. 3a): (a) 3 different mice were treated with 5 Gy of gamma radiation and sacrificed at 1, 6, or 24 h after treatment, respectively; (b) a mouse was irradiated similarly, immediately sacrificed, and the seminiferous tubules maintained in culture and subsequently processed at 1, 6, or 24 h; (c) tubules of a control mouse were irradiated while in culture and subsequently processed at 1, 6, or 24 h. The results showed that the three approaches were equally efficient to induce exogenous DNA damage, with no significant differences in the total number of induced γH2AX foci (Fig. 3b, correspondent to condition c). Moreover, the dynamics of DNA repair is comparable in all the situations, with progressive reduction of γH2AX foci at 6 and 24 h. Only a small difference could be observed 24 h after irradiation, as both cultured conditions displayed a slightly higher number of remaining γH2AX foci (Fig. 3c). Therefore, although the culture of the seminiferous tubules does not substantially perturb

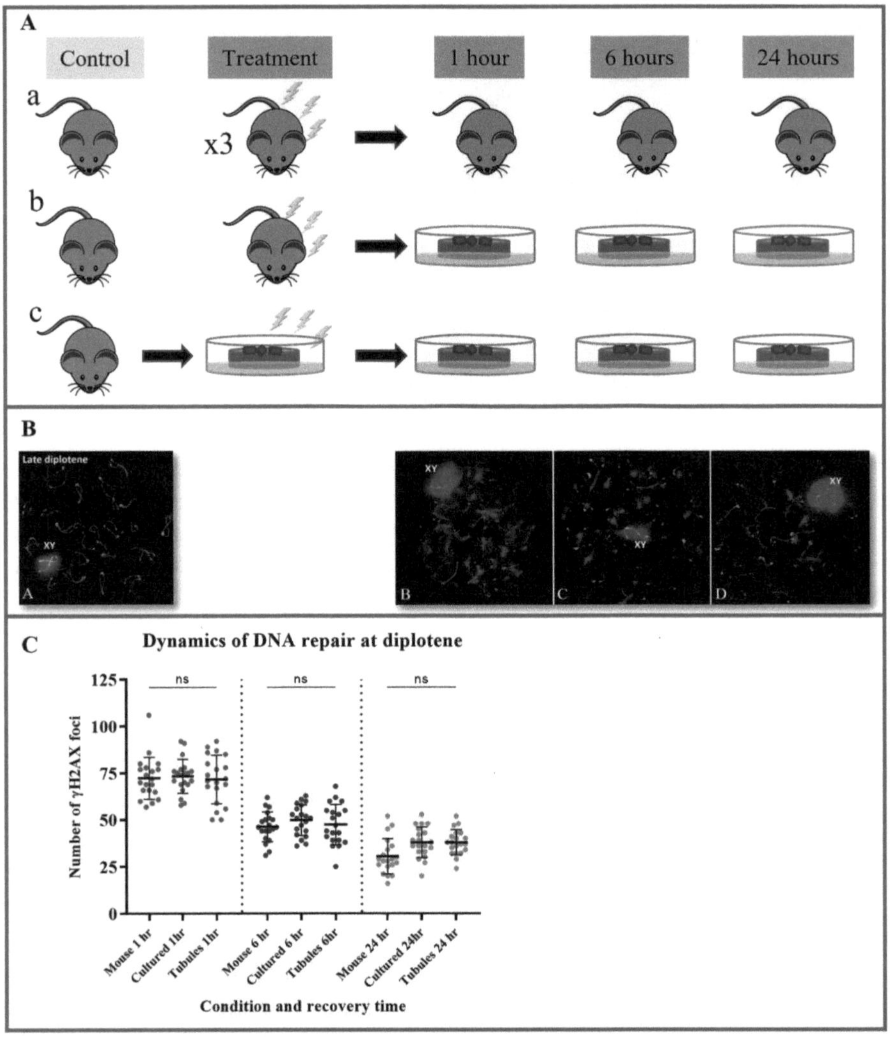

Fig. 3 Validation of the DNA damage repair in organotypic cultures of control seminiferous tubules. (**a**) Experimental approach for the comparison of the DNA damage repair in three experimental conditions (see main text): a) irradiation of alive animals and processing of fresh material (top). This required the use of four animals: one control and three irradiated mice processed at different recovery times (1, 6, and 24 hours); b) irradiation of alive animals and subsequent organotypic culture of dissected seminiferous tubules (middle). Seminiferous tubules were processed at the three different recovery times. This approach required the use of two animals: one control and one irradiated and c) irradiation of cultured fragments of seminiferous tubules, which were processed at different recovery times (bottom). This required the use of only one animal. (**b**) Representative images of the distribution of SYCP3 (green) and γH2AX (red) after immunolabeling on spread spermatocytes in control and different times of recovery post-irradiation. The images shown correspond to the irradiation of cultured seminiferous tubules (condition c). (**c**) Quantitative analysis of the dynamics of DNA damage repair identified by γH2AX foci in the three experimental conditions (a, b, and c): mouse (representing approach a), cultured (representing approach b), and tubules (representing approach c) ($n = 20$ spermatocytes for each condition)

the dynamics of DNA repair after irradiation, the different physiological environment in this condition can introduce slight differences in the absolute number of DSBs recorded. This was expectable, since previous studies have reported a delayed progression of spermatogenesis in cultured seminiferous tubules [19, 29].

4. In conclusion, the study of the level of apoptosis, synapsis progression, endogenous and exogenous DNA damage induction and repair dynamics allow us to conclude that the organotypic culture of seminiferous tubules is a trustable method to study male mouse meiosis in vitro (*see* **Note 7**).

4 Notes

1. Carry out all procedures at room temperature unless otherwise specified.
2. All instrumentation (petri dishes, well plates, razorblades, tweezes, etc.) must be handled in sterile conditions.
3. When developing the organotypic culture, fragments of the seminiferous tubules should remain over the agarose gel (not immersed in the medium) and they should not move or sink into the bottom of the well.
4. When preparing the material for the squashing technique, the slides covered with poly-L-lysine (Sigma) should be kept covered until use to avoid deposition of dust into the surface.
5. When developing the spreading of spermatocytes technique, fragments of seminiferous tubules extracted for different wells (as long as they were all kept under the same conditions) can be mixed to proceed with cell disaggregation.
6. For functional studies with the use of drugs, it has to be taken into account that the fragments of seminiferous tubules are not immersed in the medium, but rather deposited over the agarose gel absorbing the media from below, therefore requiring higher concentration when developing drug treatments.
7. Although the culture of the seminiferous tubules is a reliable method to carry out in vitro meiosis studies, undetermined deviations from normal physiological conditions could arise. In this regard, a recent study suggested that testicular explants could show dysregulation of their transcriptomic profile compared to controls for genes related to inflammation response, insulin-like growth factor, and genes involved in steroidogenesis [42]. In contrast, no major differences were found in DNA methylation between mice born after transplantation of long-term cultured spermatogonial stem cells and control, both in F1 and F2 sperm [43]. Therefore, it is advisable to optimize the

in vitro procedures at the specific conditions for each new study, and each cell type to be examined, before using this approach.

Acknowledgments

We are very grateful to José A. Suja for his advice and scientific support as some of the cited publications were partly developed and partially funded within his research group [35–38]. We thank all the students who have helped perfectionated the organotypic culture technique presented in this work (Tania Moreno, Irene Mena, Nuria García, Enrique Alfaro, Blanca Tejeda, and Inés Hidalgo).

The experiments performed in this work were funded by Ministerio de Ciencia e Innovación (Spain) grants: BFU2009-10987 to J.P. and PID2022-140364NB-I00 to J.P and R.G.; by the Ministerio de Economia y Competitividad (Spain) grant CGL2014-53106-P to J.P.; by BIOUAM02-2020 to J.P. and R. G.; and by BIOUAM05-2022 to J.P. and R.G.

References

1. Robinson NB, Krieger K, Khan FM, Huffman W, Chang M, Naik A, Yongle R, Hameed I, Krieger K, Girardi LN, Gaudino M (2019) The current state of animal models in research: a review. Int J Surg 72:9–13. https://doi.org/10.1016/j.ijsu.2019.10.015
2. Russell WMS, Burch RL (1959) The principles of humane experimental technique, vol 13. Meuthen & Co. Limited, London
3. Hess RA, Renato de Franca L (2008) Spermatogenesis and cycle of the seminiferous epithelium. Adv Exp Med Biol 636:1–15. https://doi.org/10.1007/978-0-387-09597-4_1
4. Bolcun-Filas E, Handel MA (2018) Meiosis: the chromosomal foundation of reproduction. Biol Reprod 99(1):112–126. https://doi.org/10.1093/biolre/ioy021
5. Griswold MD (2016) Spermatogenesis: the commitment to meiosis. Physiol Rev 96(1): 1–17. https://doi.org/10.1152/physrev.00013.2015
6. Handel MA, Schimenti JC (2010) Genetics of mammalian meiosis: regulation, dynamics and impact on fertility. Nat Rev Genet 11(2): 124–136
7. Kimmins S, Kotaja N, Davidson I, Sassone-Corsi P (2004) Testis-specific transcription mechanisms promoting male germ-cell differentiation. Reproduction 128(1):5–12. https://doi.org/10.1530/rep.1.00170
8. Griswold MD (1998) The central role of Sertoli cells in spermatogenesis. Semin Cell Dev Biol 9(4):411–416. https://doi.org/10.1006/scdb.1998.0203
9. Fawcett DW, Ito S, Slautterback D (1959) The occurrence of intercellular bridges in groups of cells exhibiting synchronous differentiation. J Biophys Biochem Cytol 5(3):453–460
10. Peters AH, Plug AW, van Vugt MJ, de Boer P (1997) A drying-down technique for the spreading of mammalian meiocytes from the male and female germline. Chromosom Res 5(1):66–68. https://doi.org/10.1023/a:1018445520117
11. Page J, Suja JA, Santos JL, Rufas JS (1998) Squash procedure for protein immunolocalization in meiotic cells. Chromosom Res 6(8): 639–642. https://doi.org/10.1023/a:1009209628300
12. Champy C (1920) Quelques résultats de la méthode de culture des tissues. Archives de Zoologie Expérimentale et Générale 60:461–500
13. Abu Elhija M, Lunenfeld E, Schlatt S, Huleihel M (2012) Differentiation of murine male germ cells to spermatozoa in a soft agar culture system. Asian J Androl 14(2):285–293. https://doi.org/10.1038/aja.2011.112
14. Ibtisham F, Honaramooz A (2020) Spermatogonial stem cells for in vitro spermatogenesis

and in vivo restoration of fertility. Cells 9(3). https://doi.org/10.3390/cells9030745

15. Kubota H, Avarbock MR, Brinster RL (2004) Growth factors essential for self-renewal and expansion of mouse spermatogonial stem cells. Proc Natl Acad Sci USA 101(47): 16489–16494. https://doi.org/10.1073/pnas.0407063101
16. Kubota H, Brinster RL (2006) Technology insight: in vitro culture of spermatogonial stem cells and their potential therapeutic uses. Nat Clin Pract Endocrinol Metab 2(2): 99–108. https://doi.org/10.1038/ncpendmet0098
17. Brinster RL, Zimmermann JW (1994) Spermatogenesis following male germ-cell transplantation. Proc Natl Acad Sci USA 91(24): 11298–11302. https://doi.org/10.1073/pnas.91.24.11298
18. Brinster RL, Avarbock MR (1994) Germline transmission of donor haplotype following spermatogonial transplantation. Proc Natl Acad Sci USA 91(24):11303–11307. https://doi.org/10.1073/pnas.91.24.11303
19. Sato T, Katagiri K, Gohbara A, Inoue K, Ogonuki N, Ogura A, Kubota Y, Ogawa T (2011) In vitro production of functional sperm in cultured neonatal mouse testes. Nature 471(7339):504–507. https://doi.org/10.1038/nature09850
20. Sato T, Katagiri K, Yokonishi T, Kubota Y, Inoue K, Ogonuki N, Matoba S, Ogura A, Ogawa T (2011) In vitro production of fertile sperm from murine spermatogonial stem cell lines. Nat Commun 2:472. https://doi.org/10.1038/ncomms1478
21. Sato T, Katagiri K, Kubota Y, Ogawa T (2013) In vitro sperm production from mouse spermatogonial stem cell lines using an organ culture method. Nat Protoc 8(11):2098–2104. https://doi.org/10.1038/nprot.2013.138
22. Dumont L, Rives-Feraille A, Delessard M, Saulnier J, Rondanino C, Rives N (2021) Activation of the cannabinoid receptor type 2 by the agonist JWH133 promotes the first wave of in vitro spermatogenesis. Andrology 9(2): 673–688. https://doi.org/10.1111/andr.12928
23. Hu X, Wang H, Tian GG, Hou C, Xu B, Zhao X, Zhao Y, Fang Q, Li X, He L, Chen X, Li S, Wu J (2022) Offspring production of haploid spermatid-like cells derived from mouse female germline stem cells with chromatin condensation. Cell Biosci 12(1):5. https://doi.org/10.1186/s13578-021-00697-z
24. Ibtisham F, Zhao Y, Nawab A, Wu J, Mei X, Honaramooz A, An L (2021) In vitro production of haploid germ cells from murine spermatogonial stem cells using a two-dimensional cell culture system. Theriogenology 162:84–94. https://doi.org/10.1016/j.theriogenology.2020.12.024
25. Ishikura Y, Ohta H, Sato T, Murase Y, Yabuta Y, Kojima Y, Yamashiro C, Nakamura T, Yamamoto T, Ogawa T, Saitou M (2021) In vitro reconstitution of the whole male germ-cell development from mouse pluripotent stem cells. Cell Stem Cell 28(12): 2167–2179.e2169. https://doi.org/10.1016/j.stem.2021.08.005
26. Jordan PW, Karppinen J, Handel MA (2012) Polo-like kinase is required for synaptonemal complex disassembly and phosphorylation in mouse spermatocytes. J Cell Sci 125(Pt 21): 5061–5072. https://doi.org/10.1242/jcs.105015
27. Jiang MH, Cai B, Tuo Y, Wang J, Zang ZJ, Tu X, Gao Y, Su Z, Li W, Li G, Zhang M, Jiao J, Wan Z, Deng C, Lahn BT, Xiang AP (2014) Characterization of nestin-positive stem Leydig cells as a potential source for the treatment of testicular Leydig cell dysfunction. Cell Res 24(12):1466–1485. https://doi.org/10.1038/cr.2014.149
28. Estomba H, Munoa-Hoyos I, Gianzo M, Urizar-Arenaza I, Casis L, Irazusta J, Subiran N (2016) Expression and localization of opioid receptors in male germ cells and the implication for mouse spermatogenesis. PLoS One 11(3): e0152162. https://doi.org/10.1371/journal.pone.0152162
29. Isoler-Alcaraz J, Fernández-Pérez D, Larriba E, del Mazo J (2017) Cellular and molecular characterization of gametogenic progression in ex vivo cultured prepuberal mouse testes. Reprod Biol Endocrinol 15(1): 85. https://doi.org/10.1186/s12958-017-0305-y
30. Lee SJ, Kim KH, Lee DJ, Kim P, Park J, Kim SJ, Jung HS (2023) MAST4 controls cell cycle in spermatogonial stem cells. Cell Prolif 56(4): e13390. https://doi.org/10.1111/cpr.13390
31. Feng X, Matsumura T, Yamashita Y, Sato T, Hashimoto K, Odaka H, Makino Y, Okada Y, Nakamura H, Kimura H, Fujii T, Ogawa T (2023) In vitro spermatogenesis in isolated seminiferous tubules of immature mice. PLoS One 18(4):e0283773. https://doi.org/10.1371/journal.pone.0283773
32. Hashimoto K, Odaka H, Ishikawa-Yamauchi Y, Nagata S, Nakamura H, Kimura H, Sato T, Makiyama K, Ogawa T (2023) Culture-space

control is effective in promoting haploid cell formation and spermiogenesis in vitro in neonatal mice. Sci Rep 13(1):12354. https://doi.org/10.1038/s41598-023-39323-y

33. Onen S, Atik AC, Gizer M, Kose S, Yaman O, Kulah H, Korkusuz P (2023) A pumpless monolayer microfluidic device based on mesenchymal stem cell-conditioned medium promotes neonatal mouse in vitro spermatogenesis. Stem Cell Res Ther 14(1): 127. https://doi.org/10.1186/s13287-023-03356-x

34. Saitou M, Hayashi K (2021) Mammalian in vitro gametogenesis. Science 374(6563): eaaz6830. https://doi.org/10.1126/science.aaz6830

35. Alfaro E, Lopez-Jimenez P, Gonzalez-Martinez J, Malumbres M, Suja JA, Gomez R (2021) PLK1 regulates centrosome migration and spindle dynamics in male mouse meiosis. EMBO Rep 22(4):e51030. https://doi.org/10.15252/embr.202051030

36. Gomez R, Felipe-Medina N, Ruiz-Torres M, Berenguer I, Viera A, Perez S, Barbero JL, Llano E, Fukuda T, Alsheimer M, Pendas AM, Losada A, Suja JA (2016) Sororin loads to the synaptonemal complex central region independently of meiotic cohesin complexes. EMBO Rep 17(5):695–707. https://doi.org/10.15252/embr.201541060

37. Berenguer I, Lopez-Jimenez P, Mena I, Viera A, Page J, Gonzalez-Martinez J, Maestre C, Malumbres M, Suja JA, Gomez R (2022) Haspin participates in AURKB recruitment to centromeres and contributes to chromosome congression in male mouse meiosis. J Cell Sci 135(13). https://doi.org/10.1242/jcs.259546

38. Gomez R, Viera A, Moreno-Marmol T, Berenguer I, Guajardo-Grence A, Toth A, Parra MT, Suja JA (2022) Kinase PLK1 regulates the disassembly of the lateral elements and the assembly of the inner centromere during the diakinesis/metaphase I transition in male mouse meiosis. Front Cell Dev Biol 10: 1069946. https://doi.org/10.3389/fcell.2022.1069946

39. Enguita-Marruedo A, Martin-Ruiz M, Garcia E, Gil-Fernandez A, Parra MT, Viera A, Rufas JS, Page J (2019) Transition from a meiotic to a somatic-like DNA damage response during the pachytene stage in mouse meiosis. PLoS Genet 15(1):e1007439. https://doi.org/10.1371/journal.pgen.1007439

40. Parra MT, Viera A, Gomez R, Page J, Benavente R, Santos JL, Rufas JS, Suja JA (2004) Involvement of the cohesin Rad21 and SCP3 in monopolar attachment of sister kinetochores during mouse meiosis I. J Cell Sci 117(Pt 7):1221–1234. https://doi.org/10.1242/jcs.00947

41. Hamer G, Roepers-Gajadien HL, van Duyn-Goedhart A, Gademan IS, Kal HB, van Buul PP, de Rooij DG (2003) DNA double-strand breaks and gamma-H2AX signaling in the testis. Biol Reprod 68(2):628–634. https://doi.org/10.1095/biolreprod.102.008672

42. Dumont L, Lopez Maestre H, Chalmel F, Huber L, Rives-Feraille A, Moutard L, Bateux F, Rondanino C, Rives N (2023) Throughout in vitro first spermatogenic wave: next-generation sequencing gene expression patterns of fresh and cryopreserved prepubertal mice testicular tissue explants. Front Endocrinol (Lausanne) 14:1112834. https://doi.org/10.3389/fendo.2023.1112834

43. Serrano JB, Tabeling NC, de Winter-Korver CM, van Daalen SKM, van Pelt AMM, Mulder CL (2023) Sperm DNA methylation is predominantly stable in mice offspring born after transplantation of long-term cultured spermatogonial stem cells. Clin Epigenetics 15(1): 58. https://doi.org/10.1186/s13148-023-01469-x

Chapter 10

Dynamic Live Cell Imaging of Budding Yeast Meiosis

Hoa H. Chuong, Jared M. Evatt, and Dean S. Dawson

Abstract

For over a century, major advances in understanding meiosis have come from the use of microscopy-based methods. Studies using the budding yeast, *Saccharomyces cerevisiae*, have made important contributions to our understanding of meiosis because of the facility with which budding yeast can be manipulated as a genetic model organism. In contrast, imaging-based approaches with budding yeast have been constrained by the small size of its chromosomes. The advent of advances in fluorescent chromosome tagging techniques has made it possible to use yeast more effectively for imaging-based approaches as well. This protocol describes live cell imaging methods that can be used to monitor chromosome movements throughout meiosis in living yeast cells.

Key words Budding yeast, *Saccharomyces cerevisiae*, Microfluidics, Live-cell imaging, Meiosis, Inverted microscope, Hardware triggering, Light engine, Fluorescence microscopy

1 Introduction

The advent of using fluorescent proteins to tag yeast chromosomes [1, 2] made it possible to begin visualizing chromosome movements in living yeast cells. In this method, green fluorescent protein or red fluorescent protein were fused to bacterial DNA binding proteins (LacI or TetR) which in turned localized to repeat arrays of the *lac* operator or *tet* operator. These tools could be combined with fluorescently tagged spindle components (alpha-tubulin or spindle pole body components Spc29 or Spc42 [3, 4]) that made it possible to identify meiotic stages through spindle morphology. By using fluorescent protein derivatives with non-overlapping excitation and emission profiles, for example, mVenus-lacI, mTurquoise-tetR, and Spc42-DSRed, it is possible to simultaneously

Supplementary Information: The online version contains supplementary material available at https://doi.org/10.1007/978-1-0716-3906-1_10.

Jesús A. Carballo (ed.), *Meiosis: Methods and Protocols*, Methods in Molecular Biology, vol. 2818,
https://doi.org/10.1007/978-1-0716-3906-1_10,

image the spindle poles and two homologous chromosome partners in the same experiment. Despite these advances, technical limitations have made it difficult to image individual cells throughout meiosis I, let alone all of meiosis. A first limitation revolves around keeping the cells viable and stationary for multiple hours during the imaging process. Techniques such as adhering cells to glass slides using concanavalin A or trapping them under an agarose pad are effective methods for shorter imaging experiments [5], but in our hands in longer time courses, experiments using these techniques cells tend to move on the slide and usually fail to progress through meiosis. These problems have been circumvented by the commercial availability of microfluidics chambers that trap yeast cells in place and provide temperature control and regulated media flow over the cells, locking them in place and allowing the cells to progress through meiosis without drying out or depleting the media.

A second obstacle to performing successful long time course imaging experiments is repeated exposure of the cells to fluorescent light in the multi-hour experiments that are necessary to record yeast cells proceeding through meiosis I and II. The exposure problem is twofold: (1) phototoxicity causes cell cycle arrest (mainly in prophase I) [6] and (2) repeated exposures lead to photobleaching of the fluorescent proteins [7–9]. The protocol described here reduces these issues by using new developments in imaging to reduce fluorescent light exposure. Specifically, this protocol takes advantage of controlled LED light sources, sensitive cameras that allow low-intensity light exposures, and triggering systems that bypass the use of mechanical shutters, which increase the time the cells are exposed to light for the capture of each image [10].

2 Materials

2.1 CellASIC ONIX2 Microfluidic System

- Onix2 controller.
- CellASIC ONIX2 Manifold XT.
- CellASIC ONIX2 software.
- CellAsic Y04E-01 Diploid Microfluidics Plates (Fig. 1).

2.2 Inverted Microscope System

- Nikon Plan Apo VC 60x 1.40 N.A. objective lens.
- Lumencor SpectraX light engine.
- Hamamatsu Orca-Flash4.0 CMOS camera.

A complete list of resources and materials can be found in Supplementary Table S1.

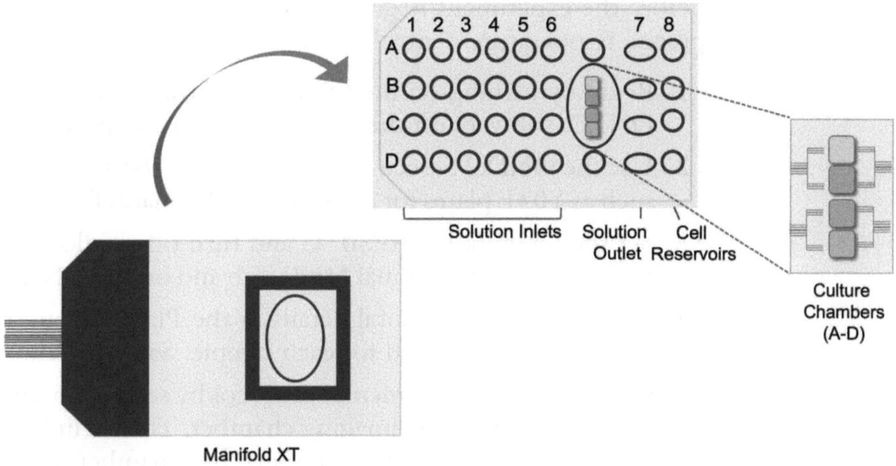

Fig. 1 Microfluidic plate and manifold. Media are pre-loaded into the solution inlet wells, and up to four cell suspensions are preloaded into cell inlet wells. The manifold is then secured over the plate. Air lines from the manifold are connected to the ONIX2 controller. Regulated air pressure flowing through the airlines sends liquid from selected wells through microchannels into the imaging chamber. The height of the imaging culture chambers (colored squares) is such that cells enter the chamber and are then wedged between the top and bottom surfaces holding them in place. The bottom surface is optical grade glass that is positioned over the objective lens of an inverted microscope. Media flowing through the imaging area is collected in the outlet wells. Controlled switches of the media that flow over the trapped cells can be programmed into the ONIX2 controller

3 Methods

3.1 Preparing Yeast Meiotic Cells

There are a variety of methods for inducing entry of yeast cultures into meiosis that are optimized for particular yeast strains or applications. In this protocol, we pre-grow yeast in YPAcetate (Yeast extract/Peptone/Acetate medium [11]) supplemented with adenine to a density of approximately 5×10^7 cells per milliliter—cells are then washed with sterile water and resuspended in 1% potassium acetate and incubated at 30 °C in baffled Erlenmeyer flasks on a rotary shaker at 250 rotations per minute [6]. See Supplementary Table S2 for a list of reagents used in this protocol.

3.2 Programming the CellAsic ONIX2 System

In the CellAsic ONIX2 system, cells and cell media are placed into wells in a plate that resembles a 96-well culture plate in which the central region has been modified to include four microfluidics chambers over optical glass (Fig. 1). A manifold fits over the plate surface. The manifold is connected to a controller by air lines that allow the controller to use air pressure to push media (and cells suspended in the media) from reservoir wells into the imaging area. The ONIX2 controller allows the investigator to control media flow rates over the cells that are trapped in the imaging area and to switch between different media. Variables such as flow rates and switch times between media are programmed into the controller before the experiment begins. Additionally, the manifold provides precise temperature control of the microfluidics imaging chamber.

1. Begin by turning on the ONIX2 controller and launching the ONIX2 software on the imaging station computer.
2. Use the ONIX2 Screen prompts to select the appropriate plate, such as Y04E plates for diploid or Y04C plates for haploid cells.
3. Set the temperature to 30 °C and turn off gas flow control by navigating to the Manual Mode Tab and on the ONIX2 screen.
4. Enter your experimental details in the Plate Layout Tab, specifying the media used for each sample. Save this information.
5. Configure the experimental protocol by setting parameters for loading cells into the imaging chamber. This includes the pressure for loading cells into the imaging chamber, the flow rate for the medium (1% potassium acetate with adenine in our case), and the overall duration of the experiment (e.g., 8 h at 2 psi or (~13.8 kPa with all well groups open)).
6. Utilize the following typical meiotic experiment program:
   - Run cell loading sequence: 10 s at 8 psi (*see* **Note 1**).
   - Run liquid priming sequence: 5 min at 5 psi (*see* **Note 2**).
   - Perfusion steps: constant flow rate pressure of 2 psi for the entire duration of the experiment (*see* **Note 3**).

3.3 Turn on Imaging System Components

1. Turn on the Nikon power supply, microscope stand, camera, and light engine.
2. Prepare the microscope for imaging by placing an empty microfluidic plate on the stage under the 10x objective, using the marks inside the chamber as a focal point for locating the viewing chamber area and focusing the microscope.

3.4 CellAsic Plate Preparation and Cell Loading

1. Select a Y04E diploid plate and open the packaging. Aspirate buffer from the wells.
2. Wash once with H_2O (200 μL per well). Aspirate the H_2O from all wells.
3. Add incubation medium (e.g., 1% potassium acetate with adenine) into the media reservoir wells—to do this, pipette 250 μL into each well 1–6 (*see* **Note 4**).
4. Ensure yeast cells are at approximately 5.0×10^6 cells/mL in 1% potassium acetate.
5. Pipette 150 μL of cell suspension into the cell reservoir—well 8 (*see* **Note 5**).
6. Place the CellAsic plate on a clean flat hard surface (for example, a clean paper towel on a clean lab bench) near the CellAsic controller. Place the CellAsic manifold (which is kept attached by the pressure air lines cable to the controller) on the CellAsic

plate, and then press the "seal" button while pressing down gently on the outer edges of the manifold. The seal button causes the controller to draw a vacuum that seals the manifold to the plate.

7. If successful, the blue "sealed" light will come on the display panel. If the sealing procedure is unsuccessful, gently clean the top of the CellAsic plate and the manifold surface with a Kimwipe moistened with 95% ethanol, and then repeat the process.
8. Place manifold sealed plate on the microscope stage. Make sure the plate is flat and properly secured on the stage.
9. Use the microscope control software to locate the chamber areas for cell entry.
10. Turn on the camera for a live view using the 10x objective lens focused on the area where cells will enter the imaging chamber from the reservoir well.
11. Run the cell load sequence by clicking the "Run cell loading sequence" while monitoring the cells flowing in.
12. Assess the number of cells coming into the chamber. For most experiments, we aim for a near monolayer of cells in a portion of the imaging chamber. This allows the simultaneous imaging of dozens to about a hundred cells in each field of view with a 60x or 100x objective lens. For other applications, it may be optimal to image well-spaced individual cells. If the desired density is reached, move on to the next step, if not repeat running cell load sequence until the preferred density is achieved.
13. To flush out untrapped cells, run a liquid priming sequence by clicking on "Run liquid priming sequence."
14. Once priming is complete, and the protocol is set on the ONIX2 (above), the system is ready for the experiment.

3.5 Setting Up the Microscope for the Experiment

1. Prepare and configure parameters in the NIS Elements software.
2. Set the illumination power of the light source at the lowest possible to prevent photobleaching (we typically set the Lumincor light engine at 5% intensity). Adjust the exposure time for each fluorescent channel to optimize signal quality using the shortest exposure times possible. Typical exposure times in our experiments are as follows: mTurquoise-TetR/tetO array, 100 milliseconds; mVenus-LacI/lacO array, 30 milliseconds; Spc42-DSRed, 30 milliseconds: DIC, 30 milliseconds (*see* **Note 6**).
3. Carefully lift the microfluidic plate off the microscope stage and set it on a clean, flat surface. Switch the objective to 60x and apply few drops of immersion oil to the objective.

4. Properly mount the microfluidic plate back on the stage, ensuring it sits flat and securely to prevent shifting during imaging.
5. Turn on live view in DIC mode. Adjust the stage position to locate cells in the chamber, identifying areas of interest for acquisition.
6. Adjust the field stop on the microscope so that only the field of view with the 60x lens is illuminated. This prevents photobleaching of areas outside of the position being imaged.
7. Set up parameters for multi-position and multi-timepoint acquisition (*see* **Note 7**).
8. Define the movie time and interval between time points.
9. Once everything is set, perform one loop run before starting the actual movie. This will allow the detection of any improper setup.
10. Initiate the flow program for the experiment in the ONIX control software.
11. Start the experiment by clicking the "Run" button in the NIS elements software.
12. After completing the movie, shut down the system properly.
13. Press the "seal" button on the manifold to unseal the plate. Carefully remove the seal and, if necessary, clean the manifold with 70% ethanol and air-dry before storage in a bag.
14. Close the program and turn off the ONIX2 system.
    - It's important to turn off the microscope first before the power supply. Ensure all lights on the front of the microscope are completely off before shutting down the power supply to prevent connection issues for the next session.
15. Export data to a server accessible from the workstation computer for subsequent analysis.
16. Utilize the NIS Elements AR Analysis software for analysis:
    - Access movie quality by navigating through time-points and z series.
    - Devolution techniques can be applied to enhance movie quality.
    - Examine fluorescent signals to evaluate chromosome segregation.

3.6 Conclusions

This protocol outlines methodology that allows for long-term live cell imaging of meiotic processes in *Saccharomyces cerevisiae* using the CellAsic ONIX2 system. This protocol addresses critical challenges associated with prolonged exposure to fluorescent light, maintaining cell viability, and achieving high-resolution imaging throughout meiosis I and II. An example of the use of this technique is shown in Fig. 2.

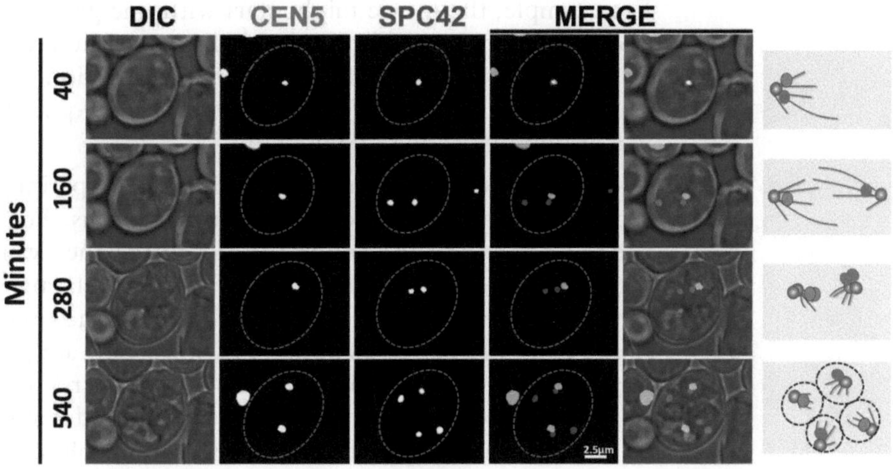

Fig. 2 Example of imaging chromosome V segregation through meiosis. A *tet* operator array was inserted adjacent to the centromere of chromosome V. Cells expressed mTurquoise-Tet and Spc42-DSRed to mark the spindle pole bodies. In this movie, images were collected every 20 min in a seven slice Z-series (spaced 200 nanometers apart). Lumincor intensity was set at 5%, mTurquoise exposures were for 100 milliseconds, and DSRed exposures were for 30 milliseconds. Maximum intensity projections were generated with NIS Elements software and images were adjusted to optimize the fluorescent signals. Selected images are shown to illustrate the progression of the cell through both meiotic divisions. The time at which each image was collected after meiotic induction (the switch to 1% potassium acetate) is indicated. The cartoon illustrates the relative positions of the tagged centromere (blue oval) of one chromosome V (CEN5) and the likely position of the other CEN5 (gray oval) and the spindle poles (red). At 40 min, the cell is in meiotic S phase or early prophase with a monopolar spindle with clustered centromeres. At 160 min, the cell is in anaphase and the tagged chromosome V homolog has segregated to one pole of the elongated spindle. At 280 min, the cell has completed anaphase I but not yet assembled the meiosis II spindle. At 580 min, meiosis II is complete and spores are forming around the segregated nuclei (DIC)

4 Notes

1. High cell density and other factors, such as debris in the culture, can cause blockage of microcapillaries in the CellAsic plate and might result in insufficient of cell loading. Thus, it is helpful to have a second monitor (in addition to the one showing the ONIX2 loading controls) to provide a live view of the plate when loading the cells. To avoid overloading or clogging of the region where cells enter the cell chamber area, it is important to visually monitor the cell loading sequence. A typical experiment would require about 10 loading pulses of 10 s each at 8 psi of loading pressure.
2. Flush out untrapped cells from the chambers by flowing in fresh medium at 5 psi.
3. At this point, it is possible to set up a program of media switches that can be timed to occur through the experiment.

For example, the movie might start with one growth medium and then after a timed interval switch to a medium containing an inducer or inhibitor of gene or protein function (e.g., rapamycin, NM-PP1, etc.). If no special media are required, load all well groups with the same medium.

4. Place the desired media in wells 1–6 for experiments that require switching between different conditions. For meiotic time courses, it is important to load the wells with pre-conditioned media harvested from sporulating cultures (see below). For movies that begin at about the time of prophase entry (1 h after meiotic induction), cells are pre-grown for 16 h in YPAcetate [11] (typically 10 milliliters) to a density of 5×10^7 per milliliter. Cells are centrifuged for 4 min at ~1000× *g*. The cell pellets are washed with a volume of sterile water equal to the culture volume, and the cells are then resuspended at a density of 1.0×10^8 cells/mL in 1% potassium acetate and adenine (20 mg/L). The resuspended cells are incubated for 1 h, shaking at 250 rpm, at 30 °C. One milliliter of the culture is removed and diluted to 5.0×10^6 cells/mL for loading in sample wells. The remaining culture is centrifuged and the supernatant is used as pre-conditioned medium that is loaded (250 μL) into wells 1–6.
5. Before loading cells, ensure the CellAsic program is set up and have the microscope 10x objective focused on the loading chamber. This allows immediate loading of the cells into the chamber before the cells settle in the cell reservoir well.
6. Set excitation light intensity as low as possible (we typically use 5%) and have exposure times set to the minimal duration that allows visualization of the fluorescent signals (50–100 millisecond exposures are typical for fluorescent proteins tethered to *lac* or *tet* operator arrays, and DSRed-SPC42 or SPC29).
7. We typically image multiple positions in a single movie. These may be in one chamber of the CellAsic plate or in some cases all four chambers. Choose positions that are as close together as is possible. This serves two purposes. First, it reduces the travel time between X-Y coordinates. In the best circumstances, it is possible to image a short Z-series (five to seven optical slices) in three colors at four X-Y coordinates and complete the cycle in about 1 min. This allows one to make four movies simultaneously with a one-minute frame rate. Second, in extended movies in which the selected X-Y positions are far apart, the immersion oil can become spread to thinly over the course of the movie, and contact between the objective lens and the glass surface of the CellAsic plate can be lost.

References

1. Straight AF, Belmont AS, Robinett CC, Murray AW (1996) GFP tagging of budding yeast chromosomes reveals that protein-protein interactions can mediate sister chromatid cohesion. Curr Biol 6:1599–1608
2. Michaelis C, Ciosk R, Nasmyth K (1997) Cohesins: chromosomal proteins that prevent premature separation of sister chromatids. Cell 91:35–45. https://doi.org/10.1016/s0092-8674(01)80007-6
3. Elliott S, Knop M, Schlenstedt G, Schiebel E (1999) Spc29p is a component of the Spc110p subcomplex and is essential for spindle pole body duplication. Proc Natl Acad Sci 96: 6205–6210. https://doi.org/10.1073/pnas.96.11.6205
4. Donaldson AD, Kilmartin JV (1996) Spc42p: a phosphorylated component of the S. cerevisiae spindle pole body (SPD) with an essential function during SPB duplication. J Cell Biol 132: 887–901. https://doi.org/10.1083/jcb.132.5.887
5. Cairo G, MacKenzie A, Tsuchiya D, Lacefield S (2022) Use of time-lapse microscopy and stage-specific nuclear depletion of proteins to study meiosis in S. cerevisiae. J Vis Exp. https://doi.org/10.3791/64580
6. Meyer RE, Kim S, Obeso D, Straight PD, Winey M, Dawson DS (2013) Mps1 and Ipl1/Aurora B act sequentially to correctly orient chromosomes on the meiotic spindle of budding yeast. Science 339:1071–1074. https://doi.org/10.1126/science.1232518
7. Shaner NC, Lin MZ, McKeown MR, Steinbach PA, Hazelwood KL, Davidson MW, Tsien RY (2008) Improving the photostability of bright monomeric orange and red fluorescent proteins. Nat Methods 5:545–551. https://doi.org/10.1038/nmeth.1209
8. Shaner NC, Campbell RE, Steinbach PA, Giepmans BNG, Palmer AE, Tsien RY (2004) Improved monomeric red, orange and yellow fluorescent proteins derived from Discosoma sp. red fluorescent protein. Nat Biotechnol 22:1567–1572. https://doi.org/10.1038/nbt1037
9. Tsien RY (1998) The green fluorescent protein. Annu Rev Biochem 67:509–544. https://doi.org/10.1146/annurev.biochem.67.1.509
10. Davis MA (2017) Hardware triggering: maximizing speed and efficiency for live cell imaging. Nat Methods 14:1223–1223. https://doi.org/10.1038/nmeth.f.403
11. Dunham M, Gartenberg M, Brown GW (2015) Methods in yeast genetics and genomics 2015 edition: A CSHL course manual. Cold Spring Harbor Laboratory Press, Cold Spring Harbor

Chapter 11

Visualization and Quantification of Rapid Chromosome Movements at Early Stages of Mouse Meiosis

Luciana Previato de Almeida, Chih-Ying Lee, Agustin Carbajal, Rodrigo O. de Castro, and Roberto J. Pezza

Abstract

Telomere-led rapid chromosome movements (RPMs) are a conserved characteristic of chromosome dynamics in meiosis. RPMs have been suggested to influence critical meiotic functions such as DNA repair and the association of the homologous chromosomes. Here, we describe a method using 3D time-lapse fluorescence imaging to monitor RPMs in Hoechst-stained mouse seminiferous tubules explants. We supplement visualization with customized quantitative motion analysis and *in silico* simulation. The ability to carry out live imaging, combined with quantitative image analysis, offers a sensitive tool to investigate the regulation of RPMs, chromosome reorganizations that precede dynamic mid-prophase events, and their contribution to faithful transmission of genetic information.

Key words Mouse gametogenesis, Spermatocyte, Chromosome dynamics, Microscopy, Telomere-led rapid prophase movements

1 Introduction

Visualization and quantification of RPMs in mice have allowed new discoveries on the machinery generating RPMs [1, 2], measuring the effect of fundamental meiotic processes in RPMs, and further understand how RPM influence dynamic complex chromosome interactions in mammals. RPMs are conserved among species and share similar structural and functional characteristics, and it is suggested to promote chromosome associations, reduce heterologous interactions, and resolve chromosome interlocks (reviewed in [3, 4]). Thus, RPMs participate in critical meiotic events required for normal progression of gametogenesis.

Since the initial identification of prophase chromosome movements in rat and Chinese hamster spermatocytes [5, 6], experiments in mammals have focused on the mouse model. The

Jesús A. Carballo (ed.), *Meiosis: Methods and Protocols*, Methods in Molecular Biology, vol. 2818,
https://doi.org/10.1007/978-1-0716-3906-1_11,

approach we describe in this chapter proved to have the sensitivity to changes provoked by deletion in components of the machinery generating RPMs, to changes in RPM characteristics during meiotic progression, and to differences in RPMs caused by defects in recombination and homologous chromosome synapsis [1].

2 Materials

2.1 Retrieving Testis and Preparing Seminiferous Tubule Explants for Visualization

1. Male mouse, 6–8 weeks old.
2. Scissors and fine tip forceps.
3. Micropipettes and tips 1–10 μL, 20–200 μL, and 100–1000 μL.
4. 1.5 mL microtubes.
5. DMEM (4.5 g/L D-glucose, L-glutamine, 25 mM HEPES, – Phenol Red; Gibco, cat. # 21063-029). Store at 4 °C.
6. MEM-α (L-glutamine, ribonucleosides, deoxyribonucleosides, Phenol Red; Gibco, cat. # 41061-029). Store at 4 °C.
7. Albumax™ I (Gibco, cat. # 11020-021).
8. Hoechst 33342 dye in DMSO (5 mg/mL).
9. Dry bath at 32 °C.

2.2 Microscopy

1. 50 mm tissue culture dish with 40 mm cover glass bottom (FluoroDish™; cat. # FD5040-100).
2. Ibidi Sticky-Slide $I^{0.2}$ Luer (cat. # 80168).
3. Ibid Perfusion set WHITE (cat. #10963).
4. Ibidi Clamp for sticky-Slides with adaptor for sitcky-Slide I Luer (cat. # 80040).
5. Cover Slips 22 × 22 mm.
6. Cover Slips 24 × 60 mm No.1 thickness.
7. Biobond Tissue Adhesive (Electron Microscopy Sciences; cat. # 71304).
8. Glass beads 150–212 microns Acid-washed (Sigma; cat. # G1145).
9. VALAP.

2.3 Equipment

1. Zeiss Axio Observer upright fluorescence microscope.
2. Zeiss Plan-APOCHROMAT 63x/1.4 Oil DIC objective (*see* **Note 1**).
3. Colibri 7 light source (Zeiss).
4. ORCA-Fusion digital camera C14440 (Hamamatsu).

5. WSB ZPiezo CAN.
6. Ibidi Air Pressure Pump 2 with one fluidic unit (Ibidi®).
7. Notebook with pump control software (Ibidi®).
8. Zen 3.2 Blue software.

3 Methods

These procedures are based on those previously developed and described by Lee et al. [1].

Before starting, set temperature of solutions in a water bath at 32 °C.

3.1 Retrieval of Testes and Seminiferous Tubules

1. Euthanize one male mice, aged about 6–8 weeks, in a CO_2 chamber and confirm dead by cervical dislocation.
2. Lay each animal on its back; sterilize the ventral surface with a 70% ethanol solution. Remove testes from the mouse abdomen with tweezers and a pair of scissors and place the pair of testes in a 35 mm Petri dish with PBS 1X at room temperature.
3. Holding the fat pad with tweezers, rinse each testis twice in PBS 1X at room temperature.
4. To obtain seminiferous tubules, hold testis from one end with a fine-tipped tweezers and make an incision in albuginea tunic with the tip of a second forceps. Pull apart the tunic albuginea by gently grasping it with two pairs of tweezers.

3.2 Seminiferous Tubule Handling and Assembling for Imaging

Here we describe two alternative methods for imaging seminiferous tubules, and experiments are performed either in rounded glass dishes (A) or using ibidi microfluidic slides (B). Although both have been shown to be proficient for measuring speed and characteristics of RPMs, the use of the Ibidi system allows keeping seminiferous tubules alive for longer time periods, and consequently extended times of visualization are possible. For both methods, nuclei spermatocyte staining with Hoechst allows visualization of centromeric heterochromatin. In each nucleus, Hoechst-stained "blobs" represent one or several chromosomal centromeres. It is important to emphasize that Hoechst 33342 staining of heterochromatin centromeric spots is pericentromeric and, in the mouse, represents half of the chromosome ends in the nucleus.

During the ~10 days those spermatocytes spend in pachytene, they move away from the seminiferous tubule wall to a location deep inside the tubule where spherical aberrations prohibit clear imaging using wide-field, high numerical aperture optics. Thus, the

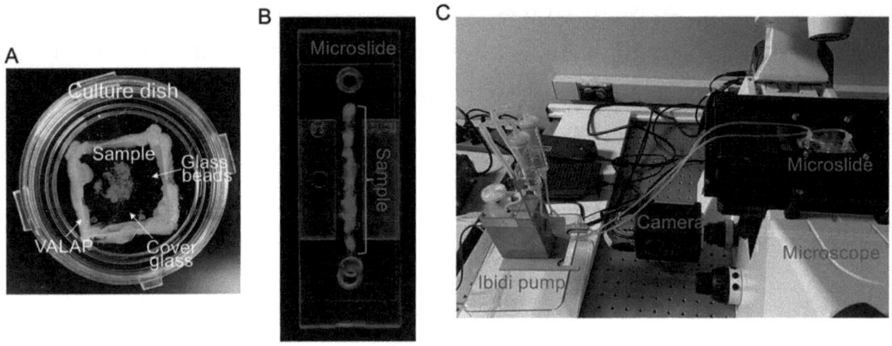

Fig. 1 Images show sample preparation and mounting before visualization. (**a**) Use of culture dish for short-term RPM measurements. (**b**) Sample preparation in a Ibidi microfluidics chamber. (**c**) Ibidi microfluidic system and inverted epifluorescent microscope used to visualize RPMs in long-term movies

meiotic prophase stages after pachytene-diplotene are not analyzed with these methods.

Four to six seminiferous tubule bundles (approximately ¾ of a testis) are transferred into Eppendorf tubes containing 1 ml MEM 1X plus 20% Albumax with added Hoechst 33342 chromatin dye (0.05 mg/mL). After 30 min of incubation at 32 °C (no shaking), the tubule bundles are rinsed twice by transferring them to Eppendorf tubes containing 1 mL MEM-Albumax (no Hoechst) and stored in this media at 32 °C until analysis (not exceeding 1 h).

For experiments performed in rounded glass bottom dish (short-term movies (10–30 min)), 10–20 seminiferous tubules (approximately 1 mm length each) were placed between a glass bottom dish (cat. # HBSt-5030, Willco Wells) and coverslip separated by glass beads (150–212 microns) in DMEM and sealed with VALAP (Fig. 1a).

Experiment using Ibidi microfluidic slide long-term movies (30 min–10 h):

1. Prior to sample mounting, we assemble the fluid unit and add 6 mL of MEM-α plus Albumax I in each reservoir holder. To eliminate bubbles in the system, it is necessary to leave the pump working for at least 5 min without any slide applying 50 mbar pressure.
2. When the pump is ready, mount seminiferous tubule bundles along the channel of the sticky-Slide $I^{0.2}$ Luer chamber (50 μm high) with sufficient media to avoid tubule dehydration and remove the protection film (Fig. 1b). Seal the chamber with a sterile glass coverslip (pretreated with biobond) and press firmly until the bottom is completely sealed. For better results, use the Ibidi Clamp for sticky-Slides with the adaptor for sitcky-Slide I Luer.

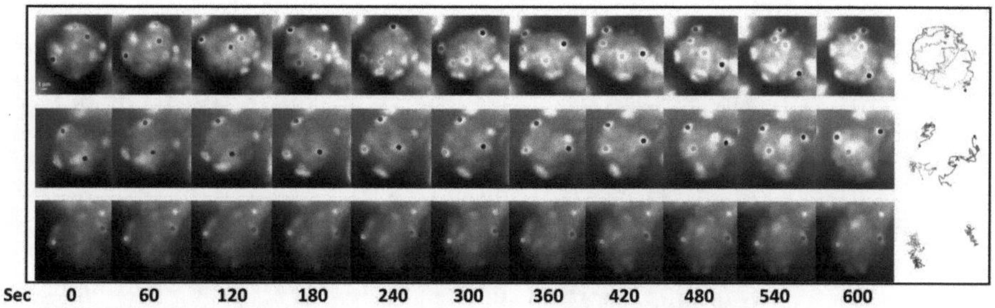

Fig. 2 Maximum intensity projections of time lapse images of wild-type zygotene spermatocytes. Red, blue, green, and yellow dots mark a defined centromere blob revealed by Hoechst 33342. Magnification bar represents 1 μm. The right panel shows the traces of spots marked in each time frame

3. Before connecting the microslide, make sure that the Ibidi pump (Fig. 1c) is turned off and use the hose clip provided with the fluid unit to block the liquid inside the tube.
4. Carefully attach the slide to the tubes in both sides and release the clip. In the pump control software change from without slide to slide-to-Slide $I^{0.2}$ Luer chamber and apply a pressure of 5 mbar.

3.3 Acquisition of Movies

1. Mounted samples were placed in the microscope stage holder (Fig. 1c).
2. 3D time course movies (stack images, 16 slices with 1 μm steps) (Fig. 2) were acquired every 10 s for a total of 10 min (short-term movies) or every 1–5 min for 1–12 h (long-term movies).

We use a Zeiss Axio Observer upright fluorescence microscope fitted with a Zeiss Plan-APOCHROMAT 63 x/1.4 Oil DIC objective, 92 HE DAPI/GFP/mCherry filter, ORCA-Fusion digital camera, and Zen 3.2 Blue acquisition software. As a light source we used the Zeiss Colibri light source 7 LED 385 nm at 5%, with 10 ms exposure and 30% shift.

3.4 Analysis of Chromosome Movements

1. To reduce the effects of noise and anisotropic resolution (0.1 μm along the X and Y axes and 1.0 μm along the Z axis), spot positions were smoothed by taking the average of the positions in a moving window over every three time-points.
2. Before starting the quantification process, save the file as .zvi in a separate folder, allowing RAW files to be saved in the selected location.
3. Open OMRFQANT [7] color movie and zvi software and open the Axiovision file (.zvi).
4. In the 20110420OMRFQANT software go to "File_Open_choose all files (.raw)." Check if the order is correct and press OK.

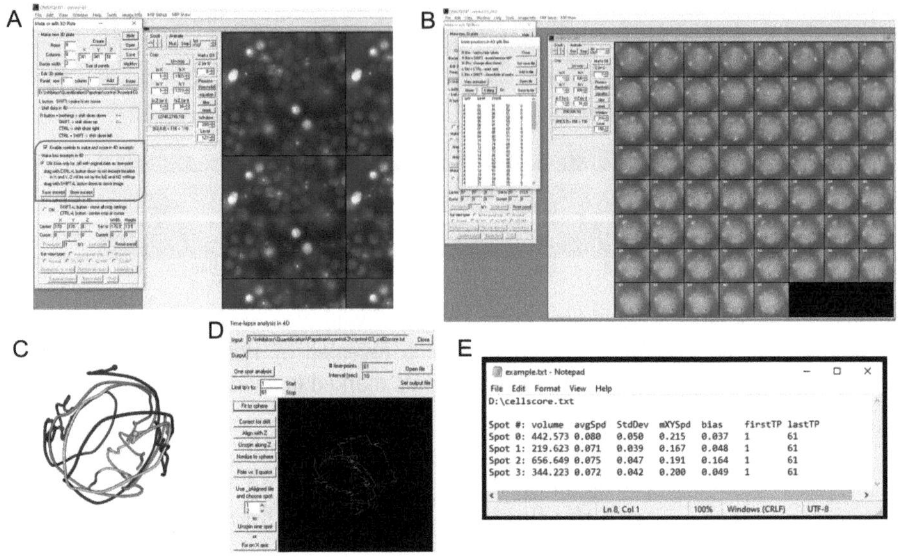

Fig. 3 Analysis of chromosomes movements using the software OMRFQANT. (**a**) Example of an image opened with OMRFQANT. (**b**) Single cell scoring. (**c**) Splines animation of four spots (red, blue, green, and yellow) scored in one cell. (**d**) Time-lapse analysis window. (**e**) Example of the final results obtained as a text file

5. Select "open as plate of thumbnails", time points should be equal X4000 Y3000; get/set image stacks settings: 1024×1024×16 and press OK. Save with the same name as the avi file.
6. Click uncrop, adjust window using loX hiX and loY hiY, and use scroll to navigate.
7. Open file in the Zen software to guide you to choose the cells. Back to OMRFQANT in the "make or edit" 3DPlate window (in the left), select "enable controls to make and score in 4D excerpts" and ON (use only for .pi8 with original data as time point) (Fig. 3a).
8. Select the cell of interest using CTRL and the left button (drag down diagonally) and click save excerpt: X = 0.103; Y = 0.103; Z = 1; T = time elapsed between stacks in seconds.
9. Click Auto and OK and save with the same name cell 1, 2, 3, etc.
10. To analyze the spots, open one cell, hit uncrop, equalize, and adjust bright/contrast in window/level.
11. Go to tool_score in 4D and set save file (save as same name cell1 score) and mark spots with CTRL+left button, scrolling up and down with mouse wheel (Fig. 3b).
12. Click add to file, follow other spots, add to file again. When finished, click view animation, Splines, save images: single tiff (Fig. 3c).
13. To analyze speed, go to Tool_XYZT analysis, and the Time analysis in 4D window will open (Fig. 3d). Open the score file and choose NO to the question in the pop-up. Set output file

(same name score analysis) and click Fit to sphere followed by One spot analysis. The results will be saved as .txt (Fig. 3e). *See* **Note 2**.

4 Notes

1. In Subheading 2.3, **step 2**, it is important to have a calibrated objective. The pixel distances can vary.
2. In Fig. 3e, the speed results are given in μM/s.

Acknowledgments

We thank Mike Dresser, a pioneer in RPMs studies, for his constant support.

This work was supported by R01-NIGMS R01 GM125803 and R21-NICHD R21 HD103562 to RP.

References

1. Lee CY et al (2015) Mechanism and regulation of rapid telomere prophase movements in mouse meiotic chromosomes. Cell Rep 11:551–563
2. Morimoto A et al (2012) A conserved KASH domain protein associates with telomeres, SUN1, and dynactin during mammalian meiosis. J Cell Biol 198:165–172
3. Link J, Jantsch V (2019) Meiotic chromosomes in motion: a perspective from Mus musculus and Caenorhabditis elegans. Chromosoma 128: 317–330
4. Alleva B, Smolikove S (2017) Moving and stopping: regulation of chromosome movement to promote meiotic chromosome pairing and synapsis. Nucleus 8:613–624
5. Parvinen M, Soderstrom KO (1976) Chromosome rotation and formation of synapsis. Nature 260:534–535
6. Ellingson DJ, Yao KT (1970) Growth and observations of Chinese hamster seminiferous epithelium in vitro. J Cell Sci 6:195–205
7. Conrad MN et al (2008) Rapid telomere movement in meiotic prophase is promoted by NDJ1, MPS3, and CSM4 and is modulated by recombination. Cell 133:1175–1187

Chapter 12

Culture of the Intact Postnatal Naked Mole-Rat Ovary: From Meiotic Prophase to Single-Cell RNASeq

Patrick T. Walsh, Ana Martínez-Marchal, and Miguel Angel Brieño-Enríquez

Abstract

Recently, we reported that, in the naked mole-rat (*Heterocephalus glaber*) ovary, there is mitotic expansion of the primordial germ cells (PGCs), and the initiation of the meiotic program occurs postnatally. This is opposite to almost all other mammals, including humans and mice, whose reproductive cycle begins very early in development. In both mouse and human, the ovaries become populated with PGCs *in utero*; these PGCs will later generate the oogonia. After mitotic proliferation, these cells will trigger the meiotic program and initiate meiotic prophase I. Given that all these processes happen *in utero*, their analysis has been very challenging; so the ability to study them postnatally and to manipulate them with inhibitors or other substances, in the naked mole-rat, opens new possibilities in the field. In this chapter, we present a comprehensive collection of protocols that permit the culture of whole naked mole-rat ovaries, followed by analysis of germ cells, from PGCs to oocytes, in meiotic prophase I, as well the obtention of single-cell suspension or single-nuclei suspension for RNASeq.

Key words Naked Mole-Rat, Ovary, Meiosis, Prophase I, Whole ovary culture, Single cell/nuclei RNASeq

1 Introduction

The adult mammalian ovary is devoid of definitive germline stem cells. A critical determinant of reproductive lifespan in female mammals is the size of the ovarian reserve, which is based on the number of germ cells in the ovary [1]. Establishment of the ovarian reserve begins with the specification of primordial germ cells (PGCs), which are the embryonic precursors of the germ cell lineage that will form the eggs and sperm required for sexual reproduction [1, 2]. To generate the ovarian reserve, human PGCs gradually proliferate during their migration to gonadal ridges at 4–5 weeks of fetal development. In the female gonads, PGCs enter asynchronously into meiosis around weeks 11–12; this process will continue

Jesús A. Carballo (ed.), *Meiosis: Methods and Protocols*, Methods in Molecular Biology, vol. 2818,
https://doi.org/10.1007/978-1-0716-3906-1_12,

up to week 22. It is estimated that, by 20 weeks of gestation, the developing human fetal ovaries contain a maximum of about five million oogonia/oocytes [3]. Oocyte numbers at birth range anywhere from half a million to one million and decline to several hundred thousand by puberty and to approximately one thousand by menopause. Because the ovarian reserve, which dictates reproductive lifespan, is established *in utero*, the reproductive aging timeline is slated before birth.

Although adult mouse or human ovaries are devoid of definitive germline stem cells, we recently reported that the naked mole-rat (NMR) does not follow that rule [4]. The NMR is well known for being the longest-lived rodent, with a maximum lifespan of >37 years [5, 6]. Female NMRs showed no decline in fertility or fecundity into their third/fourth decades of life [6, 7], whereas comparably sized mice have reduced fertility by 9 months of age [8]. Almost all the individuals in a naked mole-rat colony will never be reproductively active [9]; however, the queens (dominant breeding females) will breed until they die. Our data showed that the NMR ovary retains a population of germ cells far into postnatal life that bear markers of pluripotency as well as PGC identity and are capable of dividing in vivo, allowing the NMRs to establish and maintain an exceptionally large ovarian reserve relative to their small body size [4]. We also demonstrated, using pluripotency markers (SOX2 and OCT4) and the PGC marker BLIMP1, that PGCs persist up to postnatal day 90 (P90) alongside germ cells in all stages of female differentiation and undergo mitosis both in vivo and in vitro.

Culture of mammalian ovaries has been challenging, independently of the species. Experiments using fragments of mouse and human ovaries had shown progression of meiotic prophase and recombination [10–12]; however, these protocols do not allow the study of the ovary as a 3D structure. In this chapter, we present a compilation of protocols which have been developed to analyze the ovarian reserve, meiotic prophase I progression, and germ cell mitotic expansion in vitro and to generate single-cell suspension followed by RNASeq using whole-mount culture (Fig. 1). We also describe the protocols developed to cryopreserve the ovary after culture and the methods for performing meiotic spreads and single-nuclei suspension for single-nuclei RNASeq (Fig. 2).

2 Materials

2.1 Collection of Ovaries

2.1.1 Collection of Ovaries for Cultures

1. Warm (32 °C) phosphate buffered saline (PBS): 137 mM NaCl, 2.7 mM KCl, 8 mM Na_2HPO_4, and 2 mM KH_2PO_4; pH 7.4. PBS is certified RNase-free.

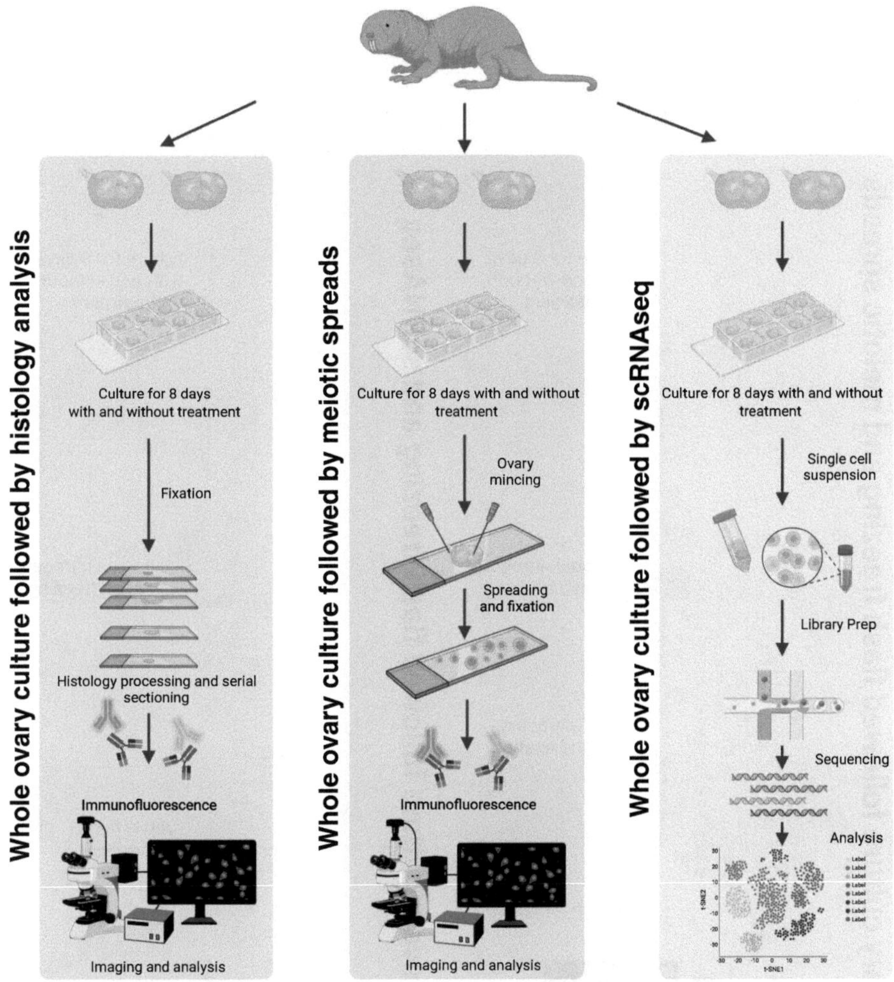

Fig. 1 Schematic diagram of the experimental design. (**a**) Histology analysis; (**b**) meiotic spread procedure; and (**c**) whole ovarian culture followed by single-cell RNASeq

2.2 Processing Ovaries in the Laboratory

2.2.1 Processing of Ovaries for Culture

1. Warm (32 °C) PBS, sterile.
2. Stereomicroscope.
3. Dumont Tweezers #5.
4. Scissors (Vanna's).
5. Forceps, curved, serrated.
6. Dissecting scissors.
7. Warm (32 °C) penicillin-streptomycin, with 10,000 units penicillin and 10 mg streptomycin per ml in 0.9% NaCl, 0.1 μm, filtered.

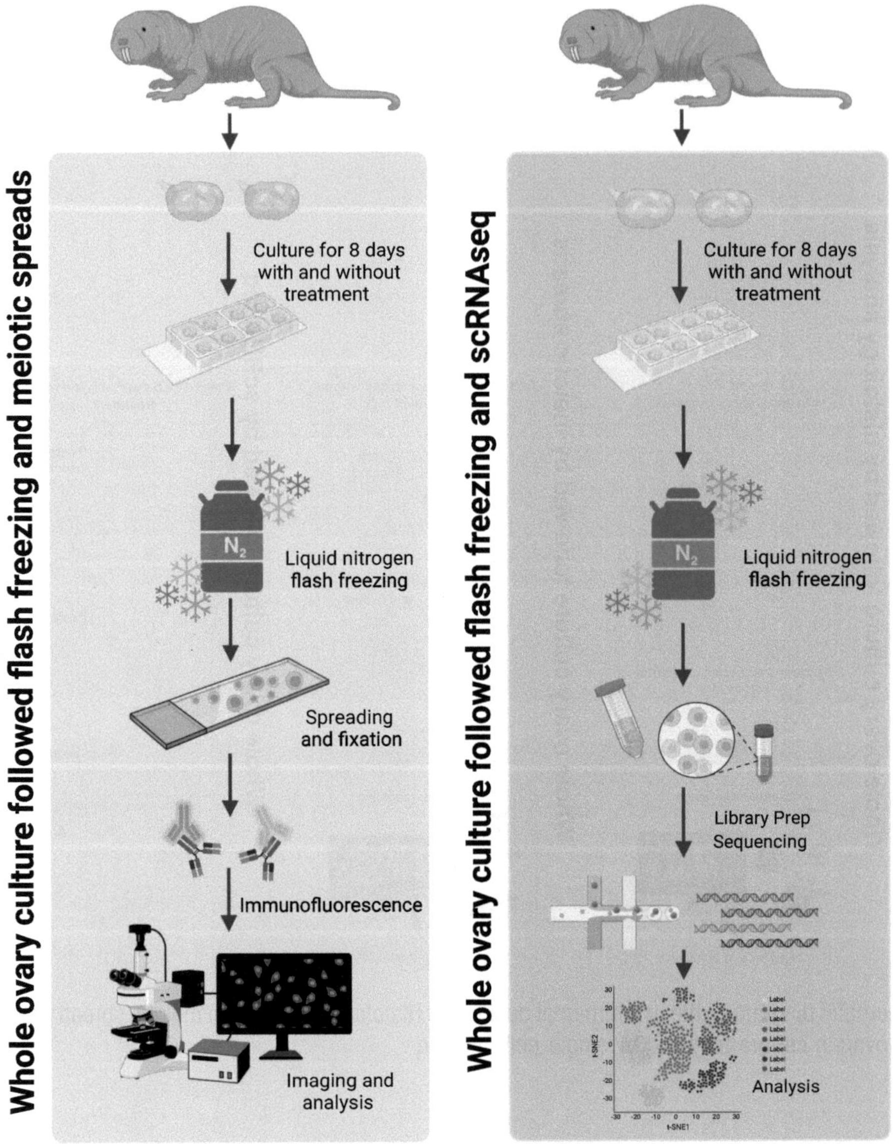

Fig. 2 Schematic diagram of the experimental design from frozen whole-ovary cultures. (**a**) Meiotic spreads from frozen whole-ovary culture; (**b**) obtention of single-cell nuclei suspension for RNASeq

8. Pre-warmed (32 °C) culture medium, D-MEM, high-glucose, GlutaMAX supplement, supplemented with 100 IU/mL penicillin (Gibco BLR), 100 μg/mL streptomycin (Gibco BLR), and 15% fetal bovine serum (FBS).
9. Nunc™ cell-culture-treated multidishes, 24 wells (polystyrene, flat, sterile).

2.3 Culture of Whole Ovaries

1. 32 °C PBS, sterile.
2. Dumont tweezers #5.
3. Pre-warmed (32 °C) culture medium, D-MEM, high-glucose, GlutaMAX Supplement, supplemented with 100 IU/mL penicillin (Gibco BLR), 100 μg/mL streptomycin (Gibco BLR), and 15% FBS.
4. Nunc™ cell-culture-treated multidishes, 24 wells (polystyrene, flat, sterile).
5. Cell culture incubator (32 °C).
6. Biological safety cabinet, Class II A2.
7. Inverted microscope.

2.4 Fixation of Whole-Ovary Cultures

1. Fixative solution: 10% paraformaldehyde, pH 9.2. Triton X 0.05% (*see* **Note 1**).
2. 70% ethanol.
3. Benchtop orbital shaker.

2.5 Clearing of Whole-Ovary Culture

1. Prepare ScaleS (0) clearing solution: 40% D-(-) sorbitol (w/v), 10% glycerol, 4.3 M urea, and 20% dimethyl sulfoxide (DMSO), pH 8.1.

2.6 Embedding of Cultured Whole Ovary in Paraffin for Histology Sections

1. Ethanol-water (v/v) 70%, 80%, and 100%.
2. Xylene.
3. Tissue infiltration medium paraffin.
4. Embedding cassettes.

2.7 Immunofluorescence on Ovarian Histology Sections

1. Xylene.
2. 70%, 80%, 90%, and 100%, ethanol-water (v/v).
3. Sodium citrate, pH 6.0.
4. 0.2% of Triton-X 100 in PBS (permeabilization solution).
5. 2.52 mg/mL glycine, 10% goat serum, 3% BSA, 0.2% Tween 20 in PBS (blocking solution).
6. 0.1% Tween 20 in PBS (PBS-T) (washing solution).
7. 0.1 μg/mL of DAPI (counterstaining solution).
8. Coplin jars.

2.8 Oocyte Spreading from Whole-Ovary Cultures

1. Tris–HCl pH 8.2 1 M (final concentration 30 mM), sucrose (final concentration 50 mM), Citrate trisodium, dehydrate (final concentration 17 mM), EDTA 0.5 (5 mM final concentration), DTT 0.5 M (0.5 mM final concentration), PMSF 10 mg/mL (final concentration 0.1 mM), milliQ water. pH 8.2. (hypotonic extraction buffer (HEB)) (*see* **Note 2**).

2. Sucrose solution (100 mM).
3. 1% paraformaldehyde, pH 9.2. Triton X 0.05% (fixative solution) (*see* **Note 1**).
4. Photoflo 0.05% in PBS (washing solution 1).
5. Photoflo 0.05% in water (washing solution 2).
6. Hydrophobic printed 8-well slides, round, 6 mm.
7. Humidity chamber.
8. Needles, 23Gx1 (0.6 mm × 25 mm), sterile.
9. Dumont tweezers #5.
10. Corning Pyrex 9-well glass spot plates.
11. Coplin jars.
12. Benchtop orbital shaker.

2.9 Meiotic Spreads Immunofluorescence from Intact Ovary Cultures

1. Bovine serum albumin (BSA, 3 g), goat serum (10 mL), 20% Triton X (1 mL), 10X PBS (10 mL), adjust with water to 100 mL (antibody dilution buffer, ADB): (*see* **Note 3**).
2. Photoflo 0.05% in PBS (washing solution 1).
3. Photoflo 0.05% in water (washing solution 2).
4. 0.005% Triton-X in PBS (permeabilization solution).
5. 10% ADB in PBS (blocking solution).
6. Antibodies: dilution of antibodies, both primary and secondary, is made in ADB.
7. Vectashield mounting medium + DAPI.
8. Parafilm.
9. 22- × 60-mm coverslips.
10. Coplin jars.

2.10 Single-Cell Suspension for Single RNASeq from Cultured Whole Ovaries

1. PluriSTEM™ Dispase-II Solution, sterile at 1 mg/mL.
2. 20 μg/mL DNase (Sigma-Aldrich).
3. Pre-warmed (32 °C) culture medium, D-MEM, high-glucose, GlutaMAX Supplement, supplemented with 100 IU/mL penicillin (Gibco BLR), 100 μg/mL streptomycin (Gibco BLR), and 15% FBS.
4. Eppendorf ThermoMixer C.

2.11 Flash Freezing Whole Ovaries After Culture

1. Warm (32 °C) PBS, sterile.
2. Ice cold PBS, sterile.
3. Liquid nitrogen.
4. 1.5 mL cryotubes.
5. Ultra-low-temperature freezer.

2.12 Meiotic Spreads from Frozen Whole-Ovary Cultures

1. Dry ice.
2. Fisher Scientific Co. Superfrost Plus Frosted Microscope Slide 3X1" White.
3. Lipsol 1%.
4. Fixative solution: 1% paraformaldehyde, pH 9.2. Triton X 0.05% (*see* **Note 2**).
5. Washing solution 1: Photoflo 0.05% in PBS.
6. Washing solution 2: Photoflo 0.05% in water.
7. Humidity chamber.
8. Needles, 23Gx1 (0.6 mm × 25 mm), sterile.
9. Dumont tweezers #5.
10. Coplin jars.
11. Benchtop orbital shaker.

2.13 Single-Nuclei Suspension from Frozen Whole-Ovary Cultures

1. Nuclei isolation to be performed with the Chromium Nuclei Isolation Kit (Genomics 10x).
2. 2.0 mL DNA LoBind tubes (Eppendorf).
3. 3.15 mL centrifuge tubes (Corning).
4. 4.10% BSA in Dulbecco's PBS (10%).
5. 5.1X PBS without calcium or magnesium.
6. Molecular-grade nuclease-free water.
7. Cell-counting nucleic acid-staining fluorescent dye: VitaStain AOPI Staining Solution.
8. Cellaca MX high-throughput automated cell counter.
9. Vortex mixer.
10. Eppendorf refrigerated centrifuge.

3 Methods

3.1 Collection of the Ovaries

All experimental procedures followed federal and institutional guidelines and were approved by the University of Pittsburgh Animal Care Committee. Naked mole-rat colonies (opposite-sex pairs) were housed in a single, medium polycarbonate cage. Naked mole-rats (NMRs) were kept on a 12:12 light/dark cycle at 28–30 °C and fed *ad libitum* with a diet consisting of sweet potato, raisins, grapes, yams, celery, ProNutro, and corn. Dissection of the ovaries can be done at room temperature, following these steps:

1. Postnatal age should be determined according to what population of cell needs to be enriched. For these protocols, we collected at postnatal day 1 (P1), but age can be selected

according to the project requirements. Female naked mole-rats can be euthanized by decapitation according to IACUC guidelines. Decapitate one animal at a time.

2. After decapitation, spray the animal with 70% ethanol and place it on a paper towel. Use tape on superior and inferior limbs to maintain the body in the correct position.
3. Ovaries will be collected in a 1.5-mL Eppendorf tube with pre-warmed PBS.
4. Make an incision that allows you to cross the skin and get into the abdominal cavity. The ovaries are in the posterior wall of the abdomen. To reach them, move the intestines out of the abdominal cavity.
5. To identify the ovaries, look for the uterine horns and, with the tweezers, follow them up. Ovaries will be at the end, semi-covered by the kidney.
6. After localizing the ovaries, pinch the uterine horn with the tweezers and pull it gently. Cut the uterine horn 2 mm from the ovary and extract the ovary and the uterine leftovers as a block. The small section of the uterus that is attached to the ovary will be used to handle the ovary, preventing its damage (*see* **Note 4**).
7. Place the ovary in pre-warmed PBS.

3.2 Culture of Ovaries

1. Keep the P1 ovaries in PBS at 32 °C.
2. Remove the residue of connective tissue and uterus, using a stereo microscope.
3. In the biological safety cabinet, class II A2, incubate the ovaries for 60 min with 500 μL of penicillin-streptomycin (10,000 units/mL penicillin and 10 mg/mL streptomycin).
4. Rinse the ovaries with pre-warmed (32 °C) sterile PBS.
5. Place the ovary on a sterile Nunc cell-culture treated 24-well multidish.
6. Add 500 μL of culture medium, D-MEM, high-glucose, GlutaMAX supplement, supplemented with 100 IU/mL penicillin (Gibco BLR) and 100 μg/mL streptomycin (Gibco BLR) and 15% FBS.
7. Remove the culture medium and replace it with fresh pre-warmed culture medium (3 times).
8. Add 500 μL of culture medium to each well and culture them at 32 °C and 5% CO_2.
9. Every other day, remove 50% percent of the culture medium and replace with fresh medium.

10. Using an inverted microscope, evaluate the cultures daily. For these purposes, ovaries were cultured for 7 days. Culture media can turn viscous; be careful (*see* **Note 5**).

3.3 Fixation of Ovarian Cultures

1. Remove all the culture media from the wells.
2. Wash the wells with sterile cold PBS, 3 times, for 10 min each, using the benchtop orbital shaker.
3. Add 1000 μL of fixative solution (1% paraformaldehyde, pH 9.2. Triton X 0.05%) and incubate overnight at 4 °C.
4. Using 1000 μL of 70% ethanol, wash the cultures three times, for 30 min each, on the benchtop orbital shaker.
5. Submit to histology core.

3.4 Oocyte Spreading from Whole Ovary for Immunofluorescence

We developed this protocol from a combination of protocols previously developed to perform mouse spreads from ovaries [13, 14] and testis [15–17], human ovarian spreads [10, 18–20], and fresh NMR spreads [4].

1. Prepare the fixative solution: 1% paraformaldehyde, pH 9.2. Triton X 0.05% (*see* **Note 2**).
2. Place the hydrophobic 8-well printed slide in the humidity chamber and 50 μL of fixative solution within each printed well (*see* **Note 5**).
3. Prepare the HEB solution and store it on ice until use (*see* **Note 1**).
4. Wash the ovaries with fresh, cold PBS for 5 min 3 times.
5. Incubate the ovary on 5 mL of HEB for 60 min.
6. Remove the ovary from HEB and place it on Corning Pyrex 9-well glass spot plates.
7. Using the sucrose solution (100 mM), wash the ovary twice.
8. Place the ovary in a new well and add 100 μL of sucrose solution.
9. Mince the ovary Using the 23G needles, creating a homogeneous solution. You should see a milky solution coming from the ovary. Culture tissue normally has a "rubbery" texture, and it is hard to get the solution. Be patient.
10. Take the milky solution and place it in a new well and remove large pieces of tissue.
11. Add 50 μL of sucrose solution. The solution needs to be slightly milky, not completely clear, so if the solution it is too milky, add another 25–50 μL of sucrose solution.
12. Be sure that the solution is fully homogenized and take 10 μL of the oocyte-sucrose solution and add it to the 50 μL of

fixative solution. Remember to preload to the hydrophobic printed 8-well slide in the humidity chamber.

13. Close the humidity chamber and let slides sit overnight at room temperature.
14. After overnight incubation in the humidity chamber, open the lid and let the slides dry for 1 h. Be careful that they do not get fully dry (*see* **Note 6**).
15. Wash the slides with washing solution 1: Photoflo 0.05% in PBS, twice, for 10 min each.
16. Wash with Photoflo 0.05% in water for 10 min two times.
17. Air-dry the slides and either process for immunofluorescence or store at −80 °C (*see* **Note 7**).

3.5 Meiotic Spread Immunofluorescence

1. Wash the slides with washing solution 1: Photoflo 0.05% in PBS for 15 min, using the benchtop orbital shaker. If frozen slides will be used, let them sit at room temperature for at least 15 min before starting the washes.
2. Incubate with permeabilization solution: 0.005% Triton-X in PBS for 30 min.
3. Using the blocking solution (10% ADB in PBS), incubate for 1 h on the benchtop orbital shaker.
4. Dilute the antibodies in ADB, following the vendor recommendations.
5. Place the slides in the humidity chamber. Keep them humid all the time.
6. If all the wells on the slide will be tested with the same antibody combination, add 100 μL of the ADB+ antibody solution to the slide and cover the slide with parafilm (the size of the parafilm should be the same as the slide). If you want to test independent antibodies, add 25 μL of the ADB+ antibody solution to each one of the hydrophobic 8-well printed slides and incubate overnight at 4 °C.
7. Using Photoflo 0.05% in PBS, wash the slides for 15 min.
8. Using the blocking solution (10% ADB in PBS), incubate for 30 min on the benchtop orbital shaker.
9. Dilute the secondary antibodies in ADB solution.
10. Add 100 μL of secondary antibody solution to each slide and cover it with parafilm. Incubate in the humidity chamber for 2 h at 37 °C.
11. Wash the slides on the orbital shaker, using washing solution 1: Photoflo 0.05%, in PBS, 3 times each, for 15 min.

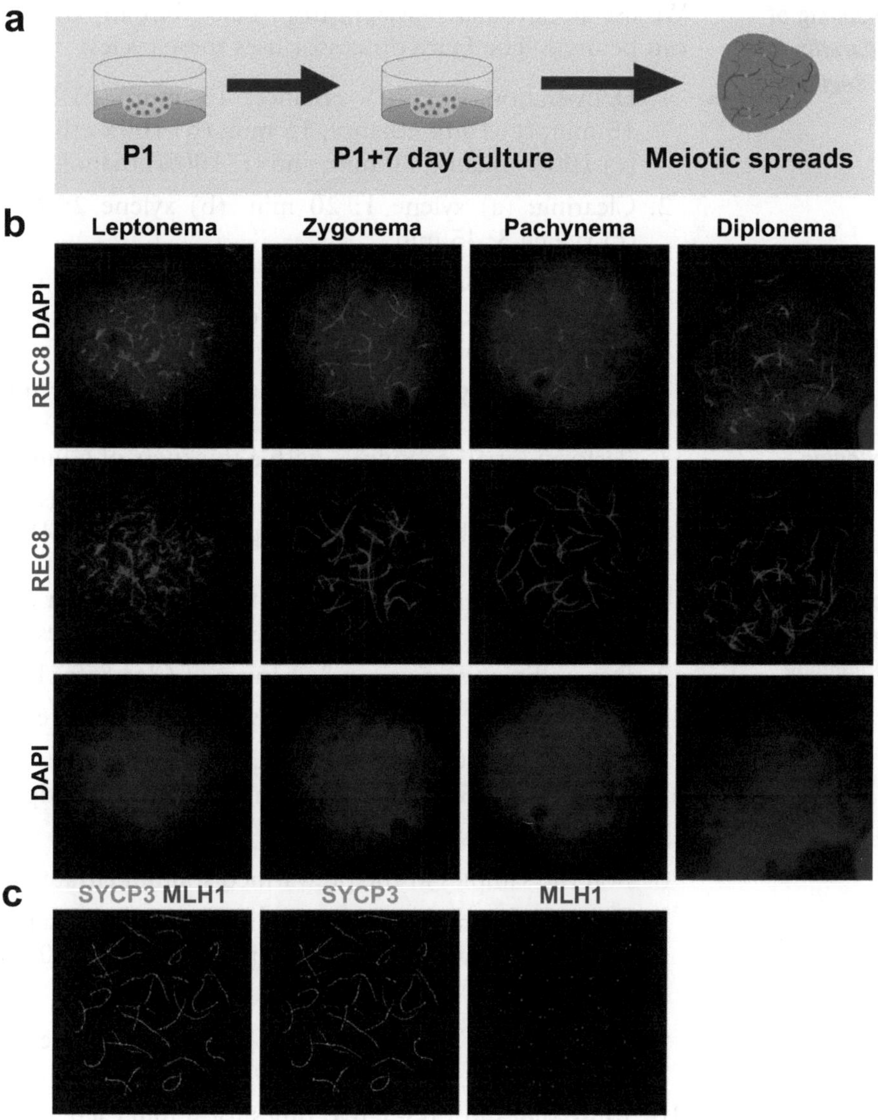

Fig. 3 Meiotic spreads. (**a**) Schematic diagram of whole ovary for meiotic spreads from P1 to P1 + 7 (equivalent to P8). (**b**) Immunofluorescence of culture of P1 + 7 ovary with DAPI (blue), REC8 (red). (**c**) Staining with SYCP3 (green) and MLH1 (red)

12. Wash the slides with Photoflo 0.05% in water for 15 min.
13. Remove the slides from the washing solution and let them dry, protecting them from light.
14. Add 25 μL of Vectashield mounting medium containing 0.1 μg/mL of DAPI. Cover with 22- × 60-mm coverslips and wait for 60 min.
15. Remove the excess DAPI and analyze. Examples of naked mole-rat meiotic spreads are shown in Fig. 3.

3.6 Embedding of Ovary in Paraffin for Histology Sections

We use an automated system from Leica, but any tissue processor can be used. The Leica processor uses these cycles:

1. Dehydration: (a) 70% ethanol, 15 min; (b) 90% ethanol, 15 min; (c) 100% ethanol, 15 min; (d) 100% ethanol, 15 min; (e) 100% ethanol, 30 min; and (f) 100% ethanol, 45 min.
2. Clearing: (a) xylene 1: 20 min, (b) xylene 2: 20 min, and (c) xylene 3: 45 min.
3. Wax infiltration: three cycles of tissue infiltration with paraffin wax are needed: first two with a duration of 30 min and the last one of 45 min.
4. Place the ovary in a mold and let it cool down for at least 24 h.

3.7 Sectioning

1. Perform serial sectioning, with a thickness of 6 μm.
2. We mounted the sections every 20th section. This is related to the large number of germ cells present in the NMR ovary.

3.8 Immunofluorescence on Ovarian Histology Sections

1. Preheat a water bath with staining dish/Coplin jar containing a sodium citrate buffer until the temperature reaches 95–100 °C.
2. Deparaffinize sections in 3 changes of ClearSafe, 10 min each.
3. Hydrate in 4 changes of 100% ethanol for 5 min each.
4. Using 95%, 80%, and 70% ethanol, hydrate the tissue. Each change is for 5 min each time and should be in triplicate.
5. Rinse 2x in distilled water.
6. Boil the samples in the prewarmed sodium citrate for 40 min.
7. Turn off steamer or water bath and return the staining dish to room temperature. Allow the slides to cool for 20 min.
8. Rinse slides in PBS-T two times for 5 min each.
9. Using 0.2% of Triton-X 100 in PBS, permeabilize the sections for 30 min.
10. Rinse slides in PBS-T two times for 5 min each.
11. Incubate the samples for 4 h with blocking solution.
12. Prepare the primary antibody in blocking solution. In Fig. 4, we show examples of tissue sections stained with DAPI, REC8, and VASA.
13. Incubate with primary antibody overnight at room temperature.
14. Rinse slides with PBS-T 3 times for 5 min each.
15. Naked mole-rat tissue can show a lot of secondary antibody background; to reduce this artifact, we use secondary antibodies with minimal cross reaction. Incubate sections with secondary antibody at appropriate dilution for 2 h at room temperature.

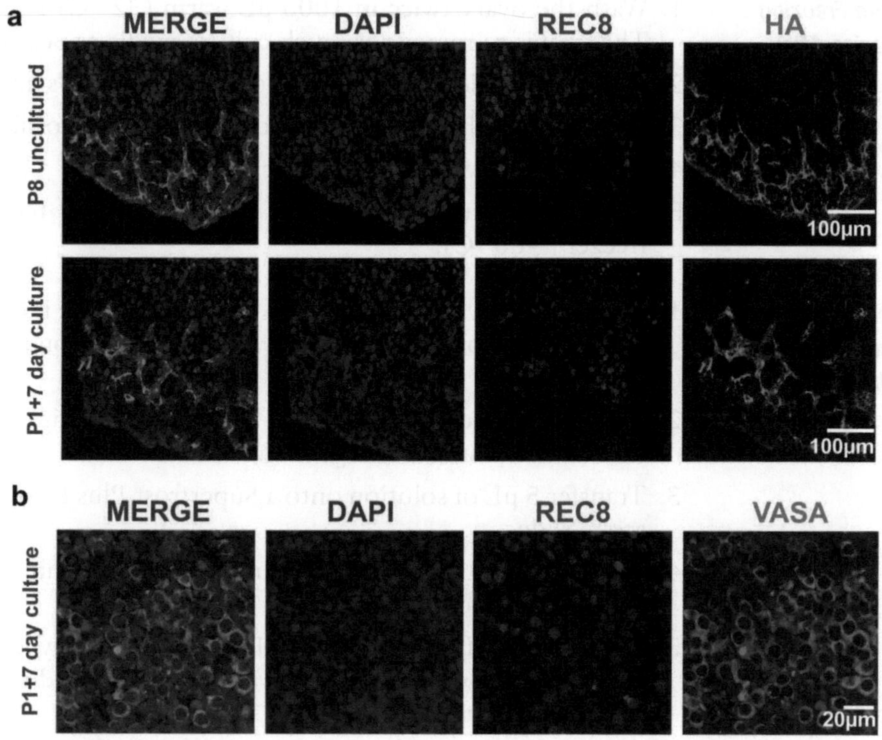

Fig. 4 Immunostaining of naked mole-rat tissue sections. (**a**) Postnatal ovary at Day 8 (P8), and P1+7 stained with REC8 (red) and DAPI (blue); (**b**) P8 ovary strained with REC8 (red), VASA (green), and DAPI (blue)

16. Rinse slides with PBS-T 3 times for 5 min each.
17. Incubate slides with 0.1 μg/mL of DAPI solution in water for 15 min at room temperature to counterstain the nucleus.
18. Rinse slides with milliQ water 3 times for 5 min each.
19. Let the slides dry for 60 min.
20. Mount with Vectashield and cover slide. See Fig. 4.

3.9 Clearing of Ovarian Cultures

The following protocol for the clearing of ovarian cultures is a modification of that previously described by Rinaldi [21], adjusted for whole cultured naked mole-rat ovaries.

1. Prepare fresh ScaleS solution.
2. Add 1000 μL of the clearing solution to each well and place them at room temperature on the benchtop orbital shaker.
3. Exchange the ScaleS solution 4 times a day until the tissue becomes transparent (approximately 4 days).
4. Mount with Vectashield and analyze.

3.10 Flash Freezing Whole Ovaries After Culture

1. Wash the ovary twice in 1000 μL warm (32 °C) PBS, sterile. The goal is to remove as much culture media as possible.
2. Wash the ovaries twice in 1000 μL ice-cold PBS, sterile.
3. Remove the PBS and place the ovary in a 1.5 cryotube.
4. Place the ovary in liquid nitrogen for 5 min.
5. Remove cryotube and keep it in an ultra-low temperature freezer (−80 °C).

3.11 Meiotic Spreads from Frozen Whole-Ovary Cultures

1. Remove the ovary from −80 °C freezer and place it on dry ice. It is possible to use the whole ovary or cut a section to perform the spreading.
2. Place the tissue in 40 μL of ice-cold PBS and mince with the needles to create a homogeneous solution.
3. Transfer 5 μL of solution onto a Superfrost Plus frosted microscope slide.
4. Add 80 μL of Lipsol 1% and incubate for 10–13 min. The time should be standardized for each condition.
5. Add 90 μL of fixative solution: 1% paraformaldehyde, pH 9.2, Triton X 0.05%, and incubate overnight in a humidity chamber at room temperature.
6. Wash the slides with Photoflo 0.05% in PBS for 15 min, twice.
7. Wash the slides with Photoflo 0.05% in water during 10 min, twice.
8. Air-dry the slides and store them at −80 °C for later analysis or start immunofluorescence as was described for spreads immediately after harvest.

3.12 Single-Nuclei Suspension from Frozen Whole-Ovary Cultures

There are various kits to perform the single-nuclei suspension. We found that the most reliable kit for our tissue is the Chromium Nuclei Isolation Kit. We follow the manufacturer's instructions with high cell yield and nuclei quality.

4 Notes

1. Fixative solution should be prepared fresh. DO NOT prepare more than 2 h before use. Keep on ice and add Tritron X after measuring the pH.
2. Do not use old (i.e., not fresh) HEB, which will prevent cells from spreading properly and create artifacts in the antibody signal.
3. Use the proper serum. For example, if secondary antibodies were raised on donkeys or other organisms, use that serum instead of goat serum.

4. Use uterine horn as a handle.
5. Naked mole-rats produce very-high-molecular-weight hyaluronan, which can turn the culture medium a bit viscous.
6. Do not let the slide become fully dry; if slides are completely dry, the secondary antibody will generate a massive amount of background.
7. Abundant proteins, such as cohesin or synaptonemal complex, are properly conserved; however, proteins that form foci, such as MLH1, are not properly preserved after a year.

References

1. Grive KJ, Freiman RN (2015) The developmental origins of the mammalian ovarian reserve. Development 142:2554–2563. https://doi.org/10.1242/dev.125211
2. Ge W, Li L, Dyce PW, De Felici M, Shen W (2019) Establishment and depletion of the ovarian reserve: physiology and impact of environmental chemicals. Cell Mol Life Sci 76: 1729–1746. https://doi.org/10.1007/s00018-019-03028-1
3. Grive KJ (2020) Pathways coordinating oocyte attrition and abundance during mammalian ovarian reserve establishment. Mol Reprod Dev 87:843–856. https://doi.org/10.1002/mrd.23401
4. Brieno-Enriquez MA et al (2023) Postnatal oogenesis leads to an exceptionally large ovarian reserve in naked mole-rats. Nat Commun 14:670. https://doi.org/10.1038/s41467-023-36284-8
5. Ruby JG, Smith M, Buffenstein R (2018) Naked Mole-Rat mortality rates defy gompertzian laws by not increasing with age. elife 7. https://doi.org/10.7554/eLife.31157
6. Buffenstein R, Jarvis JU (2002) The naked mole rat–a new record for the oldest living rodent. Sci Aging Knowl Environ 2002:pe7. https://doi.org/10.1126/sageke.2002.21.pe7
7. Edrey YH, Hanes M, Pinto M, Mele J, Buffenstein R (2011) Successful aging and sustained good health in the naked mole rat: a long-lived mammalian model for biogerontology and biomedical research. ILAR J 52:41–53. https://doi.org/10.1093/ilar.52.1.41
8. Findlay JK, Hutt KJ, Hickey M, Anderson RA (2015) How is the number of primordial follicles in the ovarian reserve established? Biol Reprod 93:111. https://doi.org/10.1095/biolreprod.115.133652
9. Jarvis JU (1981) Eusociality in a mammal: cooperative breeding in naked mole-rat colonies. Science 212:571–573. https://doi.org/10.1126/science.7209555
10. Brieno-Enriquez MA et al (2010) A new culture technique that allows in vitro meiotic prophase development of fetal human oocytes. Hum Reprod 25:74–84. https://doi.org/10.1093/humrep/dep351
11. Gonzalez-Sanz S, Barrenada O, Rial E, Brieno-Enriquez MA, Del Mazo J (2020) The antiandrogenic vinclozolin induces differentiation delay of germ cells and changes in energy metabolism in 3D cultures of fetal ovaries. Sci Rep 10:18036. https://doi.org/10.1038/s41598-020-75116-3
12. Rinaldi VD, Bolcun-Filas E, Kogo H, Kurahashi H, Schimenti JC (2017) The DNA damage checkpoint eliminates mouse oocytes with chromosome synapsis failure. Mol Cell 67:1026–1036 e1022. https://doi.org/10.1016/j.molcel.2017.07.027
13. Hwang GH, Hopkins JL, Jordan PW (2018) Chromatin spread preparations for the analysis of mouse oocyte progression from prophase to metaphase II. J Vis Exp. https://doi.org/10.3791/56736
14. Sun X, Cohen PE (2013) Studying recombination in mouse oocytes. Methods Mol Biol 957: 1–18. https://doi.org/10.1007/978-1-62703-191-2_1
15. Peters AH, Plug AW, van Vugt MJ, de Boer P (1997) A drying-down technique for the spreading of mammalian meiocytes from the male and female germline. Chromosom Res 5: 66–68. https://doi.org/10.1023/a:1018445520117
16. Pereira C et al (2022) Multiple 9-1-1 complexes promote homolog synapsis, DSB repair, and ATR signaling during mammalian meiosis. elife 11. https://doi.org/10.7554/eLife.68677
17. Holloway JK et al (2011) Mammalian BTBD12 (SLX4) protects against genomic

instability during mammalian spermatogenesis. PLoS Genet 7:e1002094. https://doi.org/10.1371/journal.pgen.1002094

18. Garcia-Cruz R et al (2010) Dynamics of cohesin proteins REC8, STAG3, SMC1 beta and SMC3 are consistent with a role in sister chromatid cohesion during meiosis in human oocytes. Hum Reprod 25:2316–2327. https://doi.org/10.1093/humrep/deq180
19. Brieno-Enriquez MA et al (2011) Human meiotic progression and recombination are affected by Bisphenol A exposure during in vitro human oocyte development. Hum Reprod 26:2807–2818. https://doi.org/10.1093/humrep/der249
20. Lenzi ML et al (2005) Extreme heterogeneity in the molecular events leading to the establishment of chiasmata during meiosis i in human oocytes. Am J Hum Genet 76:112–127. https://doi.org/10.1086/427268
21. Rinaldi VD, Bloom JC, Schimenti JC (2018) Whole mount immunofluorescence and follicle quantification of cultured mouse ovaries. J Vis Exp. https://doi.org/10.3791/57593

Chapter 13

Temporal Analysis of DSB Repair Outcome in *Caenorhabditis elegans* Meiosis

Tara Hicks, Nicola Silva, and Sarit Smolikove

Abstract

The *Caenorhabditis elegans* germline is arranged spatiotemporally and is therefore a powerful model system for the interrogation of meiotic molecular dynamics. Coupling this property with the temporal control that the auxin-inducible degron (AID) system allows can unveil new/unappreciated roles for critical meiotic factors in specific germline regions. Here we describe a widely used approach for the introduction of degron tags to specific targets and provide a procedure for applying the AID system to *C. elegans* meiotic DSB repair dynamics in the germline.

Key words Meiosis, Auxin, Crossovers, Germline, *Caenorhabditis elegans*

1 Introduction

The auxin-inducible degron (AID) system has become a priceless tool for in vivo analysis of protein function in various model systems, including *Saccharomyces cerevisiae, Caenorhabditis elegans,* and mice [1–3]. In *C. elegans* alone, the AID system has furthered understanding of many proteins since its introduction due to its easily manipulatable nature, relatively simple workflow, and rapid functionality. Furthermore, AID is easily reversible, and as a result, it can be applied for the resolution of spatiotemporal protein dynamics [4]. The underlying principle of the system is fairly simple: a 44-amino acid degron tag, sourced from *A. thaliana,* is introduced to the endogenous locus of interest via CRISPR/Cas9 in a background containing *AtTIR1.* Worms are then fed indole-3-acetic acid (i.e., "auxin") which directs *AtTIR1* to target the degron tag for polyubiquitination via the Skp1-Cullin-F-box E3 ubiquitin ligase [5]. One advantage to this is the ability to place *AtTIR1* under specific promoters, allowing tissue-specific depletion and therefore permitting deeper comprehension of process-specific protein functions and dynamics [3, 6]. This is valuable for

Jesús A. Carballo (ed.), *Meiosis: Methods and Protocols*, Methods in Molecular Biology, vol. 2818,
https://doi.org/10.1007/978-1-0716-3906-1_13,

the field of *C. elegans* meiosis, as some proteins are essential for reproduction (e.g., SPO-11 and DNA-2) and/or may play multiple roles in different regions of the germline, which makes dissecting individual functions in null backgrounds difficult. Placing *AtTIR1* under a germline-specific promoter (e.g., *pSun*-1) allows germline-specific expression and spatiotemporal analysis that can help parse shared roles of S-phase/mitosis in the pre-meiotic tip (PMT) from those in the meiotic compartment [4]. Here we show how the AID system can be used to assess the spatiotemporal dynamics of meiosis-specific proteins in the *C. elegans* germline and provide a detailed methodology for introduction of the degron tag, making of NGM/auxin plates, auxin feeding, assessment of depletion, and application to germline cytology, as well as a number of troubleshooting suggestions for successful implementation. We have applied these methods to deliver the findings reported in [4].

2 Materials

2.1 CRISPR/Cas9 Microinjection

2.1.1 Injection Mix

Resuspend tracrRNA and crRNAs in nuclease-free duplex buffer (Integrated DNA Technologies #11-05-01-12) and all repair ssODNs in nuclease-free filtered ddH_2O. Ensure all equipment and work surfaces are nuclease-free. Cas9-NLS is a purified protein (*see* **Note 1**). The *dpy-10* coinjection marker is optional and can be compensated for with nuclease-free water; however, omitting it from the reaction mix requires screening more progeny (~ 3 times more in our hands) to identify a hit.

1. 200 μM tracrRNA.
2. 100 μM target crRNA.
3. 100 μM *dpy-10* crRNA (optional).
4. 50 μM target repair ssODN.
5. 5 μM *dpy-10* repair ssODN (optional).
6. 40 μM Cas9-NLS purified protein.
7. Nuclease-free filtered ddH_2O.
8. 0.2–2 μL pipette and tips.
9. 0.2 mL (×2) PCR tubes.
10. M9 buffer (makes 1 L): Weigh 3.5 g KH_2PO_4, 6 g Na_2HPO_4, and 5 g NaCl. Add 1 mL 1 M $MgSO_4$ and bring to 1 L with ddH_2O. Sterilize by autoclave.
11. Unseeded NGM plate.12. Gloves.

2.1.2 Microinjection

1. Needle-loading capillary tubes: hold a calibrated 50 μL capillary tube at either end and bring the middle of the tube over a flame. Allow to heat until the tube is pliable in the middle,

quickly remove from heat, and pull ends straight in opposite directions in the same motion. Break the stretched middle to get two needle-loading capillary tubes with thin tapered tips.

2. Injection needles: feed a thin-walled glass 4″ capillary tube with inner filaments through an economical needle-puller. Determining optimal needle pulling settings can differ depending on different machines and the age of the heating element; refer to [7] for determining optimal settings on your needle puller.
3. Aspirator tube.
4. 3% agar pads: prepare a 3% agar solution in ddH_2O and microwave 1 min or until clear, removing the beaker to stir every ~5–10 s (watch to make sure solution does not boil over in microwave). Keep the agar solution warm in 50–55 °C water bath. With a glass Pasteur pipette, place 2–3 drops of the agar solution in the middle of a 22 × 50 mm coverslip and place a second coverslip on top for ~5 s. Gently pull the coverslips apart and place the one with the agar pad face up on a slide warmer set to 45 °C until pads dry.
5. M9 buffer.
6. Halogen injection oil.
7. Microinjection-capable microscope setup.
8. Synchronized 1-day-old adult worms of *AtTIR1* background (e.g. CA1199, available from CGC).

2.2 Auxin Plates

1. NGM media (400 mL): mix 8 g agar, 1.2 g NaCl, 1 g peptone, 400 μL 1 M $MgSO_4$, 400 μL 1 M $CaCl_2$, and bring to 400 mL with sterile ddH_2O. Sterilize by autoclave. Let cool to 55 °C and add 400 μL 5 mg/mL cholesterol in absolute ethanol and 10 mL 1 M KH_2PO_4. Pour in 60 × 15mm Petri plate using a peristaltic pump.
2. Auxin media (400 mL, *see* **Notes 2** and **3**): prepare NGM media as above and add 1 mL 400 mM indole-3-acetic acid ("auxin"). For empty vehicle control plates, omit indole-3-acetic acid and simply add 1 mL absolute ethanol. Pour in 60 × 15mm Petri plate using a peristaltic pump. Store on bench in the dark to cool, and store at 4 °C for up to 3 weeks.
3. 400 mM auxin: mix 0.7 g indole-3-acetic acid in 10 mL absolute ethanol on a nutating mixer. Store at 4 °C in the dark.
4. *Escherichia coli* OP50 cultured in 10 mL Luria Broth.
5. Glass Pasteur pipettes and bulb.

2.3 Cytological Assays

1. Sperm salts: 50 mM PIPES pH 7.0, 25 mM KCl, 1 mM $MgSO_4$, 45 mM NaCl, 2 mM $CaCl_2$, bring to volume with ddH_2O [8].

2. 37% PFA stock: Weigh 1.85 g paraformaldehyde (PFA) and top with 4.5 mL ddH_2O in a 10 mL capped glass tube. Boil in water bath for 1 min. Add 90 μL 1 N NaOH and, still in boiling water, cap and agitate by hand. After ~1 min, the solution should be nearly clear. If there is still a lot of solid PFA, add an additional 90 μL 1 N NaOH and repeat agitation in boiling water for another minute. Cool under running tap water and filter through a 0.2 μm filter. Aliquot and store stock at − 20 °C. When ready to use, melt in a 70 °C bead or water bath until solution is fully clear. Alternatively, 16% formaldehyde (FA) stock solution can be used. Once a new ampule is opened, transfer the FA solution into a 15 mL Falcon tube and store at 4 °C protected from light. Use for no longer than two weeks.
3. 4% PFA: Mix 2.23 mL 1X PBS with 270 μL 37% PFA stock. For steps which require fixation after freeze cracking, mix 500 μL 10X PBS, 400 μL 1 M HEPES pH 7.4, 8 μL 0.5 M EDTA, 540 μL 37% PFA, 8 μL 1 M $MgSO_4$, and 3.544 mL ddH_2O.
4. 18 × 18 × 1mm and 22 × 22 × 1.5 mm coverslips.
5. #10 dissection blade or 2 sharp needles for germline dissection.
6. Frosted 25 × 75 × 1mm charged slides.
7. Aluminum block on dry ice.
8. 10X PBS pH 7.4 (makes 1 L): Weigh 80 g NaCl, 2 g KCl, 2.4 g KH_2PO_4, 14.4 g Na_2PO_4. Bring to 1 L with ddH_2O. If too acidic, add 10 N NaOH to neutralize. Sterilize by autoclave.
9. 1X PBST (makes 300 mL): 267 mL ddH_2O, 30 mL 10X PBS pH 7.4, 3 mL 10% Tween-20.
10. 0.5% bovine serum albumin (BSA) in 1X PBST: weigh 0.25 g BSA and add to a Coplin jar containing 50 mL 1X PBST. Allow to dissolve ~10 min at room temperature until no flakes are present. Mix with a P1000 tip.
11. 1:10,000 5 μg/mL 4′-6-diamidino-2-phenylindole (DAPI) in 1X PBST: pipette 5 μL DAPI into a Coplin jar containing 50 mL 1X PBST. Mix with a P1000 tip and keep in dark.
12. RAD-51 antibodies: rabbit anti-RAD-51 in 1X PBST (the dilution will depend on the lab's Ab source and typically ranges 1:3000–1:30,000 [4]), 1:500 AlexaFluor donkey anti-rabbit 488 in 1X PBST.OLLAS::COSA-1 antibodies: 1:1500 rabbit anti-OLLAS tag in 1X PBST, 1:500 AlexaFluor goat anti-rabbit 488 in 1X PBST.
13. SYP-1 antibodies (optional): 1:100 rat anti-SYP-1 [4], 1:500 goat anti-rat AlexaFluor 594.
14. Vectashield anti-fade mounting medium.DAPI+Vectashield.

3 Methods

3.1 Design of Degron Tag

For designing a degron tag for the target gene, follow the instructions provided in [9]. In brief, design single-stranded oligodeoxyribonucleotides (ssODN) with 35 base pair homology arms and CRISPR RNA (crRNA, the target-specific component of the artificial small guide RNA), corresponding to the closest crRNA target to your insertion sites. It is advisable that the crRNA target site be no more than 12 base pairs away from your insertion site. Positions further away can yield successful insertion but require screening more F1s to identify a hit (see below). The 35 bases of homology arm are counted from each side of the predicted insertion site: on the side that does not have the crRNA target site the 35 bases are counted from the junction, while on the side that contains the crRNA target sequence, they are counted 3 nt upstream from the PAM site on the crRNA strand (this is the expected CRISPR cut site). If the crRNA target site does not overlap the insertion site, a silent mutation should be included in the ssODN. In our experience, a single-point mutation in one of the PAM's 2′- or 3′-G nucleotides is sufficient. If this is not possible, then design silent mutations as close as possible to the PAM. The degron sequence is 135 nt long, and with the size of homology arms, it is below the 200 nt ssODN that is cost-effectively priced to synthesize. However, since the position of the closest crRNA will vary based on the target gene, the crRNA target site may not be overlapping the insertion. This and the need to include a linker and/or additional tag (e.g., HA, 3XFLAG) will result in an overall ssODN size larger than 200 nt. In this case, it is possible either to use a longer, more costly ssODN or perform two-step insertion in which part of the tag is inserted and homozygous worms are isolated, followed by a second CRISPR editing event targeting the crRNA sequence in the inserted region (for examples, *see* **Note 4**). Use a reverse-complement ssODN if the insertion is downstream from the crRNA site.

3.2 Introduction of Auxin-Inducible Degron Tag Via Two-Step CRISPR/Cas9 Microinjection

Ensure workspace and equipment are nuclease-free before proceeding.

3.2.1 Injection Mix Preparation

1. Set thermocycler to 95 °C.
2. Mix 0.81 μL tracrRNA, 1.62 μL target crRNA, and 0.15 μL *dpy-10* crRNA in a nuclease-free 0.2 mL PCR tube.
3. Heat in 95 °C thermocycler for 5 min and then allow mix to sit at room temperature for an additional 5 min.

4. Remove 1.325 μL to a new 0.2 mL PCR tube and add 1.75 μL Cas9 to it.
5. Allow to complex at room temperature for 5 min.
6. Add 0.5 μL *dpy-10* ssODN, 0.6 μL of target ssODN, and 0.83 μL filtered sterile ddH_2O.
7. Spin down in centrifuge 14,000 rpm for 10 min.
8. While injection mix is spinning, drop 200 μL M9 buffer or ddH_2O onto the center of a clean, unseeded NGM plate.
9. Pick 40–50 synchronized young adults into liquid, being careful to minimize picking bacteria with them. Excess bacteria can cause needles to clog during microinjection procedure.

3.2.2 Microinjection and Screening for Insertion

1. Fill unbroken needle with >1 μL injection mix using a needle-loading capillary tube attached to a mouth pipette. Before proceeding, ensure injection mix migrates to the needle tip.
2. Place the loaded needle into the injection microscope's needle holder and position the needle holder on the manipulator, such that the needle tip is not touching the stage but is centered in the viewing field at the 10X objective and just out of focus.
3. Place a drop of halogen oil on the side of the injection slide and rest it on a small petri dish covered by a dissecting microscope.
4. Break the tip of your needle-loading capillary tube off and place it directly next to the oil drop. You can use your platinum pick to manipulate it on the slide. Orient it at ~45°angle for sharper needle tips. Drag some halogen oil from your droplet over the center of the capillary tube. Make sure to keep the rest of the capillary tube for recovery later.
5. Place the injection slide on the injection scope stage and center the field over the oil-coated wall of the broken capillary tip closest to your injection needle. Bring into focus so that the wall of the capillary tube is sharp and clear.
6. Manipulate the needle tip just off the side of the capillary tube wall and use the z-adjustment knob to drop the needle tip until it just touches the oil droplet. Your needle tip should *not* touch the capillary tube at this stage but should be in close proximity. Once the needle tip touches the oil, quickly use the z-adjustment knob to bring the needle just out of the oil.
7. Leaving the slide on the stage, move to the 100X objective and refocus the wall of the capillary tube using the fine focus.
8. Gently drop the needle tip back into the oil using the z-plane adjustment. Use the joystick operator to adjust as needed to bring *just* the needle *tip* into focus and to direct the needle tip to just touch the capillary tube wall. Be very careful not to make any quick movements at this stage before you're ready to break the needle tip.

9. When you are ready to break the needle tip, you can do so by using the joystick operator to bump the needle tip into the side of the capillary tube. Alternatively, you can hit the edge of the table that the microscope stands on, and the movement will cause the needle to hit the side of the capillary tube wall.
10. Make sure the needle is flowing. If flow is not sufficient, break the needle further using the same procedure. If there is no flow, try using a pulse of high pressure to push out any air bubbles (don't exceed the pressure tolerance of your microscope system), and then try rebreaking the needle if that fails.
11. Move the injection slide back to the small petri dish plate on the dissection microscope stage.
12. Collect a small droplet of halogen oil onto a platinum wire pick, scoop up one to several worms into the droplet from a bacteria-free region of the plate, and place it on the agarose pad of the injection slide.
13. Flame and cool your pick, and, being very careful not to crush the worm's pharynx/head, gently brush your pick over the length of the worm's body to facilitate adherence to the pad. If the worm does not stick, pick worms into a new drop of halogen oil on a different part of the pad.
14. Move the slide face-up to the injection scope stage. At the 10X objective, center the first worm to be injected in the field and bring it into focus. Manipulate your needle so that the needle is just off to the side of the worm body. Use the z-adjustment to bring the needle tip just into and out of the oil, as you did in step 6.
15. Leaving the slide on the stage, move to the 100X objective and adjust to bring the worm into focus so that the edges of the worm are sharp and clear. Use the fine adjustment to bring the core cytoplasm of one gonad arm into view and adjust the needle tip into the same plane using the joystick operator.
16. Using the joystick operator, apply pressure to the needle so that the cuticle depresses. Many times, this action is enough for a sharp needle to penetrate the cuticle into the syncytial gonad. If this does not occur and you can only get the cuticle depressed with your needle, you can gently move the worm into the needle tip until the tip inserts.
17. Inject your CRISPR solution. You should not need prolonged flow; a short pulse of injection mix is sufficient. The gonad should noticeably swell in both directions, and oftentimes an egg or two will pop out of the vulva with a successful injection.
18. Use the joystick operator to gently back the needle out of the worm. Move the stage away from the worm (if you do this right next to the worm, the worm will release from the pad) and check your needle flow for blockages. Clear any blockages. You

can choose to inject the other gonad from the same worm or continue on to the next worm and repeat the process. When done, move the injection slide rapidly back to the small petri dish cover on your dissection microscope stage.

19. Recover your worms by using the capillary tube you broke in step 4—attach the capillary tube to your aspirator tube and take up some M9 buffer via capillary action. Apply gentle pressure with your mouth to release the buffer on top of the oil droplet containing your injected worms. Allow the buffer to sink below the oil; worms should start moving as they try to free themselves from the pad. Once they have released from the pad, place the opening of the capillary tube near one of the worm's heads while maintaining gentle outward pressure with your mouth. Once the capillary tube is positioned to take up the worms, release the pressure with your mouth so that the capillary action takes up the buffer and the worms.
20. Drop the worms and buffer onto a seeded plate for recovery. Repeat this process (steps 11–20) for the rest of your worms. Once you are done, allow your worms to recover on the seeded plate, then move the plate to a 20 °C incubator overnight to allow worms to lay eggs that were already developing at the time of injection that will not contain your edits.
21. The next day, single injected P0s to clean seeded plates and incubate at 20 °C until the next generation of L4s arises (~3–5 days).
22. Screen P0 plates for *dpy/rol* F1s indicative of the coinjection marker—single wild-type L4 siblings only from plates that have the coinjection phenotype.
23. 1–2 days later, lyse and genotype the singled F1s for step 1 of your degron insertion. Outcross hits to eliminate any possible off-target effects and repeat injection with the same workflow.

3.3 Spreading and Picking to Auxin Plates

1. Pour *E. coli OP50* culture into a sterile conical tube and centrifuge for 30 s at 14,000 rpm.
2. Remove 9 mL supernatant so you are left with a 10X concentrated bacterial culture.
3. Using a glass Pasteur pipette, drop 3 small dots of the concentrated bacteria onto the auxin plate and allow to culture 2 days at room temperature in the dark (*see* **Note 5**).
4. Experiments rely on worms ingesting auxin, so pick synchronized worms with little to no bacteria in the middle of the 3 drops of concentrated bacteria on the auxin plate. If concerned about untreated bacterial transfer, particularly for short auxin exposures, you can pick the worms into a drop of M9 buffer on an unseeded plate to wash before picking them over to auxin plates.

3.4 Confirmation of Auxin-Inducible Functionality of Degron Tag (See Note 6)

Confirm that auxin plates are good by picking *spo-11::AID:3X-FLAG* (CA1423, available from CGC) L4s to auxin plate and allowing to grow in the dark 24 h at 20 °C. Ethanol fixes and checks diakinesis -1 for 12 univalents [4].

3.4.1 Fecundity Assay (See Notes 7–9)

1. Single 3 L4s of your degron-containing strain and 3 L4s of a wild-type control (N2) to standard NGM plates. Leave in 25 °C incubator for 5 days.
2. After 5 days, count F1 young adults and single 3 F1 L4s to auxin plates and 3 L4s to NGM plates (*i.e.* 3 mutant L4s and 3 N2 L4s per condition). Leave in 25 °C incubator for 5 days.
3. After 5 days, count F2 young adults and compare mutants to wild type.
4. Repeat the assay (steps 1–3).

3.4.2 Cytological Analysis (See Note 10)

1. Pick L4s 24 h before depletion experiment on standard NGM plates and allow to grow up overnight.
2. The next day, break synchronized young adults into three populations and plate 15–20 worms per condition:
   (a) Your NGM control population (no auxin depletion or effects from EtOH)
   (b) Your EtOH control population (no auxin depletion)
   (c) Your auxin treated population
3. 4 h after plating the three populations, pick worms into a 15 μL drop of Sperm Salts (*see* **Note 11**) on a 18 × 18 × 1mm coverslip staged on an uncharged slide on a dissection microscope, being careful not to bring over any bacteria and immediately dissect the gonads using a dissection blade or two sharp needles (*see* **Note 12**).
4. Add 15 μL 4% PFA stock in 1X PBS to the dissection buffer. Pipette up and down a few times to help the gonads extrude and remove 15 μL total solution (2% final fix).
5. Mount the coverslip on a frosted charged slide and allow to fix in a humidity chamber at room temperature for 5 min, then transfer to an aluminum block on dry ice.
6. Allow slides to set on dry ice blocks for at least 20 min.
7. Using a razor blade, carefully pop the coverslips off the frozen slides and immediately dip them into ice cold methanol for 2 min.
8. Using tweezers, dip slides in ice cold acetone for 5–10 s and move to a Coplin jar containing 50 mL 1X PBST for 10 min.

9. Repeat the 1X PBST wash in a new Coplin jar.
10. Move slides to a Coplin jar containing 0.5% BSA in 50 mL 1X PBST and allow to block for 1.5–2 h at room temperature.
11. Apply primary antibody dilution to a coverslip-sized Parafilm cutout and apply to the region of the slide where you can see worm bodies. Place slides in humidity chamber and incubate in the dark at room temperature overnight.
12. The next day, move slides to a Coplin jar containing 1X PBST for 10 min. Repeat the wash.
13. Apply secondary antibody dilution to a coverslip-sized Parafilm cutout and apply as you did in step 10. Allow it to incubate in a humidity chamber in the dark at room temperature for 2 h.
14. Wash in a Coplin jar with 50 mL 1X PBST at room temperature for 10 min. Repeat the wash.
15. Move slides to a new Coplin jar containing 5 μL DAPI and 50 mL 1XPBST. Incubate 10 min in the dark at room temperature.
16. Destain slides in 50 mL 1X PBST three times and mount 22 × 22 × 1.5 mm coverslips with antifade mounting medium (*e.g.*, Vectashield). Aspirate excess liquid around the edges of the coverslip with Kimwipe strips.
17. Seal the coverslips with clear nail polish and proceed to imaging. When imaging, ensure exposure time is consistent across slides to avoid analytical errors.
18. *Before you start the analysis:* Pick a single z-plane in Fiji ImageJ to work with that is in the middle of most nuclei in the image. Record the z-plane and take a screenshot of your image; mark which nuclei you will measure. Using the draw tool, trace the borders of the nuclei in the DAPI channel, then move to your antibody channel and measure the fluorescence (CTRL+M). Record the data output (CTRL+A, CTRL+C) in your Excel file and repeat this process for all the selected nuclei.
19. When done measuring the nuclear fluorescence, measure the cytoplasmic background intensity by using the draw tool to select at least 5 regions that do not contain nuclei in the image (i.e. the cytoplasm). Measure the antibody fluorescence of these 5 regions and mark them as controls; this will constitute your background signal later.
20. In your Excel file, calculate the raw intensity divided by the area for each data point. This gives you the raw intensity density per area ("RDPA") and corrects for differently-sized nuclei. For cytoplasmic ("control") measurements, you want an average of your five measurements, so take the sum of all of the raw intensity densities and divide by the sum of all the areas.

21. To correct the fluorescent values of your nuclear stain against the cytoplasmic background, subtract the average control RDPA from each RDPA of your nuclear readings. The output of this is the corrected fluorescence intensity of the data point.
22. Compare fluorescence intensity measurements to determine if detectable cytological depletion has occurred.

3.5 Specific Applications to Dynamics in C. elegans Germlines (See Note 13)

Nuclei move at a rate of ~1 nuclear row per hour [10, 11] and one-day-old adult germlines typically contain ~70 rows of nuclei, ~48 of which are from meiotic entry (transition zone) to late pachytene. Therefore, it is possible to calculate backwards from the region in which an outcome (*e.g.*, RAD-51, COSA-1) is analyzed, to identify which germline region the nuclei were depleted of your protein of interest. This can be done by multiplying the row number by the time of depletion. For example, using this rationale to analyze the time window for crossover formation via COSA-1 foci, ~20 h of auxin exposure will examine the effect of depletion of your protein of interest from mid to late pachytene, while depletion of 24 h followed by recovery of 60 h on standard NGM plates will test the effect of mitotic depletion on diakinesis DAPI bodies [12].

3.5.1 RAD-51

1. As described in "Cytological analysis," pick L4s 24 h before short depletion experiments on standard NGM plates. For longer depletion times (e.g. 24 h+), transfer L4s directly to auxin and control empty vehicle plates unless experiment dictates otherwise.
2. Allow worms to sit on plates in the dark at 20 °C for the desired time. If recovering, pick back to standard NGM plates for the desired recovery time; otherwise, proceed to antibody staining as described in "Cytological Analysis" above.
3. Acquire intact gonads and divide them into seven equal zones from the mitotic tip to the end of pachytene stage. As a general indication, Zone 7 should not include nuclei in Diplotene stage, and therefore it should end before the gonadal bend.
4. Nuclei with obvious apoptotic features, such as hyper condensed chromatin or reduced nuclear size (pycnotic nuclei), should not be included in the counts.
5. Set z-stacks at 0.25 μm and only one half of the gonad should be imaged. Generally, the two halves of the germ line are not stained with equal efficiency, and therefore the one showing better staining should be acquired.
6. Before proceeding with the RAD-51 foci quantification, performing image deconvolution will improve sharpness and noise/signal ratio. Different softwares can be employed for deconvolution and parameters should be set accordingly (see **Note 14**). For representative results, see Fig. 1.

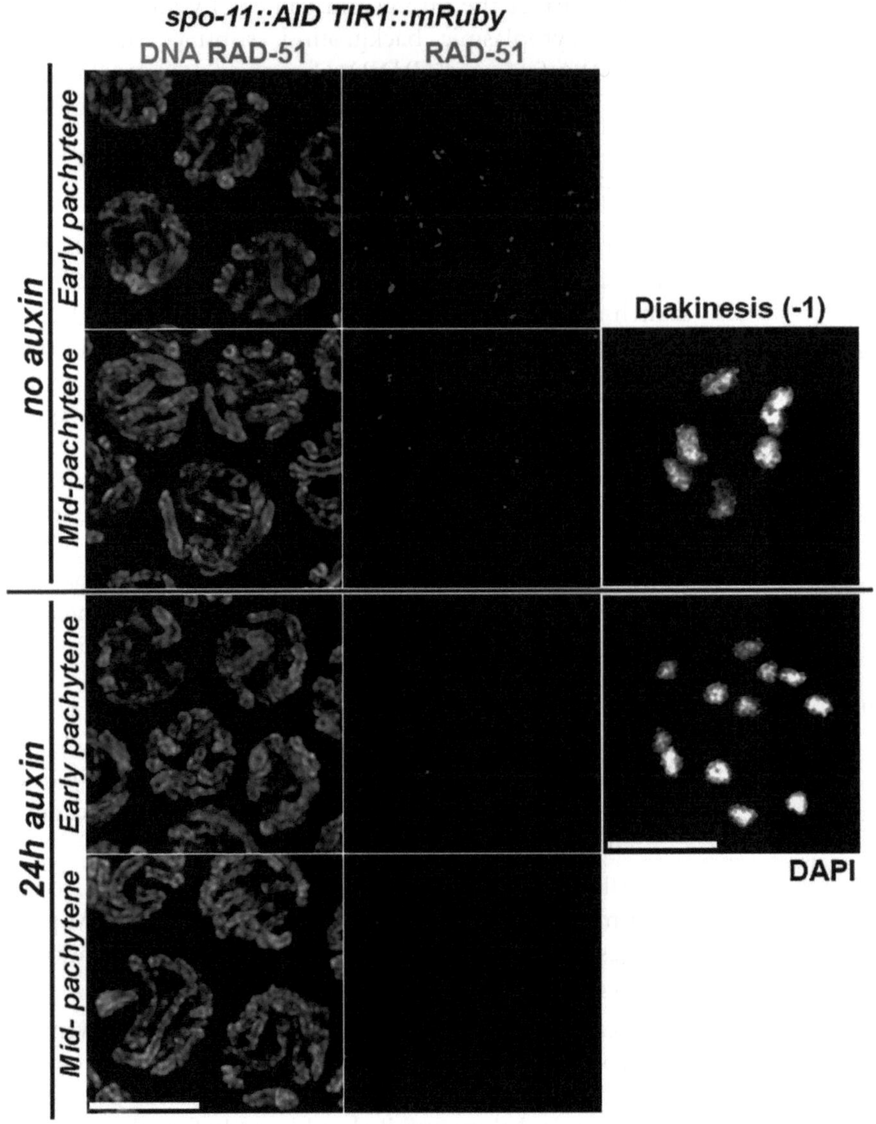

Fig. 1 Nuclei of the indicated genotype, stage, and conditions of exposure to auxin, stained for RAD-51 and counterstained by DAPI. Scale bar: 5 μm

7. Three to four gonads per genotype should be acquired in order to obtain roughly one hundred nuclei for each of the seven zones and, therefore, obtaining robust statistical comparisons.
8. The number of RAD-51 foci should not be counted on whole projected stacks since individual foci on different focal planes might appear as singlets or doublets, thus affecting quantification outcomes. Once RAD-51 foci are detected in a nucleus, scoring their number through the z-stacks will provide accurate counts.

9. Given that some background might occur, only the foci clearly localizing along the chromatin should be included in the quantification.

3.5.2 DAPI Bodies in Diakinesis -1

1. Drop 8 μL of M9 buffer onto the surface of an uncharged slide, and put another drop of the same volume on the slide in close proximity to the first.
2. Pick 10–25 worms into one of the drops of M9 and use a worm pick to slowly drag the edge of the drop containing the worms toward the empty drop to clean them.
3. Use a triangle of filter paper to remove as much M9 as possible without sucking up any worms or letting them dry out.
4. Add 10 μL of 100% ethanol to the top of the worms and let them dry. While the ethanol is drying, you can quickly use a platinum worm pick to straighten any worms that might overlap, but don't do so once the liquid is gone.
5. Mount a 22 × 22 × 1.5 mm coverslip with DAPI+Vectashield and seal with clear nail polish. Store in the dark at 4 °C overnight to let the DAPI penetrate (*see* **Note 15**). Proceed to imaging the next day. For representative results, see Fig. 1.
6. Analysis of DAPI bodies should be done in 3D. DAPI bodies can touch each other but still constitute individual chromosomes/chromosomal fragments, if one is falling out of focus while the other one is not.

3.5.3 COSA-1

1. At the end of the desired depletion time course, pick 10–15 worms into a 15 μL drop of M9 buffer on a 18 × 18mm coverslip staged on an uncharged slide on a dissection microscope.
2. Dissect gonads immediately using either a dissection blade or two sharp needles, as preferred.
3. Once all worms on the slide are dissected, mount the coverslip on a frosted charged slide and immediately transfer to an aluminum block on dry ice to freeze. Allow slides to sit on frozen block at least 20 min. Alternatively, you can dip the slide +coverslip in liquid nitrogen for 30 s.
4. Freeze-crack the coverslip with a razor blade. Regardless of which freezing method is preferred, removal of the coverslip should be very fast (while slide is still frozen).
5. Immediately dip into ice cold methanol for 1 min.
6. Rapidly dry just the back of the slide using a Kimwipe, transfer to a humidity chamber, and apply 800 μL 4% paraformaldehyde fix solution directly to gonads on the slide. Gently cover with a slide cover-sized Parafilm cutout and incubate in the humidity chamber for 30 min to fix.

7. Wash in 1X PBST for 10 min, and repeat wash step once more.
8. Proceed with labeling as described in steps 9–16 of "Cytological analysis" (3.D.B.). At this step, you can add SYP-1 antibody (or any other antibody against synaptonemal complex component) for clear visualization of late pachytene nuclei, but this is not required for the analysis.
9. COSA-1 undergoes a dynamic loading between mid and late pachytene, as it reaches full localization pattern (six foci/nucleus) gradually. To avoid overestimations in the number of nuclei with absent or reduced COSA-1 foci formation, only the last 7–10 cell rows preceding gonadal bend are usually included in the quantifications, as they reproducibly show full COSA-1 foci complement in control animals. Like RAD-51 counts, COSA-1 foci quantification should exclude nuclei with aberrant morphology and only foci clearly localizing along the chromatin should be included in the quantification. For representative image of COSA-1 see Fig. 2.
10. Nuclei in Diplotene should also be excluded from the counts, since COSA-1 starts to be released from the chromatin and therefore the foci number cannot be considered as reliable at this stage.

    An antibody directed against endogenous COSA-1 is currently unavailable in the literature, however functional tagged lines obtained by biolistic bombardment or more recent CRISPR-tagged lines are available, which all show comparable numbers [13, 14].

4 Notes

1. Avoid freeze-thaw cycles for all CRISPR components. It is advised to aliquot frequently used components (tracrRNA, Cas9) into 10 μL batches if not supplied as aliquots. The shelf life of Cas9 is several years at -20 °C. Avoid bubbles from pipetting and ensure all solutions are ejected fully from your pipette as you make the injection mix to ensure proper complexing.
2. Auxin is light-sensitive. Store plates in a dark container. Auxin plates are good for up to 3 weeks when stored at 4 °C in the dark.
3. Make auxin/ethanol solution 5–10 min before ready to pour media. Wrap tube in aluminum foil to prevent light exposure and set on nutating mixer at room temperature to allow auxin to fully dissolve.

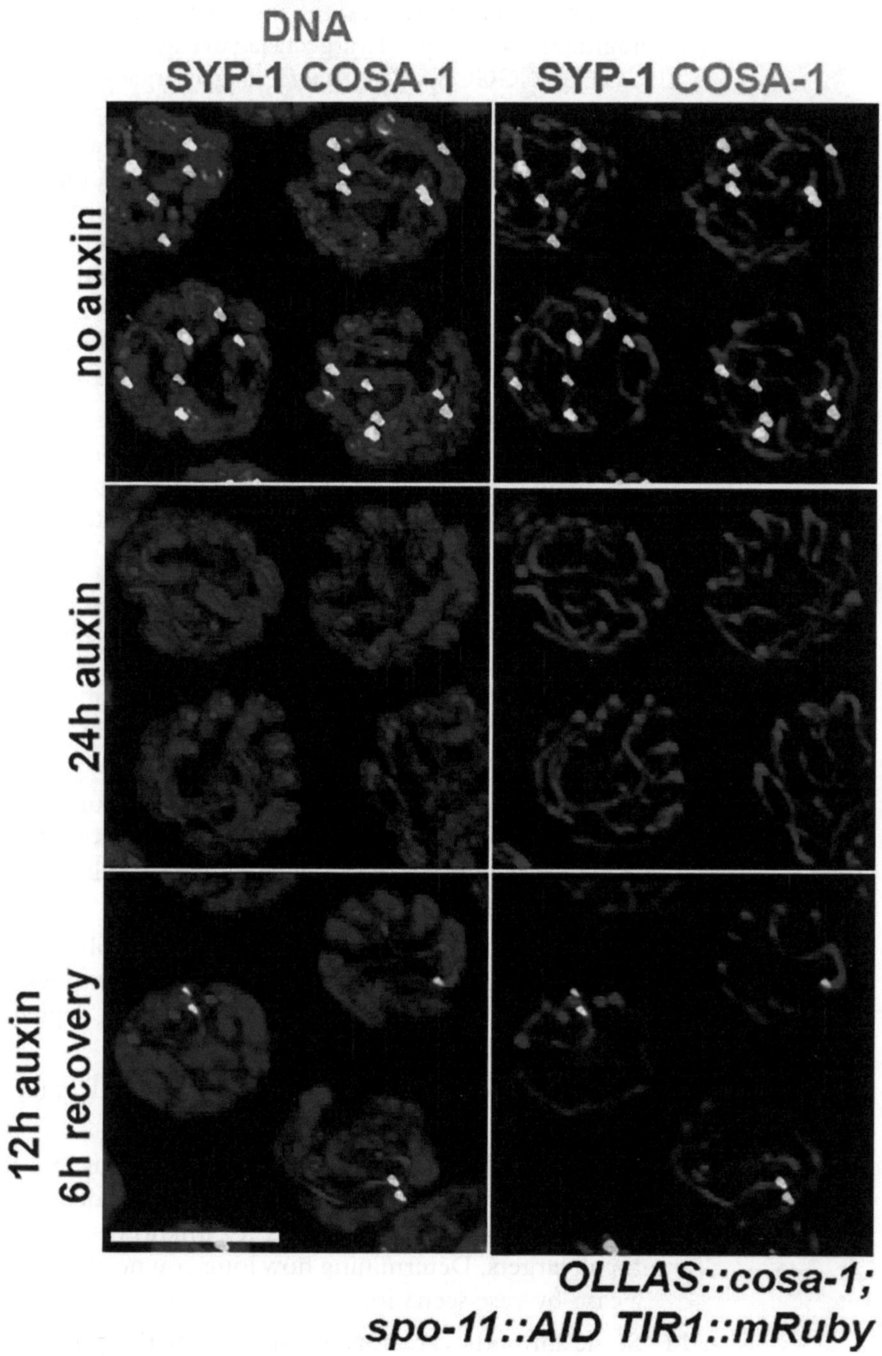

Fig. 2 Late pachytene nuclei of the indicated genotype and exposure conditions to auxin were stained for SYP-1 and OLLAS::COSA-1, and counterstained by DAPI. Scale bar: 5 μm

4. We have successfully used the insertion of a degron-linker-flag fragment as step 1.(atgcctaaagatccagccaaacctccggccaaggca GGTGGCGGCAAATTCgactacaaagaccatgacggtgattataaagatca tgaTatcgaTtacaaggatgacgatgacaag) followed by screening and isolation of a homozygous insertion. This was followed by a second CRISPR injection using step 2 crRNA (aaacctccggccaa ggcaGG) and ssODN. (atgcctaaagatccagccaaacctccggccaaggca caagttgtgggatggccaccggtgagatcataccggaagaacgtgatggtttcctgcc aaaaatcaagcggtggcccggaggcggcggcgttcgtgaagGGTGGCGGC AAATTCgactacaaagaccatgacgg) to obtain a line that has a degron-linker-flag C′ insertion. Note that the sequences here do not contain the homology arms' sequences which will be gene-specific. We follow the guidelines of [9].
5. Auxin plates inhibit bacterial growth; for short exposures you can do an unconcentrated OP50 lawn if you spread it the night before rather than dropping it on 2 days before your experiment.
6. AID strains can undergo silencing of the TIR1 transgene over time; it is advisable to use fresh stocks roughly every four weeks. Silencing may not be readily detected since it will first affect the penetrance of the phenotype upon short rather than longer exposures. Therefore, strains should be frozen as multiple tubes and new stock thawed every month.
7. To ensure worms still have food throughout the fecundity assay, check plates every day. Add more *E. coli* OP50 as needed. Due to the length of the assay, this is true for all plate types, not just auxin plates that inhibit bacterial growth.
8. It is important to wait 2 generations to see observable effects on fertility because even though degron works quickly, P0s will already have eggs stacked that have not been depleted during development.
9. Can compare phenotypes to null mutants if available as an alternative method of assessment.
10. For troubleshooting; if depletion isn't obvious in 4 h, plate as L4s and allow to deplete 24 h. In our hands, detectable depletion occurs within 2 h, however it's unknown if this is the case for all targets. Determining how long you need to deplete for is a case-by-case scenario.
11. Some antibodies are not compatible with this protocol. If this is true, you can follow the antibody staining procedure outlined in section 3.E.C ("COSA-1").
12. If using a GFP-tagged degron line, you can simply do an ethanol fixation (as outlined in 3.E.B "DAPI bodies in diakinesis -1") or skip antibody staining steps.

13. All analyses done by eye may be influenced by the user. Therefore, it is essential to include controls in the same dataset, blind the data before counting, and conduct all analysis of a particular experiment by the same individual.
14. Our samples are acquired with a fully motorized widefield upright microscope Zeiss AxioImager.Z2, equipped with a monochromatic camera Hamamatsu ORCA Fusion, sCMOS sensor, 2304 × 2304 pixels, 6.5 × 6.5 μm size. Deconvolution is performed with ZEN Blue Software, using Batch Deconvolution set with the "Constrained Iteration" mode and at maximum strength. Alternatively, we have successfully used the DeltaVision imaging system to acquire similar data. It is likely that any widefield upright microscope of good quality or any spinning disk confocal will suffice. From our experience, standard confocal microscopes are not suitable for this analysis since they bleach the sample in the step and z-stack thickness required. To test if your microscope system is suitable, perform the analysis of wild type and confirm it matches published data (for example: [4]).
15. Slides mounted with DAPI+Vectashield keep for about a week when stored at 4 °C in the dark before DAPI signal becomes weak.

References

1. Yesbolatova A, Saito Y, Kitamoto N et al (2020) The auxin-inducible degron 2 technology provides sharp degradation control in yeast, mammalian cells, and mice. Nat Commun 11:5701
2. Ashley GE, Duong T, Levenson MT et al (2021) An expanded auxin-inducible degron toolkit for Caenorhabditis elegans. Genetics 217:iyab006
3. Zhang L, Ward JD, Cheng Z et al (2015) The auxin-inducible degradation (AID) system enables versatile conditional protein depletion in C. elegans. Development 142:4374–4384
4. Hicks T, Trivedi S, Eppert M et al (2022) Continuous double-strand break induction and their differential processing sustain chiasma formation during Caenorhabditis elegans meiosis. Cell Rep 40:111403
5. Tan X, Calderon-Villalobos LI, Sharon M, Zheng C et al (2007) Mechanism of auxin perception by the TIR1 ubiquitin ligase. Nature 446:640–645
6. Divekar NS, Davis-Roca AC, Zhang L et al (2021) A degron-based strategy reveals new insights into Aurora B function in C. elegans. PLoS Genet 17:e1009567
7. Ghanta KS, Ishidate T, Mello CC (2021) Microinjection for precision genome editing in Caenorhabditis elegans. STAR Protoc 2: 100748
8. Raices M, Bowman R, Smolikove S et al (2021) Aging negatively impacts DNA repair and bivalent formation in the C. elegans germ line. Front Cell Dev Biol 9:695333
9. Paix A, Folkmann A, Goldman DH et al (2017) Precision genome editing using synthesis-dependent repair of Cas9-induced DNA breaks. Proc Natl Acad Sci USA 114: E10745–E10754
10. Hillers KJ, Jantsch V, Martinez-Perez E et al (2017) Meiosis. WormBook 2017:1–43
11. Jaramillo-Lambert A, Ellefson M, Villeneuve AM et al (2007) Differential timing of S phases, X chromosome replication, and meiotic prophase in the C. elegans germ line. Dev Biol 308:206–221
12. Hicks T, Koury E, McCabe C et al (2022) R-loop-induced irreparable DNA damage evades

checkpoint detection in the C. elegans germline. Nucleic Acids Res 50:8041–8059

13. Yokoo R, Zawadzki KA, Nabeshima K et al (2012) COSA-1 reveals robust homeostasis and separable licensing and reinforcement steps governing meiotic crossovers. Cell 149: 75–87

14. Janisiw E, Dello Stritto MR, Jantsch V et al (2018) BRCA1-BARD1 associate with the synaptonemal complex and pro-crossover factors and influence RAD-51 dynamics during Caenorhabditis elegans meiosis. PLoS Genet 14:e1007653

Chapter 14

Rapamycin-Induced Translocation of Meiotic Nuclear Proteins in *Saccharomyces cerevisiae*

Jinsha Padmarajan, Abhijna Krishnan Edilyam, and Vijayalakshmi V. Subramanian

Abstract

Conditional depletion of proteins is a potential strategy to elucidate protein function, especially in complex cellular processes like meiosis. Several methods are available to effectively deplete a protein in a conditional manner. Conditional loss of a protein function can be achieved by depleting it from its region of action by degrading it. A conditional loss of protein function can also be achieved by sequestering it to a functionally unavailable compartment inside the cell. This chapter describes anchor away, a conditional depletion tool that can deplete a protein both temporally and spatially by translocation. It utilizes the affinity of FRB to bind FKBP12 in the presence of rapamycin for a quick and efficient translocation of the protein to a designated location. Anchor away is a reliable tool for the study of meiotic proteins, as only small quantities of rapamycin are required to efficiently and rapidly translocate the protein of interest without compromising meiotic progression.

Key words Anchor away, FRB tag, FKBP12 tag, Rapamycin, Conditional sequestration, Conditional depletion, Nuclear proteins, Translocation

1 Introduction

Sexually reproducing eukaryotes maintain ploidy over generations by halving their genome through a specialized cell division called meiosis [1]. This highly programmed cellular division involves regulation by several proteins at each step [2, 3]. A study of the functional significance of these proteins will contribute to a better understanding of regulation of the meiotic process.

One approach to study a protein is to cull its activity and analyze the functional consequences with reference to the wild-type background. This can be achieved by introducing point mutations in the protein of interest (POI) that would affect its cellular

Jinsha Padmarajan and Abhijna Krishnan Edilyam contributed equally with all other contributors.

Jesús A. Carballo (ed.), *Meiosis: Methods and Protocols*, Methods in Molecular Biology, vol. 2818,
https://doi.org/10.1007/978-1-0716-3906-1_14,

roles or to delete the gene [4]. However, under certain circumstances, deletion of genes can have severe consequences in cell functions such as multiple phenotypes for pleiotropic genes or even lethality in case of some genes, thus making such functional characterization difficult [5]. This problem can be overcome by a conditional depletion of proteins at any desired stage of cell growth. Methodologies for conditional depletion include the use of temperature sensitive (ts) mutants where the protein activity is impaired at the non-permissive temperature. However, the stress response induced by the temperature-shift and severe defects in processes including meiosis progression limits its applicability [6, 7].

An innovative method is the use of chemical genetic engineering, wherein a small-molecule inhibitor can specifically bind to and conditionally inhibit only the genetically altered allele. Analog-sensitive alleles can be created for conditional inactivation of specific protein activities, in particular, kinases. This method allows the use of non-hydrolyzable analogs of ATP that can bind to the altered ATP-binding pockets of the kinase to make it catalytically inactive at the desired time point [8]. However, this method is limited to only a subset of proteins, and designing the analog-sensitive allele is also not trivial. Furthermore, potential adverse effects from the introduction of a gatekeeper amino acid mutation and the search for effective inhibitors for different kinases are added challenges to this approach [9].

Auxin-inducible degron (AID) system is a robust method that allows the inducible degradation of AID-tagged protein by proteasome machinery in the presence of auxin. Although this is a fast and efficient technique to study the function of essential genes, auxin-independent leaky degradation of proteins and the need to use increased concentration of auxin, especially in meiosis, are potential drawbacks of this approach [10–13].

Haruki et al. in 2008 developed a method for altered protein localization called the anchor-away technique. This method has the capacity to deplete the POI in the presence of rapamycin and uses protein tagging technique instead of mutation[14]. Importantly, rapamycin is required in small amounts and does not interfere with meiotic progression [15]. Here we describe the use of anchor-away technique to deplete meiotic proteins from the budding yeast nucleus to effectively create a nuclear-null phenotype during meiosis.

Anchor away is a conditional protein depletion technique developed on the basis of a strong interaction between FKBP12 and FRB (FKBP12-rapamycin binding domain) in the presence of the antibiotic rapamycin [16]. The FRB domain is part of human FRAP (FKBP12-rapamycin-associated-protein), while FKBP12 is a human protein which can bind to the drug FK506. Affinity of FKBP12 toward FRB increases by 2000-fold upon its binding to

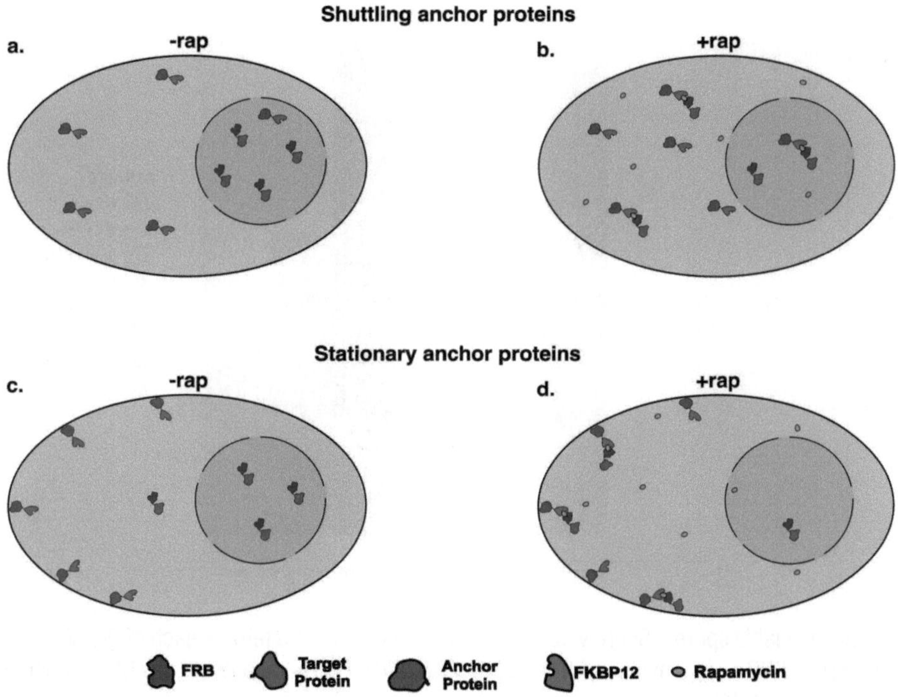

Fig. 1 Schematic of the anchor-away strategies for nuclear depletion of proteins. (**a**) Anchor proteins shuttling between nucleus and cytoplasm do not bind to target protein in the absence of rapamycin, (**b**) but sequester the target protein from the nucleus upon addition of rapamycin. (**c**) Interaction between a stationary anchor and a shuttling target protein is not observed in the absence of rapamycin. (**d**) The target protein gets sequestered at the plasma membrane upon rapamycin-mediated interaction with stationary anchor protein. -rap denotes absence of rapamycin; +rap denotes presence of rapamycin

rapamycin [17]. This interaction creates a stable complex between FRB-rapamycin-FKBP12. Anchor away utilizes this strong interaction to deplete or translocate the protein of interest (POI) from its region of action.

Since the development of anchor away in *Saccharomyces cerevisiae* by *Haruki et al.* in 2008 [14], it has also been successfully implemented in the study of meiotic proteins [15, 18]. Meiotic POI is tagged with 11kD FRB domain, while a suitable anchor is tagged with two copies of the 12kD FKBP12 tag. Upon addition of rapamycin, the FRB domain in the POI interacts with the FKBP12 leading to its depletion from the target location by sequestering it to anchor protein [14] (Fig. 1). The depletion can be confirmed by also tagging the target with a fluorophore (e.g., GFP) in addition to FRB to visualize the translocation or by comparing the phenotype with a null mutant [15, 19]. Anchor away of Zip1 and Spo11 led to reduced spore viability, similar to the respective null mutants (Fig. 2).

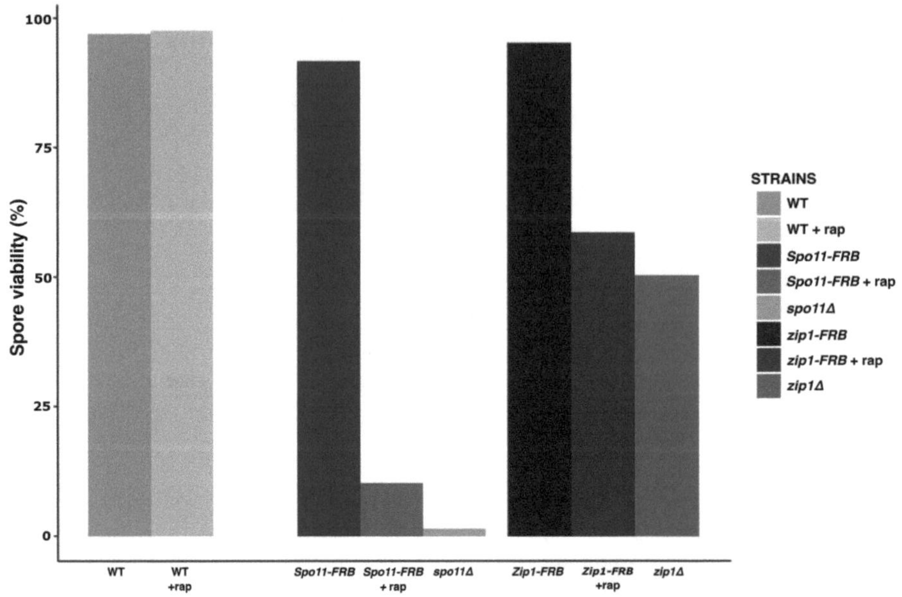

Fig. 2 Percentage of viable spores. Spore viability was analyzed in $n = 66$ tetrads each of WT, WT + rapamycin, *SPO11-FRB*, *SPO11-FRB* + rapamycin, *spo11*Δ, *ZIP1-FRB*, *ZIP1-FRB* + rapamycin, *zip1*Δ. (Dataset used from Subramanian et al. 2016 [15])

2 Materials

2.1 Strain Background

Addition of rapamycin causes lethality due the presence of Tor1 in *S. cerevisiae*. Thus, *tor1–1* allele with a missense mutation that alleviates this sensitivity is used as the background strain. In addition to the *tor1–1* mutant, the endogenous FRB gene, *FPR1*, is deleted. This prevents Fpr1 from competing with the FRB-tagged POI for interaction with FKBP12 [14].

2.2 Selection of Anchor Protein

Anchor protein aids in the sequestration of POI to the target location when rapamycin is added to the media. Efficient depletion of POI would require that the anchor protein be abundant in comparison to the POI. Importantly, both proteins should also be transiently present in the same compartment. Choice of the anchor protein would depend on the location the target needs to be sequestered to (Table 1) [14].

2.3 Media

1. 2X YEP: Sterilize 12 g Bacto™ Peptone, 6 g yeast extract, and 0.09 g L-tryptophan in 300 mL of distilled water.
2. 2% YPD broth: Sterilize 300 mL of 1X YEP. Add sterile glucose (2% final concentration) and sterile adenine (0.6 mM final concentration) to 1X YEP and mix well.
3. BYTA: Sterilize 10.2 g potassium phthalate-buffered, 10 g potassium acetate, 20 g Bacto™ Tryptone, and 10 g yeast extract in 1 L distilled water.

Table 1
Anchor proteins for POI sequestration to different cellular compartments

Organism	Anchor location	Signal/anchor protein	Target protein
S. cerevisiae	Plasma membrane	Pma1 [14, 19]	Shuttling nuclear proteins
S. cerevisiae	Nucleus	Htb2 [19]	Shuttling cytoplasmic proteins
S. cerevisiae	Nucleus	NLS [28]	Cytoplasmic proteins
S. cerevisiae	Cytoplasm	NES [28]	Nuclear proteins
S. cerevisiae	Cytoplasm	RPL13A [14, 15, 26]	Nuclear proteins
S. cerevisiae	Mitotic kinetochore	Mif2 [29]	Proteins functionally associated with kinetochore

4. Sporulation media (SPO): Sterilize 3 g potassium acetate in 1 L of distilled water with 250 μL of 0.5% acetic acid.
5. Media plates:
   (a) Prepare and sterilize 300 mL of 2X YEP and 4% agar (12 g in 300 mL).
   (b) Add the following solutions to sterile 2X YEP and mix well.
      (i) YPG plates: Add sterile glycerol (2% final concentration) and sterile adenine solution (0.6 mM final concentration).
      (ii) 4% YPD plates: Add sterile glucose solution (4% final concentration) and adenine solution (0.6 mM final concentration).
      (iii) 2% YPD plates: Add sterile glucose (2% final concentration) and adenine solution (0.6 mM final concentration).
      (iv) G418 (drug) plates: Add sterile 40% glucose (2% final concentration) and adenine solution (0.6 mM final concentration), 120 mg geneticin (final concentration: 0.2 mg/mL).

   Mix 2X YEP and sterile agar solution and pour into sterile petriplates. Allow the media to solidify and label the plates.

2.4 Amplification and Tagging

1. Plasmid: pFA6a-FRB-KanMX6 [14].
2. Forward and reverse primers with 45 bp homology to the region flanking insert site.

   PAGE- or HPLC-purified primers for C-terminal tagging:

   Forward primer: 5′ (45 bp before stop codon) + cggatccccgggttaattaa 3′

Reverse primer: 5′ (45 bp after stop codon) + gaattcgagctcgtttaaac 3′

3. Recipient strain with *tor1–1* and *fpr1Δ* alleles and RPL13A-(2×)FKBP12 for cytoplasmic translocation upon nuclear depletion (for anchoring to different compartments, please *see* Table 1).
4. Ethanol precipitation: 3 M sodium acetate, pre-chilled ethanol.
5. 2% YPD broth.
6. 1 M Lithium acetate (LiAc).
7. Polyethylene glycol (PEG) 50%.
8. Sonicated single-stranded salmon sperm DNA (ssssDNA).
9. Sterile MilliQ water.
10. G418 selection plates.

2.5 Synchronous Meiosis

1. YPG plates.
2. 4% YPD plates.
3. 2% YPD broth.
4. BYTA (50 mM potassium phthalate-buffered, 1% yeast extract, 2% Bacto™ Tryptone, and 1% acetate).
5. Sporulation media (SPO).
6. 3 M Sodium citrate.
7. RNase A (30 mg/mL).
8. Proteinase K (20 mg/mL).
9. SYTOX™ green.

2.6 Chromosome Spreads

1. Coplin jars.
2. Ethanol cleaned glass slides.
3. Cover slips (rectangular, W × L 22 mm × 50 mm).
4. 10% Sodium azide stock.
5. 200 mM Tris–HCl pH 7.5.
6. 1 M Dithiothreitol.
7. 2 M Sorbitol stock.
8. 2% Potassium acetate, 1 M sorbitol mix.
9. 10 mg/mL Zymolyase 100 T.
10. Ice-cold MES-Sorbitol (0.1 M MES-NaOH pH 6.4, 1 mM EDTA, 0.5 mM $MgCl_2$, 1 M sorbitol).
11. Fixative (3% paraformaldehyde, 3.4% sucrose; store at 4 °C).
12. 1% Lipsol detergent.
13. 0.4% Photoflo (Kodak).

14. 1X TBS (Tris buffered saline) (20 mM Tris–HCl, pH 7.5/ 150 mM NaCl).
15. Blocking Buffer (1× TBS, 0.2% gelatine, 0.5% BSA; store at −20 °C as 1 mL aliquots).
16. Primary antibody (diluted in blocking buffer).
17. Secondary antibody (diluted in blocking buffer).
18. Vectashield with DAPI.

2.7 Depletion of Target with Rapamycin

1. Rapamycin (stock concentration: 100 mM; dissolved in DMSO).

2.8 Sporulation Efficiency Assay

1. Sporulation media (SPO).
2. Clean glass slides.
3. Cover slips.

2.9 Spore Viability Assay

1. 2% YPD plates.
2. 10 mg/mL Zymolyase.

3 Methods

3.1 Tagging Protein of Interest (POI) with FRB

1. Amplify FRB along with a selection marker from pFA6a-FRB-KanMX6 using PCR in up to 400 μL reactions (*see* **Note 1**).
2. Confirm amplification and check the size of the amplicon by agarose gel electrophoresis.

3.2 Concentrating the PCR Product for Transformation

1. Add 1/10th the volume of 3 M sodium acetate and twice the volume of chilled ethanol.
2. Store at −20 °C for overnight precipitation.
3. Centrifuge the sample for 5 min at max speed (~ 18,400 g) and air-dry the pellet. Resuspend the pellet in 74 μL of sterile MilliQ water.

3.3 Introducing the PCR Product into Recipient Strain by Transformation

1. Inoculate the recipient strain in 3 mL of 2% YPD broth and incubate at 30 °C with agitation at 200 rpm for 16 h.
2. Measure OD_{600} of the sample and inoculate 20 mL of 2% YPD broth to a final OD_{600} of 0.3. Incubate for 2 h at 30 °C with agitation at 200 rpm.
3. Harvest the cells by centrifugation at 1240 g for 5 min in a tabletop centrifuge with swinging bucket rotors. Discard the supernatant carefully without disturbing the pellet.
4. Wash the cells in 1 mL of 0.1 M LiAc and transfer to a 1.5 mL microcentrifuge tube.

5. Harvest the cells by centrifugation at 845 g in a microcentrifuge for 5 mins and aspirate the supernatant.
6. Resuspend the cells in 4 volumes of 0.1 M LiAc.
7. Boil the sonicated single-stranded salmon sperm DNA (ssssDNA) at 100 °C for 5 mins and immediately keep on ice.
8. Prepare 360 μL of transformation mix as below.

   Transformation mix (one reaction): 240 μL 50% PEG, 36 μL 1 M LiAc, 10 μL ssssDNA, 74 μL PCR product. Mix the contents by vortexing (*see* **Note 2**).
9. Add 50 μL of cells suspended in 0.1 M LiAc to the transformation mix and vortex it.
10. Subject the cells to heat shock at 42 °C for 40 mins.
11. Keep the cells at room temperature for 10 mins.
12. Harvest the cells by centrifugation at 845 g in a microcentrifuge for 5 mins and aspirate the supernatant.
13. Resuspend the cell pellet in 200 μL sterile MilliQ water.
14. Spread the cells on 2% YPD plate and incubate for 24 h at 30 °C.
15. On the following day, replica the cells on selection media (here G418 plates). Incubate at 30 °C for 2 days.
16. Streak out the transformed colonies obtained onto the selection media (G418) plate to obtain single colonies. Incubate for 2 days at 30 °C.
17. Patch out the positive single colonies onto a fresh YPD plate.
18. Isolate genomic DNA from the selected colonies followed by checking PCR to confirm the site of tagging.
19. Generate homozygous diploids in the required background including *tor1–1*, *fpr1Δ*, FRB-tagged POI, and FKBP12-tagged anchor.
20. Make a glycerol stock of the positive clone and diploids for long-term storage at −80 °C.

3.4 Synchronous Meiosis

Day 1

1. Patch out the stored diploids onto a YPG plate from −80 °C. Incubate overnight at 30 °C.

Day 2

2. Patch the strain onto a 4% YPD plate and incubate overnight at 30 °C.

Day 3

3. Scoop out cells using sterile loop and inoculate 15 mL of 2% YPD broth in a 250 mL sterile conical flask.

4. Incubate the culture at room temperature (~ 23 °C) with agitation at 200 rpm for 24 h.

Day 4

5. Check the OD_{600} (1:50 dilution).
6. Dilute culture into 50 mL BYTA to a final OD_{600} of 0.3 in a 500 mL sterile flask and incubate for 16 h at 30 °C with agitation at 200 rpm.

Day 5

7. Check the OD_{600} (1:10 dilution) and calculate the volume of cells required to get OD_{600} of 2.0 in the final volume.
8. Harvest the required volume by centrifugation at 1240 g for 5 mins in a tabletop centrifuge with swinging bucket rotors. Discard supernatant, vortex the pellet, and wash twice in sterile distilled water.
9. Resuspend cells in 100 mL sporulation media (0.3% potassium acetate, 0.00125% acetic acid) in a 1 L sterile flask at an OD_{600} of 2.0 for inducing meiosis. Incubate at 30 °C with agitation at 200 rpm.
10. To assess synchrony of meiotic progression, collect 150 μL of culture every hour for five hours and fix the cells in 350 μL 100% ethanol. Harvest the cells, resuspend in 500 μL of 50 mM sodium citrate +0.7 μL RNase A (30 mg/mL), and incubate overnight at 50 °C to digest RNA in the sample. Treat with 5 μL proteinase K (20 mg/mL) and keep at 50 °C for 1–2 hrs to degrade cellular proteins. For staining the cells, add an additional 500 μL of 50 mM sodium citrate +0.1 μL SYTOX green. Sonicate the cells for 5 s at the lowest output. Subject the cells to flow cytometry to measure frequency of cells with respect to DNA content as a readout for synchrony of cells.
11. To assess sporulation efficiency and spore viability, collect 1 mL of sample after 24 h and store at 4 °C (*see* **Note 3**).

3.5 Depletion of Target with Rapamycin

1. Add 1 μM of rapamycin to the media at the required time points (T = 0 h at the time of meiosis induction or any appropriate time point for your experiment).
2. Confirm conditional depletion of proteins using immunofluorescence of chromatin-bound proteins on spread meiotic chromosomes or compartmental translocation by whole cell immunofluorescence. Depletion can also be assessed by assaying sporulation and spore viability defects or other assays that mimic the null phenotype (*see* **Note 4**).

3.6 Visualization of Chromatin-Bound Proteins in Meiotic Cells by Immunofluorescence on Spread Chromosomes

1. Soak glass slides in ethanol overnight. Air-dry before use.
2. Collect 2 mL of culture at required time points and kill the cells by treating with 0.1% sodium azide.
3. Harvest the cells at 455 g for 2 min. Resuspend the pellet in 500 μL of 200 mM Tris–HCl pH 7.5 and 10 μL 1 M DTT.
4. Harvest the cells at 455 g for 2 min.
5. Spheroplast the cells by treating with 65 μg zymolyase (6.5 μL of 10 mg/mL) in 500 μL of 2% potassium acetate and 1 M sorbitol. Incubate at 30 °C for 30 min on a rotating rack.
6. Assess the efficiency of spheroplasting by treatment with detergent. Spheroplasts will lyse in the presence of detergent.
7. Harvest the spheroplasts at 541 g for 3min.
8. Wash the spheroplasts in 2 mL of ice-cold MES-sorbitol (0.1 M MES pH 6.4/ 1 mM EDTA/0.5 mM $MgCl_2$/1 M sorbitol) by gentle tapping. Keep the samples on ice.
9. Harvest the spheroplasts at 541 g for 3 mins. Add 55 μL of MES-sorbitol and resuspend by tapping. Keep on ice.
10. Add 20 μL of cells on ethanol-cleaned slides.
11. Add 40 μL fixative (3% para-formaldehyde and 3.4% sucrose) to the cells to fix them, followed by 80 μL of 1% lipsol to simultaneously lyse the cells.
12. Mix the contents by gentle swirling and tilting the slide without spilling over the sample.
13. Spread the chromosomes by rolling a clean glass rod over the liquid layer without touching the slide.
14. Let the slides dry on the bench. Repeat the steps for all the slides. Wait for 10 mins for the last slide to dry.
15. Soak the dried slides in 0.4% Photoflo for 2 min and dry overnight at room temperature.
16. Store the slides at −80 °C or treat the slides for immunofluorescence.
17. Rehydrate the slides in 1X TBS in a Coplin jar with gentle shaking for 10 min.
18. Add blocking buffer to the slides and incubate under coverslip for 2 h at room temperature in a humid chamber followed by gentle wash in 1X TBS.
19. Incubate in primary antibody overnight at 4 °C in a humid chamber.
20. Wash the slides for 10 min in 1X TBS. Repeat three times.
21. Incubate in secondary antibody in a humid chamber for 4 h at room temperature.
22. Wash the slides in 1X TBS for 10 min. Repeat three times.

23. Put a drop of Vectashield (+ DAPI, 1 μg/mL) and lay the cover slip. Dry the slides gently with a tissue. Seal the cover slip using nail polish.
24. Visualize slides with a fluorescence microscope.

3.7 Sporulation Efficiency Assay

1. Collect 1 mL of meiotic culture and store at 4 °C.
2. Check for sporulation efficiency by visualizing tetrads under light microscope on a clean glass slide.
3. Count the number of monads, dyads, tetrads, and non-sporulating cells from a random pool of 200 cells. Calculate the percentage of sporulation efficiency and compare with a control strain.

   Sporulation efficiency = Total number of sporulated cells/Total number of cells counted
4. Proceed with tetrad dissection to assess spore viability if a good number of tetrads are present.

3.8 Spore Viability Test

1. Prepare 1 mg/mL zymolyase in 1 M sorbitol.
2. Add 20 μL of zymolyase to 20 μL of sporulated culture in a 1.5 mL microcentrifuge tube and incubate for 20–30 min at 37 °C.
3. Add 200 μL of sterile MilliQ water to the zymolyase treated culture (*see* **Note 5**).
4. Take a dry and even 2% YPD plate and draw a straight line along the diameter.
5. Drop 20 μL of culture onto a corner of the line and let it flow along the line. Collect the excess drop at the other end and let the plate dry briefly.
6. Dissect the tetrads and incubate the plate for 48 h at 30 °C (*see* **Note 6**).
7. Count the number of viable spores per tetrad.

3.9 Concluding Remarks

Anchor-away method establishes a method to conditionally deplete protein of interest from its cellular compartment. This method has also been used in various other systems including fission yeast [20], *Drosophila*, [21] and human cell lines [22]. While most of the applications involve translocation of nuclear proteins to cytoplasm, this method can also be used for sequestering proteins to specific cellular compartments by selecting the right anchor protein or for forcing a direct interaction between proteins [22–24]. In HeLA cell lines, protein AP-2 was depleted by sequestering it to mitochondria where it became functionally inactive [22]. Similarly, the anchor protein can also be directed for nuclear export or import by adding a NES/NLS signal to it [25]. Cytoplasmic proteases were successfully targeted to the nucleus by using Htb2 as the anchor [19]. Therefore, this method is not limited to just depletion of nuclear proteins.

In yeast meiosis, this method has been used to deplete proteins like Zip1, Mer2, Spo11[15] with direct depletion and Pch2 by depleting Rpo11[26]. The amount of rapamycin required for each protein may vary depending on the cellular abundance of the protein. For instance, Zip1 depletion required 1 μM rapamycin, while depletion of Spo11 and Mer2 required 2 μM rapamycin [15].

Despite being part of the synaptonemal complex, Zip1 was efficiently depleted from the nucleus [15]. Hence, this method can also be used for depleting proteins with multiple interacting partners or proteins that are part of a large complex. Additional advantages of the method are the short period of time required for depletion [14] and the long retention time of rapamycin in the cell [25]. One setback of this approach is the requirement for building a complex strain with two additional mutations (*fpr1Δ*, *tor1-1*) besides the two tags. Regardless, the efficiency and the flexibility of its usage render anchor away as a promising tool to understand intricate roles of meiotic proteins.

4 Notes

1. Notes for primer designing:
   (a) For C-terminal tagging, the forward primer should include 45 bp upstream of POI stop codon and reverse primer should include 45 bp sequence downstream of the gene but skipping the stop codon [27].
   (b) The POI can be tagged with GFP, in addition to FRB, to visualize the protein translocation.
   (c) For N-terminal tagging, the primers have to be designed to incorporate an appropriate promoter.
2. Gently pipette the PEG as it is very viscous.
3. Thoroughly shake the culture before collection of samples.
4. If depletion is not complete, repeat the experiment with a higher concentration of rapamycin and/or increase the copies of anchors in the genome.
5. Do not tap or pipette the cells as it will separate the spores in tetrads.
6. Mark positions on the plate where tetrad dissections were incomplete.

Acknowledgments

This work was supported in part by Ramalingaswami reentry fellowship (BT/RLF/Re-entry/04/2019), DBT-Wellcome Trust India Alliance intermediate fellowship (IA/I/21/1/505647), SERB CRG grant (CRG/2021/007107/BHS) to V.V.S., and CSIR-JRF fellowship (09/1178(12250)/2021-EMR-I) to A.K.E.

References

1. Villeneuve AM, Hillers KJ (2001) Whence meiosis? Cell 106:647–650
2. Sanderson ML, Hassold TJ, Carrell DT (2008) Proteins involved in meiotic recombination: a role in male infertility? Syst Biol Reprod Med 54:57–74
3. Grey C, de Massy B (2021) Chromosome organization in early meiotic prophase. Front Cell Dev Biol 9:688878
4. Alberts B, Johnson A, Lewis J et al (2002) Studying gene expression and function. Garland Science
5. Miklos GL, Rubin GM (1996) The role of the genome project in determining gene function: insights from model organisms. Cell 86:521–529
6. Varadarajan R, Nagarajaram HA, Ramakrishnan C (1996) A procedure for the prediction of temperature-sensitive mutants of a globular protein based solely on the amino acid sequence. Proc Natl Acad Sci 93:13908–13913
7. Harris SD, Pringle JR (1991) Genetic analysis of *Saccharomyces cerevisiae* chromosome I: on the role of mutagen specificity in delimiting the set of genes identifiable using temperature-sensitive-lethal mutations. Genetics 127:279–285
8. Blethrow J, Zhang C, Shokat KM, Weiss EL (2004) Design and use of analog-sensitive protein kinases. Curr Protoc Mol Biol. Chapter 18:Unit 18.11
9. Lopez MS, Kliegman JI, Shokat KM (2014) The logic and design of analog-sensitive kinases and their small molecule inhibitors. Methods Enzymol 548:189–213
10. Nishimura K, Fukagawa T, Takisawa H et al (2009) An auxin-based degron system for the rapid depletion of proteins in nonplant cells. Nat Methods 6:917–922
11. Tang S, Wu MKY, Zhang R, Hunter N (2015) Pervasive and essential roles of the Top3-Rmi1 decatenase orchestrate recombination and facilitate chromosome segregation in meiosis. Mol Cell 57:607–621
12. Kaur H, Ahuja JS, Lichten M (2018) Methods for controlled protein depletion to study protein function during meiosis. Methods Enzymol 601:331–357
13. Copsey A, Tang S, Jordan PW et al (2013) Smc5/6 coordinates formation and resolution of joint molecules with chromosome morphology to ensure meiotic divisions. PLoS Genet 9: e1004071
14. Haruki H, Nishikawa J, Laemmli UK (2008) The anchor-away technique: rapid, conditional establishment of yeast mutant phenotypes. Mol Cell 31:925–932
15. Subramanian VV, MacQueen AJ, Vader G et al (2016) Chromosome synapsis alleviates Mek1-dependent suppression of meiotic DNA repair. PLoS Biol 14:e1002369
16. Chen J, Zheng XF, Brown EJ, Schreiber SL (1995) Identification of an 11-kDa FKBP12-rapamycin-binding domain within the 289-kDa FKBP12-rapamycin-associated protein and characterization of a critical serine residue. Proc Natl Acad Sci USA 92:4947–4951
17. Banaszynski LA, Liu CW, Wandless TJ (2005) Characterization of the FKBP·rapamycin·FRB ternary complex. J Am Chem Soc 127:4715–4721
18. Vincenten N, Kuhl L-M, Lam I et al (2015) The kinetochore prevents centromere-proximal crossover recombination during meiosis. elife 4:e10850
19. Tsuchiya H, Arai N, Tanaka K, Saeki Y (2013) Cytoplasmic proteasomes are not indispensable for cell growth in *Saccharomyces cerevisiae*. Biochem Biophys Res Commun 436:372–376
20. Ding L, Laor D, Weisman R, Forsburg SL (2014) Rapid regulation of nuclear proteins by rapamycin induced translocation in fission yeast. Yeast 31:253–264
21. Bosch PS, Pepperl J, Basler K (2020) Anchor away—a fast, reliable and reversible technique to inhibit proteins in *Drosophila melanogaster*. G3 Genes|Genomes|Genetics 10:1745–1752

22. Robinson MS, Sahlender DA, Foster SD (2010) Rapid inactivation of proteins by rapamycin-induced rerouting to mitochondria. Dev Cell 18:324–331
23. Pyatnitskaya A, Andreani J, Guérois R et al (2021) The Zip4 protein directly couples meiotic crossover formation to synaptonemal complex assembly. Genes Dev 36:53–69. https://doi.org/10.1101/gad.348973.121
24. Andersen TG, Nintemann SJ, Marek M et al (2016) Improving analytical methods for protein-protein interaction through implementation of chemically inducible dimerization. Sci Rep 6:1–11
25. Geda P, Patury S, Ma J et al (2008) A small molecule-directed approach to control protein localization and function. Yeast 25:577–594
26. Cardoso da Silva R, Villar-Fernández MA, Vader G (2020) Active transcription and Orc1 drive chromatin association of the AAA+ ATPase Pch2 during meiotic G2/prophase. PLoS Genet 16:e1008905
27. Longtine MS, McKenzie A 3rd, Demarini DJ et al (1998) Additional modules for versatile and economical PCR-based gene deletion and modification in *Saccharomyces cerevisiae*. Yeast 14:953–961
28. Xu T, Johnson CA, Gestwicki JE, Kumar A (2010) Conditionally controlling nuclear trafficking in yeast by chemical-induced protein dimerization. Nat Protoc 5:1831–1843
29. García-Rodríguez LJ, Kasciukovic T, Denninger V, Tanaka TU (2019) Aurora B-INCENP localization at centromeres/inner kinetochores is required for chromosome bi-orientation in budding yeast. Curr Biol 29:1536–1544.e4

Part III

Automated Computational Tools for Meiosis Analysis

Chapter 15

synapsis: A Bioconductor Package to Automate the Analysis of Meiotic Double-Strand Break Repair and Crossover Formation

Lucy McNeill, Vanessa Tsui, and Wayne Crismani

Abstract

Immunofluorescent staining is commonly used to generate images to characterize cytological phenotypes. The manual quantification of DNA double-strand breaks and their repair intermediates during meiosis using image data requires a series of subjective steps, from image selection to the counting of particular events per nucleus. Here we describe "synapsis," a bioconductor package, which includes a set of functions to automate the process of identifying meiotic nuclei and quantifying key double-strand break formation and repair events in a rapid, scalable, and reproducible workflow, and compare it to manual user quantification. The software can be extended for other applications in meiosis research, such as incorporating machine learning approaches to categorize meiotic substages.

Key words Meiosis, Double-strand break, Crossover, Immunofluorescence, Automated image analysis, R package, Bioconductor

1 Introduction

Meiosis is a special type of cell division that is necessary to make sperm and eggs [1]. During meiosis, chromosomes are recombined and their numbers halved through highly regulated processes. Two key steps in the meiotic process are studied heavily and can be visualized with immunofluorescent (IF) techniques: (1) the reorganization of chromosome structure is unique at meiosis because homologous chromosomes pair and align in a process known as synapsis, and this can be appreciated when visualizing the synaptonemal complex [2–4]; (2) DNA double-strand breaks (DSBs) form in high numbers and are repaired by meiosis-specific recombination pathways. This is a homeostatic process, which lends itself to visualization as many of the proteins involved form foci [5–8]. Most meiotic immunofluorescent localization analyses currently rely on subjective manual counting or ImageJ macros, the latter requiring

Jesús A. Carballo (ed.), *Meiosis: Methods and Protocols*, Methods in Molecular Biology, vol. 2818,
https://doi.org/10.1007/978-1-0716-3906-1_15,

significant user intervention to run. To the best of our knowledge, "synapsis" is the first open access software package designed for the analysis of meiotic immunofluorescent image data that has been optimized with samples from different taxonomic kingdoms. Below we describe how to use the software.

2 Materials

This protocol describes the use of example mutichannel immunofluorescent images of meiotic prophase I that come with "synapsis."

2.1 Installing Rstudio

Rstudio is a user-friendly way to use R and can be downloaded freely from https://www.rstudio.com/.

2.2 Installing Synapsis

"synapsis" is free and an open-source software that can be downloaded for Windows, Mac OS X, or Linux.

From Rstudio, Install "BiocManager":

```
if (!requireNamespace("BiocManager", quietly = TRUE))
install.packages("BiocManager")
```

Then install "synapsis":

```
BiocManager::install("synapsis")
```

(*see* **Note 1**).

3 Methods

The user should note that the parameter `'annotation = "on"'` results in the output of particular `'synapsis'` functions being far more descriptive than if it were off. Examples with this parameter setting are included below. Novice R users who are more interested in the biology, than the computational aspects of the package, should consider running this tutorial in R markdown with `'annotation = "on"'` in order to visualize output (i.e., "decisions" being made by `'synapsis'`). In the case of running these functions as an R script, instead of R markdown, extra descriptions will be output to the R studio console and the R studio plots window.

3.1 Loading Synapsis

To load `'synapsis'` in R, run the following command:

```
library(synapsis)
```

3.2 Checking Documentation

You can access the documentation for functions in ‘`synapsis`‘, which explains their purposes and how to use them. For example,

```
??auto_crop_fast
```

This will provide information on the function ‘auto_crop_fast‘ in the “help” window in Rstudio.

3.3 Data Preparation

Here, we will use the example images which come with ‘synapsis.‘.

```
path = paste0(system.file("extdata",package = "synapsis"))
```

3.3.1 Synaptonemal Complex Channel

Inspect images in the SYCP3 channel. An example of what to expect is shown in Fig. 1.

```
library(EBImage)

file_SYCP3 <- paste0(path,"/MLH3rabbit488_SYCP3mouse594_
fancm_fvb_x_fancm_bl6_724++_slide01_006-SYCP3.tif")

image_SYCP3 <- readImage(file_SYCP3)

display(2*image_SYCP3)
```

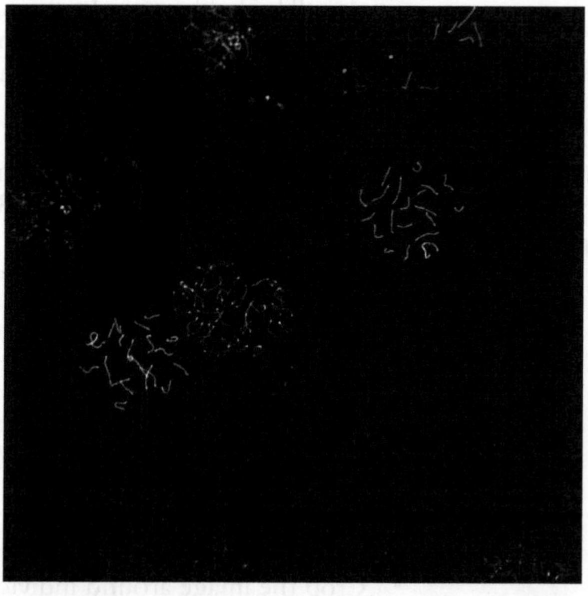

Fig. 1 Immunofluorescent image that visualizes SYCP3, a component of the mouse synaptonemal complex

3.3.2 Foci Channel

Inspect images in the MLH3 channel.

```
file_MLH3 <- paste0(path,"/MLH3rabbit488_SYCP3mouse594_
fancm_fvb_x_fancm_bl6_724++_slide01_006-MLH3.tif")

image_MLH3 <- readImage(file_MLH3)

display(2*image_MLH3)
```

3.3.3 Comment on Resolution and Size

Many of the input parameters in synapsis are in pixel units. It is good to get a feel of pixel unit dimensions before you start analyzing these digital images. You can inspect the dimensions, in pixels, of an image by running the following command:

```
cat("dimension of image:", dim(image_MLH3)[1], " x ", dim
(image_MLH3)[2], sep = " ")
## dimension of image: 750 x 750
```

You can inspect the size of an image by running the following command:

```
mb = 1e06
cat("file size in mb:", file.size(file_MLH3)/mb, sep = " ")
## file size in mb: 2.250206
```

The image is 750 × 750 pixels and 2.25mb. If we recall the SYCP3 channel image, keeping in mind that it is 750 pixels wide, a single cell looks like it might measure roughly 1/8th of the width of the image, i.e., 125 pixels. Keeping this approximate *diameter of a cell* in *pixels* will come in handy when determining input parameters for your own images.

3.4 Calling 'synapsis' Functions on a Sample Image

Here we will demonstrate the functionality of synapsis. Given an image whose *foci* channel (MLH3) is stored in the file:

```
MLH3rabbit488_SYCP3mouse594_fancm_fvb_x_fancm_bl6_724++_sli-
de01_006-MLH3.tif
```

and *synaptonemal complex* channel (SYCP3) stored in:

```
MLH3rabbit488_SYCP3mouse594_fancm_fvb_x_fancm_bl6_724+
+_slide01_006-SYCP3.tif
```

We can:

- Crop the image around individual nuclei.
- Determine whether these cells are in the pachytene phase (optional).
- Count the number of foci per nucleus.

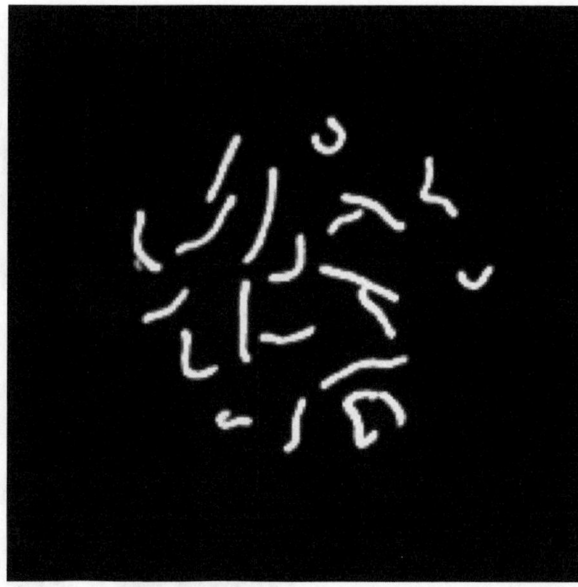

Fig. 2 A nucleus identified by the 'auto_crop_fast' function

These are currently the three main features of 'synapsis'. Some notable features mentioned in the tutorial are:

- Segmenting cells which are proximally close or overlapping
- Setting function parameters so that images have foci counts that are comparable to manual quantification

3.4.1 Cropping Routine

Without Nuclei Separation

We begin analyzing an image by using the 'auto_crop_fast' function. An example of a cropped meiotic nucleus from the example image is shown in Fig. 2.

```
auto_crop_fast(path, annotation = "on", max_cell_area =
30000, min_cell_area = 7000, file_ext = "tif")
```

Here we call 'path,' plus other optional parameters (that would otherwise take on default values). But only 'path' is essential. This is because 'auto_crop_fast' has built-in default values that are assumed when the user doesn't specify. These default values assume that there is a low density of nuclei in an image. Below we deal with higher nuclei densities.

A 'crops' folder with three channels per detected nuclei should have been generated inside the folder where these images are kept, i.e., in 'path.'.

With Nuclei Separation Using Watershed

Often multiple nuclei will be "touching" in an image. It is at the user's discretion as to whether these nuclei should be included in any quantification. Here we show how this can be achieved with well-established "watershed" functions. We will run the

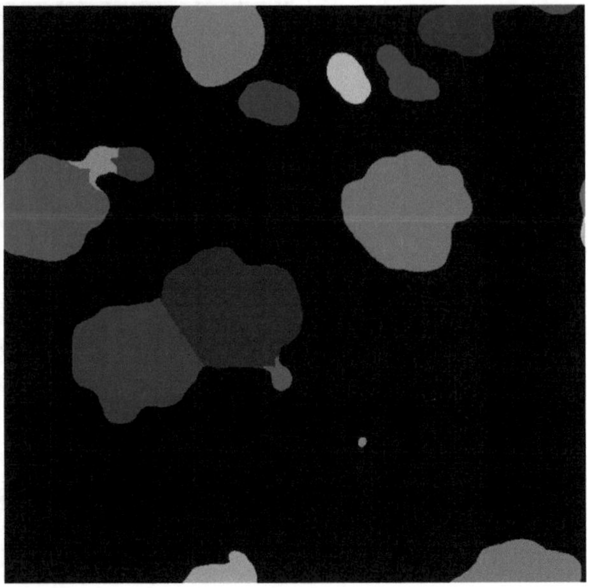

Fig. 3 An intermediate step of the 'auto_crop_fast' function where the parameter 'crowded_cells' is set to 'TRUE,' which makes use of watershed methods

'auto_crop_fast' function again with different settings, so we will delete the 'crops' folder. An example of an intermediate processing step that shows the output of a watershed method is presented in Fig. 3.

```
unlink(paste0(path,"/crops-RGB/"), recursive = TRUE)
unlink(paste0(path,"/crops/"), recursive = TRUE)
```

'synapsis' can separate nuclei by making use of the 'EBImage::watershed' and 'EBImage::distmap' functions. If we set 'crowded_cells = TRUE', then.

```
auto_crop_fast(path, annotation = "on", max_cell_area = 30000, min_cell_area = 7000, file_ext = "tif",crowded_cells = TRUE)
```

Instead of one nucleus, which was identified previously in Subheading 3.4.1.1, the output of 'auto_crop_fast ' is now three nuclei, due to setting the parameter 'crowded_cells' to 'TRUE'.

3.4.2 Getting Pachytene

The following function was optimized for meiotic surface spreads for mice and can be used to identify nuclei that are likely in pachytene. This function may prove useful for researchers who are specifically interested in quantifying features of meiocytes in pachytene.

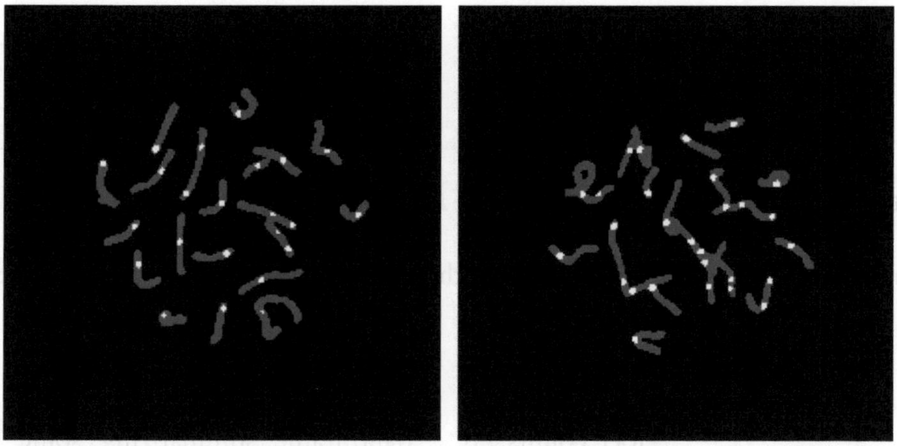

Fig. 4 Binary output images that represent coincident SYCP3 (red) and MLH3 (white) signal as determined by the 'count_foci' function. These two nuclei can be seen in Fig. 1

```
SYCP3_stats <- get_pachytene(path,ecc_thresh = 0.8, area_-
thresh = 0.04, annotation = "on", file_ext = "tif")
```

SYCP3_stats is a data frame summarizing some features of the cells classified as pachytene. The data frame is populated with information such as the image file name where the data comes from. This will be more useful when you have many images in a folder. Then there are columns of output from the cells that were identified. You will notice a column for genotype. The default settings assume that '++' or '—' are present in the file name. However, these can be modified with the parameters 'WT_str' and 'KO_str'.

3.4.3 Counting Foci

Now that we have the cropped images, and those which have been selected as our stage of interest (pachytene), we can quantify the number of foci per nucleus.

Note the option 'crowded_foci = FALSE,' because for our antibodies we only typically expect a couple dozen foci per cell. If you are dealing with an antibody that typically has in excess of ~100 foci per cell, e.g., DMC1 or RAD51, it is recommended to use 'crowded_foci = TRUE.' An example visualization of an overlap of binary masks from a nucleus for detected SYCP3 and MLH3 signal is shown in Fig. 4.

```
foci_counts <- count_foci(path,offset_factor = 8, brush_size
= 3, offset_px = 0.6, brush_sigma = 3, annotation = "on",
stage = "pachytene",file_ext = "tif", watershed_stop= "on",
crowded_foci = FALSE, C1 = 0.03, disc_size_foci = 9)
```

	filename	cell_no	genotype	stage	foci_count	sd_foci	mean_foci	median_foci
1	/Users/wcrismani/Library/R/x86_64/4.2/library/syna...	1	+/+	pachytene	23	3.58621242263058	6.96666666666667	8
2	/Users/wcrismani/Library/R/x86_64/4.2/library/syna...	2	+/+	pachytene	28	4.53763044882349	10.5135135135135	11

mean_px	median_px	percent_on	sd_px	lone_foci	comments	verdict	C1
0.0192701445994088	0.0117647058823529	0.880382775119617	0.0241255328238543	7	meets crisp-ness criteria	keep	0.0214486388913312
0.0289901843781464	0.0196078431372549	0.794344473007712	0.048167318981632	9	meets crisp-ness criteria	keep	0.0148827521147477

Fig. 5 The data frame output from the '`count_foci`' function applied to nuclei in pachytene

Reading the Data Frame

We can now inspect the output of '`count_foci`.' Example output is shown in Fig. 5.

```
print(foci_counts)
```

'foci_counts' is a data frame summarizing some features (including foci counts) of the cells classified as pachytene.

The "verdict" column recommends keeping or discarding an image based on how it compares to the "crispness criteria" C1, a measurement of how uniformly sized the foci identified are, which is empirically indicative of a high-quality image with coincident foci and low background.

This is currently how we can identify (and hence ignore) cells with low signal/high background in the foci channel which passed through the automated pipeline.

The "lone_foci" column indicates how many of the detected foci were not coincident with the SYCP3 signal.

We will now clean up the crops folders that we just generated.

```
unlink(paste0(path,"/crops-RGB/"), recursive = TRUE)
unlink(paste0(path,"/crops/"), recursive = TRUE)
```

3.5 Summary

In this tutorial, we have demonstrated the main functions of the synapsis package:

1. Cropping images around cell candidates, where 'synapsis' can identify individual nuclei even when they are close/touching, after inputting the location of your raw images (which have already been separated into foci and synaptonemal complex channels).

The output is stored in a folder called "crops" with square images (from all channels) of the nuclei candidates. Note that a crops-RGB folder is also generated for you to look at the overlapping channels in color, for your own troubleshooting purposes.

2. Identifying nuclei in the pachytene phase. This is optional and was made for niche biological problems that are relevant to this meiotic substage.

If you use this, a folder called "pachytene" inside the folder "crops" is generated with the pachytene nuclei candidates. Make sure you specify this for the counting function (3) with stage = pachytene, so that `'synapsis'` knows to look in this folder.

3. Counting coincident foci on synaptonemal complexes

The output is a data frame, which includes a measure of the "crispness" of the foci channel image. Related to this is a recommendation to keep or discard the cell candidate.

The processes outlined above may be second nature to biologists who work with microscopy, and some nuclei that pass through this automated pipeline would be discarded by a real person much earlier. However, with an automated package, these decisions are recorded along with a quantification of real signal vs background/ noise. The user can also change these criteria to be, e.g., less strict and allow more noisy images (in the foci channel) to be kept.

3.6 Next Steps

Before trying to repeat the same process on your own images, let's note some things. It is recommended to use high-resolution images (e.g., more than 4 MB and 1000 × 1000 pixels) so that underlying smoothing operations do not erase small foci that you may aim to detect. This may require some testing on a small number of images to determine the appropriate balance between computational efficiency (if images are excessively large) and throughput. To run a function with a small number of images, you can set the parameter test_amount within the function auto_crop_fast.

Next, make sure you are aware of your file extensions, and when you call functions on images that are not .jpeg, to specify, e.g., file_ext = tif.

Finally, there are many optional inputs to the functions. If the default settings do not generate the desired results, progressively try and alter different parameters by 10s of percent at a time. The annotation = "on" option is designed for you to knit a notebook which steps you through all the decisions being made, so that you can troubleshoot parameter values.

4 Notes

1. Availability and implementation: synapsis can be freely downloaded and installed via:

   http://bioconductor.org/packages/release/bioc/html/synapsis.html. R functions and further information can be found at https://gitlab.svi.edu.au/drr-public/synapsis.

   This tutorial has been adapted from material that we first published with an MIT license at: http://bioconductor.org/packages/release/bioc/vignettes/synapsis/inst/doc/synapsis_tutorial.html. Contact: wcrismani@svi.edu.au

References

1. Hunter N (2015) Meiotic Recombination: The Essence of Heredity. Cold Spring Harb Perspect Biol 7. https://doi.org/10.1101/cshperspect.a016618
2. Bisig CG, Guiraldelli MF, Kouznetsova A et al (2012) Synaptonemal complex components persist at centromeres and are required for homologous centromere pairing in mouse spermatocytes. PLoS Genet 8:e1002701. https://doi.org/10.1371/journal.pgen.1002701
3. Capilla-pérez L, Durand S, Hurel A et al (2021) The synaptonemal complex imposes crossover interference and heterochiasmy in Arabidopsis. Proc Natl Acad Sci U. S. A 118:1–11. https://doi.org/10.1073/pnas.2023613118
4. Voelkel-Meiman K, Moustafa SS, Lefrançois P et al (2012) Full-length synaptonemal complex grows continuously during meiotic prophase in budding yeast. PLoS Genet 8:e1002993. https://doi.org/10.1371/journal.pgen.1002993
5. Brown MS, Grubb J, Zhang A et al (2015) Small Rad51 and Dmc1 complexes often co-occupy both ends of a meiotic DNA double strand break. PLoS Genet 11:1–22. https://doi.org/10.1371/journal.pgen.1005653
6. Cole F, Kauppi L, Lange J et al (2012) Homeostatic control of recombination is implemented progressively in mouse meiosis. Nat Cell Biol 14:424–430. https://doi.org/10.1038/ncb2451
7. Kurzbauer M-T, Uanschou C, Chen D, Schlögelhofer P (2012) The recombinases DMC1 and RAD51 are functionally and spatially separated during meiosis in Arabidopsis. Plant Cell 1–14:2058–2070. https://doi.org/10.1105/tpc.112.098459
8. Reynolds A, Qiao H, Yang Y et al (2013) RNF212 is a dosage-sensitive regulator of crossing-over during mammalian meiosis. Nat Genet 45:269–278. https://doi.org/10.1038/ng.2541

Chapter 16

Automated Quantification of Meiotic Recombination Foci Position and Intensity

John Fozard and Chris Morgan

Abstract

During meiosis, homologous chromosomes reciprocally exchange segments of DNA via the formation of crossovers. However, the frequency and position of crossover events along chromosomes are not random. Each chromosome must receive at least one crossover, and the formation of a crossover at one location inhibits the formation of additional crossovers nearby. These crossover patterning phenomena are referred to as "crossover assurance" and "crossover interference," respectively. One key method for quantifying meiotic crossover patterning is to immunocytologically measure the position and intensity of crossover-associated protein foci along the length of meiotic prophase I chromosomes. This approach was recently used to map the position of a conserved E3 ligase, HEI10, along *Arabidopsis* pachytene chromosomes, providing experimental support for a novel mechanistic "coarsening model" for crossover patterning. Here we describe a user-friendly method for automatically measuring the position and intensity of recombination-associated foci along meiotic prophase I chromosomes that is broadly applicable to studies in different eukaryotic species.

Key words Meiosis, Synaptonemal complex, Crossover interference, Quantitative cytogenetics

1 Introduction

Meiosis is a vital step in eukaryotic sexual reproduction and promotes genetic diversity within offspring via two major mechanisms: 1. the independent assortment of homologous chromosomes into gametes and 2. DNA exchange between maternal and paternally inherited chromosomes. The latter, homologous recombination, is initiated early during prophase I and, in most cases, is a necessary step in facilitating the balanced, independent segregation of homologs during anaphase I. Homologous recombination starts with the programmed formation of numerous DNA double-strand breaks (DSBs) along the lengths of the chromosomes. In most species, the majority of these DSBs are repaired as non-crossovers, with a smaller proportion being repaired as crossovers (COs) [1]. For example, in *Arabidopsis,* ~200 DSBs are thought to occur per

Jesús A. Carballo (ed.), *Meiosis: Methods and Protocols*, Methods in Molecular Biology, vol. 2818,
https://doi.org/10.1007/978-1-0716-3906-1_16,

meiosis, but only ~10 of these are repaired as COs [2]. The final number and position of DSBs that progress to form COs during meiosis are shaped by various well-described CO patterning phenomena, including CO interference and CO assurance [3–5]. CO interference was first described over a century ago and refers to the observation that COs tend to be more distantly spaced than would be expected by chance [6]. CO assurance describes the observation that each pair of homologous chromosomes must receive at least one CO, the obligate CO, which plays a mechanistic role in the segregation of homologous chromosomes [7].

Recently, a mathematical model, coined the "coarsening model," was derived to mechanistically explain a host of CO patterning features, including CO interference and CO assurance [8–14]. In this model, the conserved pro-crossover protein HEI10 is initially loaded along the length of the synaptonemal complex (SC), a protein scaffold that assembles between homologous chromosomes during late prophase I, and at the sites of partially repaired DSBs (recombination intermediates) [8, 15]. HEI10 molecules can then preferentially diffuse through the SC along synapsed pairs of homologous chromosomes (bivalents) and reversibly bind into protein foci at recombination intermediate sites. As more HEI10 molecules accumulate at an individual focus, the escape rate of HEI10 molecules out of that focus, back to the SC compartment, decreases. These binding and diffusion dynamics then drive a coarsening process, whereby large, well-separated foci grow over time at the expense of smaller, closely-spaced foci. By late-prophase I, the majority of HEI10 molecules will have accumulated into a small number of distantly-spaced foci where they trigger CO designation [8].

In order to refine the mathematical parameters of the coarsening model and to test certain model predictions, it was necessary to quantitatively compare model simulation outputs with experimental measurements of HEI10 protein localization and relative abundance along late-prophase I chromosomes. To accomplish this, we designed a bespoke image analysis pipeline for quantifying HEI10 immunofluorescent intensity on individual bivalents in partially spread pachytene nuclei imaged using super-resolution 3D-structured illumination microscopy (3D-SIM) [8, 10]. This work was performed on the model eukaryote *Arabidopsis thaliana*. Using this approach, it was found that the mathematical coarsening model could explain a host of CO patterning features in *Arabidopsis*, including CO interference, CO assurance, CO homeostasis, heterochiasmy, various HEI10 focal intensity measurements, HEI10 over−/under-expression data, the timing of CO designation, and CO patterning in SC mutants [8, 10].

Measuring the number or position of recombination-associated protein foci is a common approach in meiosis research [16]. However, it is also common for these cytological measurements to be obtained manually without clear classification criteria, meaning that this approach can be somewhat subjective and potentially inconsistent between studies. Therefore, one of the major advantages of the imaging pipeline described above is that it allows transparent and reproducible reporting of experimental immunocytological data. For example, different researchers are, in principle, able to perform identical analysis and obtain the exact same results using publicly available imaging data and computer code. However, one drawback is that a certain level of literacy in computational coding is required to run this open-access code in its raw form. Therefore, we have sought to develop a user-friendly version of this analysis software as a simple online tool for use by the wider meiosis community. We have named this online tool "FociMapper" and in this chapter, we describe how to use this software to measure the position and intensity of axis/SC associated recombination foci. We demonstrate the utility of this software for measuring the position and intensity of SC-associated HEI10 foci in late-pachytene nuclei of the model plant species *Arabidopsis thaliana* ($n = 5$). We have also tested and validated this software using HEI10 imaging data from another plant species, *Arabidopsis arenosa* (diploid, n = 8), and from mouse (*M. musculus*) pachytene nuclei stained for MLH1 and SYCP3 (mouse images courtesy of S. Gray).

2 Materials

1. A computer with internet access and an up-to-date version of the FIJI image processing package installed [17]. The SNT plugin for FIJI must also be installed (currently distributed through Fiji's Neuroanatomy Update Site [18]) (*see* **Note 1**).
2. Images of appropriately prepared and immunolabeled prophase I meiotic nuclei, preferably captured as z-stacks in 3-dimensions using fluorescence microscopy (*see* **Note 2**). The image file will require at least two color channels: one channel labeling a component of the meiotic axis, or SC, and one channel labeling recombination foci (*see* **Note** 3).

3 Methods

3.1 Segmenting Prophase I Bivalents or Chromosomes Using SNT

1. Open the image file of your immunolabeled prophase I nucleus using FIJI.
2. For multichannel images, split the image into separate color channels (*Image › Color › Split Channels*) and save the axis/SC channel (e.g., ZYP1 in Fig. 1a, b) and recombination foci channel (e.g., HEI10 in Fig. 1a, b) as separate single-color TIFF files.

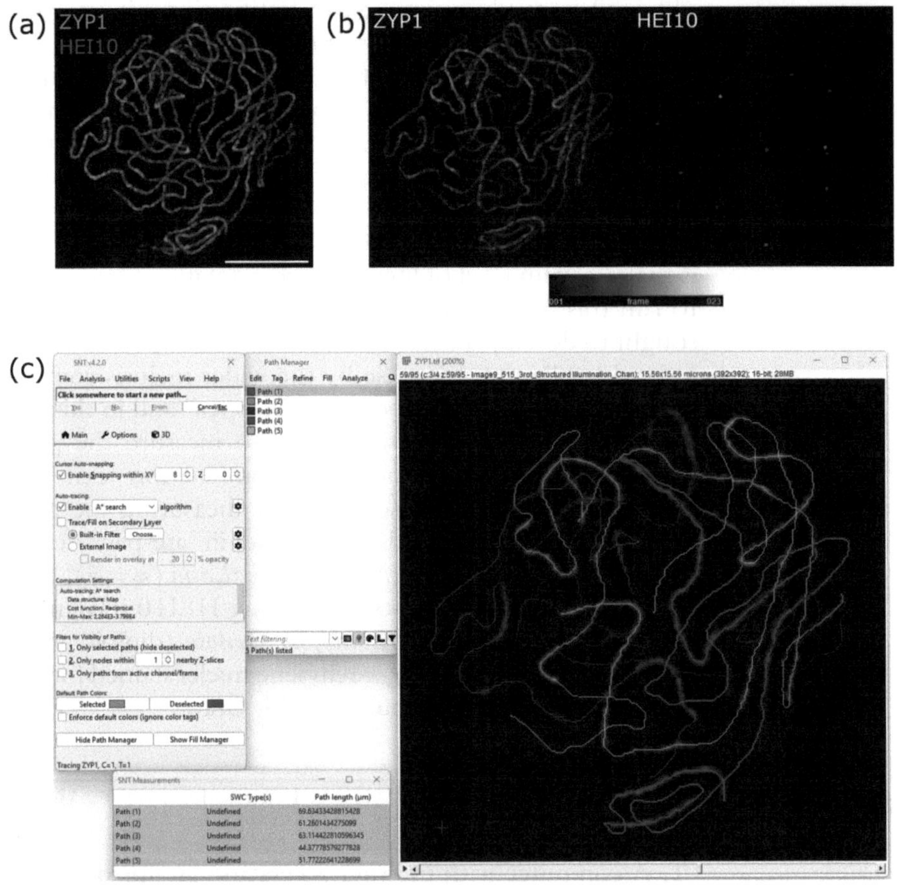

Fig. 1 Segmenting *Arabidopsis* pachytene bivalents using SNT. (**a**) A maximum-intensity projection of a late-pachytene nucleus from *Arabidopsis*, labeled for ZYP1 (green) and HEI10 (red) and imaged with 3D-SIM. Scale bar = 5 μm. (**b**) Single color channels from (a) with color coded z-depth. (**c**) Screenshot of using the SNT FIJI plugin to trace the paths of the five bivalent chromosomes using the ZYP1 channel from the nucleus in (**a**)

3. Run the SNT plugin (*Plugins* › *NeuroAnatomy* › *SNT*) using the axis/SC color channel TIFF as the image file.
4. Trace the path of each individual bivalent or chromosome within the nucleus from end-to-end using SNT's semi-automated tracing tool, following the path of the discrete SC or axis-associated protein marker, respectively (Fig. 1c) (*see* **Note 4**). Take care to avoid introducing branches or hairpins into your SNT traces, as this may cause downstream problems with path orientation in "FociMapper."
5. Once the paths of all bivalents or chromosomes within a nucleus have been traced, save the SNT tracings as a .TRACES file.

3.2 Using "FociMapper" Online Tool to Measure Recombination Foci Position and Relative Intensity

1. Open the "FociMapper" online tool at: https://huggingface.co/spaces/JFoz/FociMapper.
2. Enter an image name in the Cell ID textbox (Fig. 2a).
3. Upload the recombination foci TIFF (from 3.1.2) file into the "Input foci image" upload box (Fig. 2b).
4. Upload the SNT .TRACES file (from Subheading 3.1.5) into the "SNT traces file" upload box (Fig. 2d).
5. Complete the "Additional options" input fields (Fig. 2e) using values that are appropriate for your data type and analysis (*see* **Note 5**).
6. Click the "Process" button (*see* **Note 6**).
7. Download the Output Data file (Fig. 2j). (*see* **Note** 7).

4 Notes

1. SNT is a FIJI plugin designed and developed for the tracing, visualization, quantitative analyses, and modeling of neuronal morphology in light microscopy images [18]. It is a comprehensively updated version of the original Simple Neurite Tracer plugin [19]. Despite being developed for tracing neurons, SNT's software is capable of tracing 3-dimensional filaments in any imaging data. To this end, SNT has been widely used in the meiosis field in different organisms for segmenting and measuring the lengths of unsynapsed or synapsed meiotic chromosomes in meiotic leptotene or pachytene nuclei, respectively (e.g., [20, 21]).
2. In *Arabidopsis*, we routinely use 3D-SIM super-resolution microscopy for imaging prophase I nuclei. Due to the particularly tangled nature of plant chromosomes, the enhanced resolution in x, y, and z orientations afforded by 3D-SIM microscopy facilitates the accurate segmentation of overlapping bivalents. For this analysis, it is also important to ensure that the microscope has been correctly calibrated to ensure correct channel alignment between different imaging channels. A comprehensive protocol for the preparation and imaging of *Arabidopsis arenosa* nuclei using 3D-SIM can be found in [22]. Additionally, while this software is compatible with 2D images, 3D z-stacks are preferable as they avoid issues associated with minor z-depth variations within spread nuclei that could affect the quality or reliability of results.
3. To image the meiotic axis in leptotene nuclei, common proteins to detect are *Saccharomyces cerevisiae* HOP1 homologs (e.g., ASY1 in *Arabidopsis*, HORMAD1 in mammals, or HIM-3 in *Caenorhabditis elegans*). To image the SC in

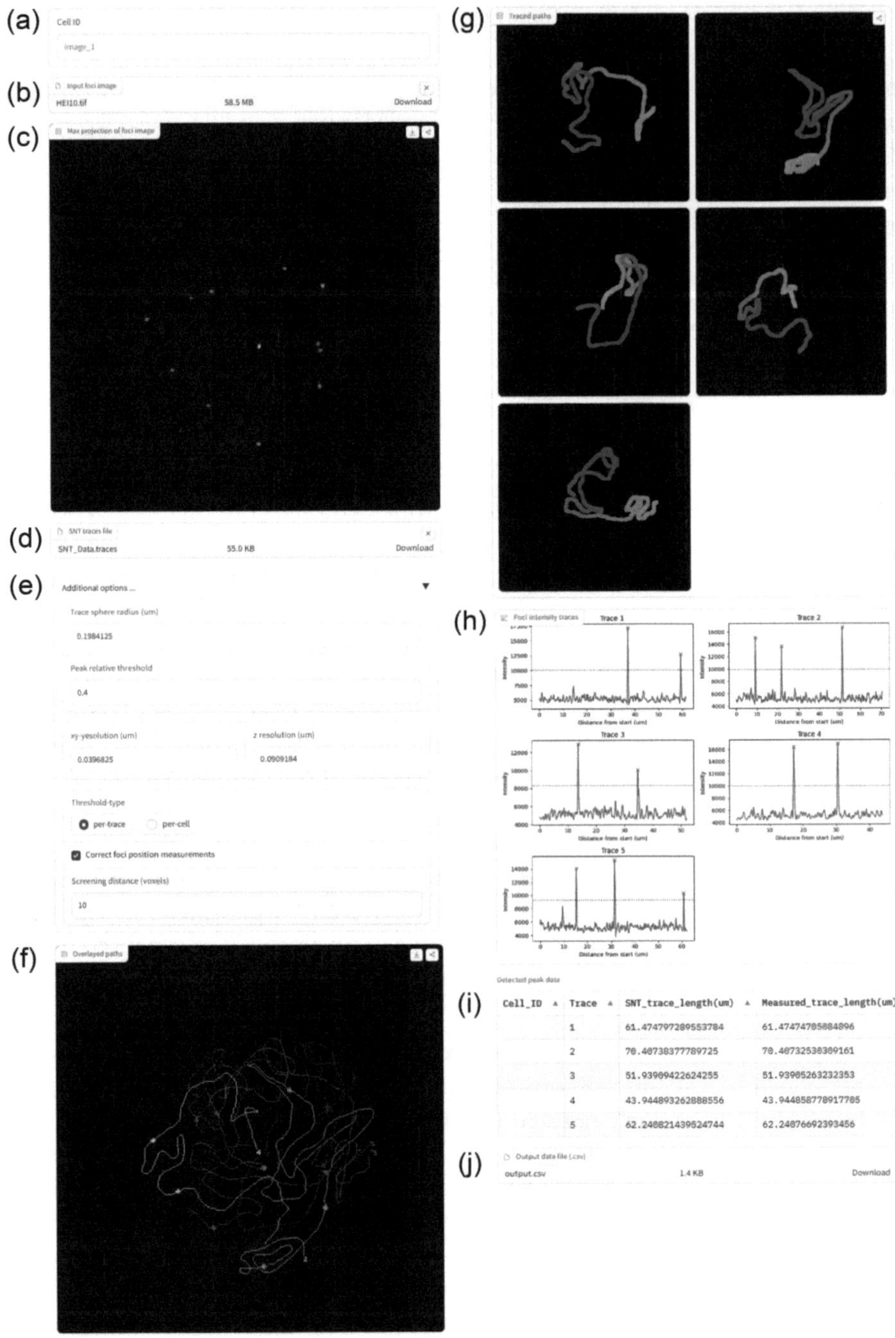

Fig. 2 The FociMapper image analysis tool. (**a**) Input—Cell ID textbox. (**b**) Input—Recombination foci TIFF file upload box. (**c**) Output—Maximum intensity projection of uploaded TIFF file. (**d**) Input—SNT .TRACES file upload box. (**e**) Input—Additional options input fields. (**f**) Output—Image of traced paths (from SNT .TRACES file) with identified foci peaks (colored crosses) overlayed onto maximum intensity projection. If present, grey lines mark distances between screened peaks and nearby higher-intensity peaks. (**g**) Output—Individual traced paths within the nucleus orientated from "start" (green) to "finish" (red). (h) Output—Plots showing the intensity of recombination foci along traced bivalents. Red crosses show the location of identified foci peaks and red dashed-lines show the input threshold value. If present, gray circles indicate foci peaks screened by nearby higher-intensity peaks. (**i**) Output—table showing details of individual path length, focus position, and focus intensity. (**j**) Output—results download box

pachytene nuclei, common proteins to detect are *S. cerevisiae* ZIP1 homologs (e.g., ZYP1 in *Arabidopsis*, SYCP1 in mammals and SYP-1 in *C. elegans*). REC8 homologs can also be used to image leptotene or pachytene nuclei. By labeling different recombination proteins, the number of foci associated with different stages of recombinational repair can be quantified. For example, homologs of *S. cerevisiae* SPO11 and RAD51 proteins are regularly used to measure the number of meiotic double-strand breaks. Homologs of *S. cerevisiae* MLH1 or ZIP3 (e.g., HEI10 in *Arabidopsis*, HEI10/RNF212 in mammals, and ZHP-3 in *C. elegans*) are regularly used to measure the number and position of COs.

4. As a quick quality control step, after tracing the paths of all bivalents or chromosomes within the nucleus, make sure that the final number of paths traced equals the expected number and that the path lengths fall roughly within the expected range. This number and range will vary from species to species, depending on karyotype. For example, in a spread of *Arabidopsis* pachytene nuclei, you would expect to measure five complete paths (one for each of the five *Arabidopsis* bivalents), with lengths ranging between ~30 and 70 μm (Fig. 2h) [8].

5. The values required for the "Additional options" input fields should be tailored to your individual experiment and imaging set-up. The default values are based on those used for our analysis of *Arabidopsis* pachytene nuclei imaged using 3D-SIM microscopy. "Trace sphere radius (μm)" determines the radius of the intensity detection spheres centered around trace-path voxels (*see* **Note 6**). In other words, this value determines how far away foci can be from the traced paths in 3D space to be detected by the software. In *Arabidopsis*, we find that a radius of 0.2 μm is sufficient for detecting HEI10 foci co-localizing with the synaptonemal complex. "Peak relative threshold" determines the threshold value for foci peak-detection relative to the normalized intensity of the brightest focus on the same trace or within the same cell. Using a lower threshold value will mean that a greater number of dimmer foci are detected, and vice versa, using a higher value will mean that only the few brightest foci are detected. While this value is arbitrary and based on the user's personal judgment, using the same threshold value in experiments across different replicates and samples will ensure consistency in quantification. Note that for peak detection, the per-trace mean intensity is subtracted from the normalized intensity, for consistency with our earlier work. "xy-resolution (μm)" and "z-resolution (μm)" values will vary depending on the image acquisition parameters of your microscope. This x, y and z resolution information should be contained within the TIFF file metadata

and can be looked up in FIJI (*Image › Show Info*... see "Voxel size"). For 2D image analysis, an input value is still required for z-resolution, but this value will not affect the analysis. "Threshold-type" can be set as either "per trace" or "per cell." "Per-trace" sets the threshold on each trace separately, relative to the normalized intensity of the brightest focus on that individual trace. "Per-cell" sets the same absolute intensity threshold on all the traces, relative to the normalized intensity of the brightest focus detected across all traces. Different threshold types may be suitable for different analyses. For analyses in which CO assurance may be affected (i.e. CO recombination foci may be absent on some bivalents), then we suggest using the "per-cell" threshold, as the "per-trace" threshold-type will always detect *at least* one focus on every trace relative to the brightest point on that trace. The "Correct foci position measurements" checkbox determines whether foci positions in the output data will be determined by measuring the sum of the distances between successive voxel centers from the start of the trace, rather than counting the proportion of voxels along the trace and multiplying this by the trace length measured by SNT (the approach in our original work). Note that both approaches are approximate. "Screening distance" is the distance (measured in voxels) within which a peak, on the same or different trace, inhibits the detection of another peak with lower (absolute) intensity. This parameter is useful for resolving situations where multiple traces pass close to a single focus, ensuring that the focus is only assigned to a single trace.

6. The FociMapper analysis tool operates using the same analysis principles described in detail in [8]. In brief, starting from an arbitrary end, SNT trace skeleton voxels are ordered from start to finish. Recombination focus intensities on the trace are then extracted using the mean intensity from a spherical region centered at each voxel. The radius of the spherical region can be modified using the "Trace sphere radius (μm)" input field. Peaks in the intensity traces are identified using the scipy find_-peaks criterion [23]. Peaks are suppressed by other, more intense peaks, within a distance of "Screening distance" voxels. This resolves difficulties that occur when multiple paths pass close to a single focus. Peaks are then identified as foci if their normalized intensity is more than a certain focus intensity threshold of the largest peak (either the largest peak "per trace" or "per cell"). This intensity threshold can be modified using the "Peak relative threshold" input field. Focus peak intensities are measured using the trace median as an estimate of the background fluorescence along the trace and normalized relative to the average intensity of all foci intensities within the same cell.

7. The output data can be downloaded as a .csv file. Within this file are: the total length measurements for each trace, either calculated by SNT (SNT_trace_length(μm)) or using our software (Measured_trace_length(μm)); the median intensity of the imaging channel along the entire length of the trace (Trace_median_intensity), which is used as a measure of 'background' signal; the user-inputted detection sphere radius value (Detection_sphere_radius(μm)); the user-inputted screening distance (Screening_distance(voxels)); the user-inputted threshold value (Foci_ID_threshold); the distance of the first identified focus from one end of the trace (Foci_1_position(μm)); the absolute intensity of the first identified focus (Foci_1_absolute_intensity); the relative intensity of the first identified focus (Foci_1_relative_intensity), which is the absolute intensity, minus background, normalised by the mean intensity of all foci in the same cell; as well as position and intensity measures for all other foci on the same trace. Users can easily use this information to calculate metrics such as inter-foci distances and to investigate relationships between focus intensity and focus number per bivalent. Note that differences in the length values measured by our software (Measured_trace_length(μm)) and from the SNT software (SNT_trace_length(μm)) likely indicate that the xy or z resolution values inputted by the user are incorrect.

Acknowledgments

We thank Martin Howard and Matthew Hartley for their support in developing the original version of this analysis pipeline, Chris Franklin for supplying antibodies, Eva Wegel for microscopy support, and Steven Gray for providing sample images from *M. musculus* for software testing. This work was supported by a BBSRC Discovery Fellowship to C.M. (BB/V005774/1).

References

1. Gray S, Cohen PE (2016) Control of meiotic crossovers: from double-strand break formation to designation. Annu Rev Genet. https://doi.org/10.1146/annurev-genet-120215-035111
2. Mercier R, Mézard C, Jenczewski E et al (2015) The molecular biology of meiosis in plants. Annu Rev Plant Biol 66:297–327. https://doi.org/10.1146/annurev-arplant-050213-035923
3. Zickler D, Kleckner N (2016) A few of our favorite things: pairing, the bouquet, crossover interference and evolution of meiosis. Semin Cell Dev Biol. https://doi.org/10.1016/j.semcdb.2016.02.024
4. von Diezmann L, Rog O (2021) Let's get physical—mechanisms of crossover interference. J Cell Sci 134. https://doi.org/10.1242/jcs.255745
5. Pazhayam NM, Turcotte CA, Sekelsky J (2021) Meiotic crossover patterning. Front Cell Dev Biol 9:1940. https://doi.org/10.3389/fcell.2021.681123

6. Sturtevant AH (1915) The behavior of the chromosomes as studied through linkage. Z Indukt Abstamm Vererbungsl. https://doi.org/10.1007/BF01792906
7. Darlington CD, Dark SOS (1932) The origin and behaviour of chiasmata, II. Cytologia (Tokyo) 3:169–185. https://doi.org/10.1508/cytologia.3.169
8. Morgan C, Fozard JA, Hartley M et al (2021) Diffusion-mediated HEI10 coarsening can explain meiotic crossover positioning in Arabidopsis. Nat Commun 12:4674. https://doi.org/10.1038/s41467-021-24827-w
9. Girard C, Zwicker D, Mercier R (2023) The regulation of meiotic crossover distribution: a coarse solution to a century-old mystery? Biochem Soc Trans 51:1179–1190. https://doi.org/10.1042/BST20221329
10. Fozard JA, Morgan C, Howard M (2023) Coarsening dynamics can explain meiotic crossover patterning in both the presence and absence of the synaptonemal complex. elife 12: e79408. https://doi.org/10.7554/eLife.79408
11. Lloyd A (2022) Crossover patterning in plants. Plant Reprod. https://doi.org/10.1007/s00497-022-00445-4
12. Rafiei N, Ronceret A (2023) Crossover interference mechanism: new lessons from plants. Front Cell Dev Biol 11:1156766
13. Durand S, Lian Q, Jing J et al (2022) Joint control of meiotic crossover patterning by the synaptonemal complex and HEI10 dosage. Nat Commun 13:5999. https://doi.org/10.1038/s41467-022-33472-w
14. Zhang L, Stauffer W, Zwicker D, Dernburg AF (2021) Crossover patterning through kinase-regulated condensation and coarsening of recombination nodules. bioRxiv:2021.08.26.457865. https://doi.org/10.1101/2021.08.26.457865
15. Chelysheva L, Vezon D, Chambon A et al (2012) The Arabidopsis HEI10 is a new ZMM protein related to Zip3. PLoS Genet. https://doi.org/10.1371/journal.pgen.1002799
16. Kim H, Choi K (2022) Fast and precise: how to measure meiotic crossovers in Arabidopsis. Mol Cells 45:273–283. https://doi.org/10.14348/molcells.2022.2054
17. Schindelin J, Arganda-Carreras I, Frise E et al (2012) Fiji: an open-source platform for biological-image analysis. Nat Methods 9: 676–682. https://doi.org/10.1038/nmeth.2019
18. Arshadi C, Günther U, Eddison M et al (2021) SNT: a unifying toolbox for quantification of neuronal anatomy. Nat Methods 18: 374–377. https://doi.org/10.1038/s41592-021-01105-7
19. Longair M, Baker D, Armstrong J (2011) Simple neurite tracer: open source software for reconstruction, visualization and analysis of neuronal processes. Bioinformatics 27:2453–2454. https://doi.org/10.1093/bioinformatics/btr390
20. Woglar A, Yamaya K, Roelens B et al (2020) Quantitative cytogenetics reveals molecular stoichiometry and longitudinal organization of meiotic chromosome axes and loops. PLoS Biol 18:e3000817
21. Crichton JH, Dunce JM, Baarends WM et al (2022) Parallel recruitment pathways contribute to synaptonemal complex assembly during mammalian meiosis. bioRxiv:2022.04.14.488335. https://doi.org/10.1101/2022.04.14.488335
22. Morgan C, Wegel E (2020) Cytological characterization of Arabidopsis arenosa Polyploids by SIM. In: Pradillo MHS (ed) Plant meiosis. Methods in molecular biology, vol 2061. Humana, New York
23. Virtanen P, Gommers R, Oliphant TE et al (2020) SciPy 1.0: fundamental algorithms for scientific computing in python. Nat Methods 17:261–272. https://doi.org/10.1038/s41592-019-0686-2

Chapter 17

Computational Tools for the Analysis of Meiotic Prophase I Images

James H. Crichton and Ian R. Adams

Abstract

Prophase I is a remarkable stage of meiotic division during which homologous chromosomes pair together and exchange DNA by meiotic recombination. Fluorescence microscopy of meiotic chromosome spreads is a central tool in the study of this process, with chromosome axis proteins being visualized as extended filaments upon which recombination proteins localize in focal patterns.

Chromosome pairing and recombination are dynamic processes, and hundreds of recombination foci can be present in some meiotic nuclei. As meiotic nuclei can exhibit significant variations in staining patterns within and between nuclei, particularly in mutants, manual analysis of images presents challenges for consistency, documentation, and reproducibility. Here we share a combination of complementary computational tools that can be used to partially automate the quantitative analysis of meiotic images. (1) The segmentation of axial and focal staining patterns to automatically measure chromosome axis length and count axis-associated (and non-axis associated) recombination foci; (2) Quantification of focus position along chromosome axes to investigate spatial regulation; (3) Simulation of random distributions of foci within the nucleus or along the chromosome axes to statistically investigate observed foci-axis associations and foci-foci associations; (4) Quantification of chromosome axis proximity to investigate relationships with chromosome synapsis/asynapsis; (5) Quantification of and orientation of focus-axis distances. Together, these tools provide a framework to perform routine documentation and analysis of meiotic images, as well as opening up routes to build on this initial output and perform more detailed analyses.

Key words Meiosis, Bioimage analysis, Segmentation, Axis, Recombination, Chromosome

1 Introduction

During meiotic prophase I, homologous chromosomes are organized into chromatin loops, anchored into an axial pattern by cohesion complexes [1]. Further proteins assemble along the chromosome axis, including HORMAD proteins [2] and the axial element proteins of the synaptonemal complex [3–5]. Meiotic

Supplementary Information: The online version contains supplementary material available at https://doi.org/10.1007/978-1-0716-3906-1_17.

Jesús A. Carballo (ed.), *Meiosis: Methods and Protocols*, Methods in Molecular Biology, vol. 2818,
https://doi.org/10.1007/978-1-0716-3906-1_17,

recombination is initiated by double-strand DNA break formation early in meiotic prophase I, with the chromosome axis playing a central role in its regulation as both a hub around which recombination events physically organize [6] and as a regulator influencing the distribution of recombination events [5]. Repair of these DNA double strand breaks drives pairing and synapsis between homologous chromosomes, which in turn is accompanied by completion of synaptonemal complex assembly. Fully assembled synaptonemal complex then holds homologous chromosomes in close proximity along the length of their axes while meiotic recombination progresses and exchanges genetic material between homologous chromosomes [5].

Fluorescence microscopy is a particularly valuable tool in the study of meiosis. The progressive formation of chromosome axes and pairing of homologous axes is used to define substages of meiotic prophase I progression, and the quantification of axis length can be used to indicate variation in chromatin loop/axis organization [7]. Recombination foci, which are typically associated with chromosome axes, are often abundant, exceeding 200 per nucleus [5], and are commonly counted to monitor the progression of recombination. Such scoring is typically performed manually and can be subject to subjective scoring biases. As recombination intermediates can follow different repair pathways and result in different repair outcomes, variation between and within nuclei can be difficult to capture. More complex analyses, such as quantifying the position of recombination intermediates relative to each other along chromosome axes, typically to measure crossover interference, are almost exclusively restricted to low-abundance recombination intermediates partly due to it being impractical to measure the positioning of more abundant foci [8, 9]. More automated alternatives have, however, recently been developed that begin to accelerate this process [10].

Advances in microscope resolution have dramatically improved the ability to visualize intricate events taking place during meiotic prophase I, particularly enabling the resolution of the components of the synaptonemal complex, and the relative positioning of axis-associated proteins [11–14]. These gains in resolution present many new opportunities for quantitative analysis of images of meiotic cells, but the extraction of this data requires the development of computational tools that can be used reproducibly. Several such tools have been developed, particularly for the analysis of single molecule localization microscopy images (SMLM) [11–13]. However there remains a lack of generally applicable tools to address many research questions.

Here we share a toolkit of open-source computational programs that can be used in the analysis of meiotic prophase I chromosome spread images. These tools provide an efficient and effective framework for recording image data, improving the speed, documentation, and reproducibility of analysis, and generating a system of metadata that can be further explored, enabling greater insight into the complex processes taking place along chromosome axes during meiotic prophase I.

2 Materials

The analytical tools described are designed to run in Fiji (ImageJ) 1.53f [15] and Python 3.10.9, both of which can be freely downloaded (https://fiji.sc/, https://www.python.org/downloads/). Code was written on the Windows operating system. Fiji scripts require the installation of the "IJPB-plugins" [16], "Neuroanatomy" [17], and "SCF MPI CBG" plugins, which can be added by selection within the Help≫Update≫Manage Update Sites setting in Fiji. All code described for the analysis of meiotic images can be downloaded from https://github.com/JamesCrichton/Meiotic-Image-Analysis-Toolkit. as a zip file and extracted to the desktop (*see* **Note 1**). All the code herein is designed to process stacked 2D tiff images of chromosome spreads. These preparations are beneficial for analyzing chromosomes and associated proteins, as nuclei are typically spread out and quite flat. Where z-stacks are taken to capture signals in 3D, we reduce images to 2D by maximum intensity projection before analyzing. Example images and metadata are included in the zip file. To run the Image Segmentation macro, first open Fiji, then simply drag and drop this file into the Fiji toolbar. A stack image to analyze should be opened before running the script; this can similarly be dragged and dropped into the toolbar (test image provided: "Example_Pachytene_DAPI_RPA_SYCP3.tif").

Positioning of focal staining on chromosome axes, quantification of paired axis proximity, and positioning of foci relative to axial pairs require tracing of axes using SNT (in the Neuroanatomy plugin). Guidance on using the SNT plugin can be found here: https://imagej.net/plugins/snt/step-by-step-instructions. The subsequent interpretation of saved .traces files in our scripts uses the TracePy package by Ziwei Huang. This should be downloaded from https://github.com/huangziwei/TracePy as a zip file and similarly extracted to the desktop (*see* **Note 1**). Example scripts using python functions to batch process files are written with the assumption that these repositories are located on the user's desktop. Python scripts will require modifications to direct toward users' specific images. We edit and run these scripts using the Spyder IDE (integrated development environment) downloaded with Anaconda (https://www.anaconda.com/). Measurements generated from all images are not scaled, but given in pixel values. Images analyzed in the data presented are previously studied synapsed pachytenes (*Tex19.1*$^{+/+}$, *Tex19.1*$^{+/-}$, *Tex19.1*$^{-/-}$) [18] images by widefield microscopy, and structured illumination microscopy images of *Sycp1*$^{+/-}$ and *Sycp1*$^{-/-}$ spermatocytes from other publications [19, 20] (Slides kindly provided by Dr. Willy Baarends [21]).

3 Methods

3.1 Axis Length Measurement and Axial Focus Counting

Cell staging, counting of axial foci, and measurement of axis length are common analyses conducted on meiotic prophase I images. Here we share an Fiji macro that guides users to record the prophase I substage of their image and any comments. The macro then segments and measures the chromosome axes, segments and counts foci, and finally measures the proximity of foci to the axes.

Begin by opening an image to be analyzed (*see* **Note 2**). This should be 2D and contain a minimum of two channels (axis and foci staining). The example image in the "Widefield_sample_data" folder, "Example_Pachytene_DAPI_RPA_SYCP3.tif" is a mouse pachytene spermatocyte with stained axial elements (SYCP3), recombination foci (RPA), and nucleus (DAPI). Nuclear staining is not essential for the script to run but provides a useful guide to selecting the region of an image to analyze. This image was captured on a diffraction-limited widefield microscope, so the separate paired axial elements cannot be resolved.

Once you have your image open, run the "Image Segmentation.ijm" macro in Fiji (Fig. 1, Video 1). The user will be guided through a dialogue to define the channel of the stack image corresponding to the axial and focal staining to be analyzed. Distinct axis and foci staining patterns are expected to be in different channels in the image to be analyzed. If needed, the user can open the Fiji Channels tool at this point to inspect the different channels in the image. If an image has previously been analyzed then segmented staining patterns can be reapplied, enabling the addition and comparison of further focal or axial channels (*see* **Note 3**). To facilitate this type of analysis, the user can select the "Pre-existing mask" and "Pre-existing labelmap" options for the axis and foci channels, respectively. The user also has an option to customize the titles of their analyses, which will be used in the output filenames. However, unless the user is re-analyzing or performing additional analysis on a previously analyzed image, we recommend keeping these titles as their defaults.

The user will be asked to define the destination of the analysis output. This output will consist of a folder of metadata named in the format "*imagename*_Output". We recommend arranging the images to be analyzed for an experiment or condition into one folder before starting the analysis, then selecting this folder as the output destination for these new metadata folders generated. This organization of images and metadata folders is assumed in subsequent analytical pipelines. If the image has been previously

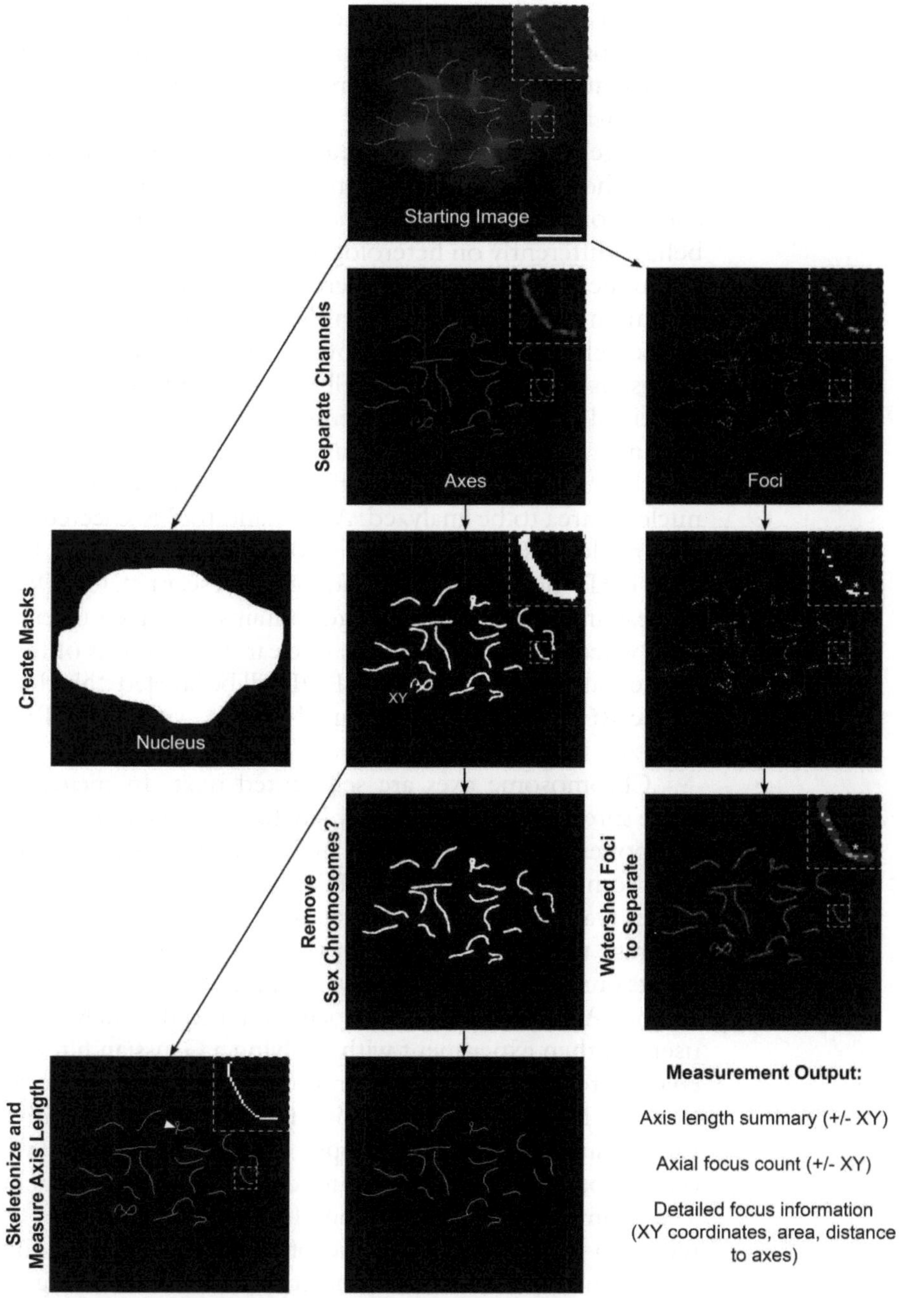

Fig. 1 Segmentation pipeline. Beginning with a fluorescence image, nuclear, axial, and focal binary masks are generated. Sex chromosomes (marked "XY" on Axes Mask image) can be erased to create an autosomal mask. Axial masks are skeletonized to create single-pixel thick lines, the length of which is automatically measured. Occasionally, proximal axes can become fused during this process (arrowhead), underestimating length measurements. Adjacent points fused together in the focus mask (asterisk) can be separated using a maxima watershed function. 10 μm scale bar

analyzed and the data has been saved in this format, segmented regions of interest (ROIs) and an "Image_Summary.csv" file will be opened automatically. If the image has not been analyzed before, a dialog will open asking the user to define the cell's meiocyte sub-stage, add any comments, and define whether the sex chromosomes should be separated from analysis using the "Separate Autosome Scoring" checkbox. In mammals, recombination and synapsis behave differently on heterologous XY sex chromosomes than they do on the autosomes [22]. Therefore it can be important to analyze recombination and synapsis on the autosomes independently from the sex chromosomes (*see* **Note 4**). If this option is chosen,then analysis is run on all axes, as well as just the autosomal axes (with the identified sex chromosomes removed).

Image segmentation begins with the nucleus. The user should use the "Selection tools" in the Fiji toolbar to draw around the nucleus/area to be analyzed. A freehand tool is selected by default. The nucleus is typically visualized in meiotic samples by staining with DAPI or other DNA-binding fluorescent stains. Defining the nuclear area is useful if an image contains multiple nuclei and is also used when generating random nuclear distributions of foci for that image (Subheading 3.4). This ROI will be labeled "Nucleus_Mask" in the ROI manager. Only signals found within this ROI will be analyzed by the macro from this point on.

Chromosome axes are segmented next. In meiotic chromosome spreads, axes are typically labeled with synaptonemal complex components such as SYCP3 in mammals, though chromosome core components such as SMC3 cohesin could also be used (*see* **Notes 5** and **6**). The macro first asks the user to adjust brightness/contrast to visualize the axes (these brightness/contrast adjustments purely to help the user visualize their image. Changes should not be "Applied" to avoid propagation into the analysis steps). The user can then experiment with applying a Gaussian filter to the axis channel to reduce noise and improve the quality of the segmentation (*see* **Note 7**). The Gaussian filter will accept positive integer values, and a 0 value will apply no filtering. Images with more off-axis background staining and discontinuities/fragmentation of axis staining may benefit from higher Gaussian filter values to reduce both fragmentation and off-axis staining being interpreted as chromosome axes. The macro records the settings used for Gaussian filtering in the "Image_Summary" output to allow the user to reproduce analyses from the original image. The user then generates a binary axial mask by setting a threshold above which the macro will consider staining to represent bone fide axis staining. This threshold is somewhat subjective, though there are auto-threshold methods available within the Fiji thresholding tool to reduce subjectivity. The resulting mask from the axes will be recorded in the ROI Manager and named with the given Axes analysis title ("Axes_Mask" by default).

If the user previously selected the "Separate Autosome scoring" option, the user will then have the option to remove the sex (XY) chromosomes from the axes mask using the Eraser Tool (in Drawing Tools). During this erasure process, the mask is overlayed on the original axial channel for reference, and the two can be toggled on/off with the Channels Tool. The new axes mask will have an "Autosome_" prefix in the ROI manager. The macro will count foci in the original mask and in the autosome mask. If the user is choosing to reuse a previously generated mask, they will be invited to select this in the ROI Manager (*see* **Note 8**).

The thickness of the chromosome axis staining will depend on the antibody used, the intensity of the immunostaining signal, and the image capture conditions. Therefore, reducing the axis mask to a single pixel thickness line along the middle of the axis can facilitate some types of analysis. To do this, the axis masks are skeletonized using an algorithm that removes pixels from the circumference of each axis in the mask until a continuous single-pixel line remains. The quality of this process can vary depending on the axial mask used, therefore, users are given the opportunity to manually refine the axis skeleton by removal or addition of necessary connections using the eraser and/or pencil drawing tools (set width to 1px) found in the Fiji toolbar (*see* **Note 9**). Users can also comment on the quality of these skeletons in a subsequent dialogue for reference. The skeletonized axis mask will be labeled "Axes_Mask_Skeleton" in the ROI Manager.

A similar process of segmentation is followed for the foci channel. The macro will display an image with the axial mask outline overlayed on the foci for reference to aid foci segmentation. The macro first asks the user to adjust brightness/contrast to visualize the foci and identify areas with weak/closely spaced foci to monitor foci segmentation (these brightness/contrast adjustments are lost when the user proceeds; they are not propagated into the analysis steps). The user then generates a binary focal mask by setting a threshold above which the macro will consider staining to represent bone fide focal staining. This threshold is somewhat subjective, though there are pre-set auto-thresholding methods available within the Fiji thresholding tool to reduce subjectivity. In images with numerous foci, it is often impossible to separate the foci by thresholding alone, so a maximum watershed step is included (Fig. 1). The macro will ask users to set a maxima prominence, which represents the minimum difference in intensity between surrounding pixels used to identify distinct peaks in regions with merged/overlapping signals. By default, the resulting focal masks will be labeled "Watershed_Foci_Total" in the ROI Manager, the positions of the foci maxima will be labeled "Watershed_Foci Maxima," and a labelmap image of these foci will be saved as "Watershed_Foci.tif," which can be used in other analyses discussed later. These default names and related measurement titles will alter depending on the "Foci analysis title" set within the macro. The title of the axes mask will default to "Axes_Mask" and will similarly apply to the related files, ROIs, and measurements but can also be modified in the macro under "Axes analysis title." Measurements relating to these individual foci are then automatically generated: area, circularity, centroid X/Y coordinates, distance from the nearest point on the axial mask, and distance from the nearest point on the autosomal mask (if selected). A basic summary of measurements is also recorded in the "Image_Summary.csv" file, including a count of foci on axes, off axes, and on autosomes (if selected) and the Euclidean length of the skeleton masks produced (NB: All distance measurements produced are in pixel units). The mask ROIs in the ROI Manager are also saved as a RoiSet.zip file. If images are being reanalyzed, the summary metadata is appended to the pre-existing "Image_Summary.csv" file each time, whereas a new file is generated for the measurements of each focal and axial pattern analyzed.

3.2 Validating Counts from Segmentation Pipeline

To assess the reliability of the measurements generated using the "Image_Segmentation" macro, we first investigated its ability to measure chromosome axis length. The autosomal axis length of a library of pachytene spermatocyte images was first measured using the SNT plugin [17]. This is a plugin, which allows users to trace axial signals and is widely used for chromosome axis tracing in meiosis research [14, 23]. The image library consists of wild type, *Tex19.1*$^{+/-}$, and *Tex19.1*$^{-/-}$ mouse pachytene spermatocytes pooled together from a study into pachytene progression in this mutant [18]. The same images were also processed using the "Image_Segmentation" macro, generating automated axis length measurements. Fitting a linear model to the data for each image generated using these two methods gave an R^2 of 0.93, demonstrating that the automated measurements capture a large amount of the variation in the manual measurements of axis length and perform to a very high standard. Drawbacks to the use of skeletonization to measure axis length include the systematic removal of pixels from object ends and the potential fusion of adjacent regions of axes during thresholding, resulting in multiple axes being reduced to single skeleton traces. Both of these caveats reduce the total axis length measured (Figs. 1 and 2a). This reduction likely accounts for the slight underestimate of length using the automated method (coefficient = 0.87) and should be considered during the segmentation process.

Similarly, the performance of the automated focus counting feature of the macro was assessed by counting autosomal RPA foci in a library of pachytene spermatocytes manually [18] and automatically using the macro. Initially, the segmented focus masks were analyzed directly, without watershedding. Comparison with the manual counts gave an R^2 of 0.82 (Fig. 2b), showing that a large amount of the variation identified by manual counting was

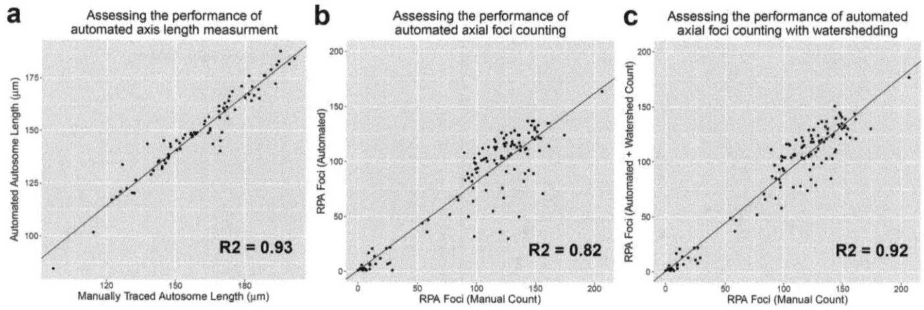

Fig. 2 Automated measurement validation. (**a**) Scatter plot comparing total autosomal lengths for a library of images measured by manual tracing using SNT and automated measurement using the Meiosis-Toolkit Image_Segmentation macro ($n = 78$, $R^2 = 0.93$, coefficient $= 0.87$, intercept $= 9.57$). Pixel output of analysis is converted to microns using the original image-capture metadata. (**b**) Scatter plot comparing total autosomal RPA foci in a library of images measured by manual counting [18] and automatically, using the Meiosis-Toolkit Image Segmentation macro, without using watershedding to separate foci ($n = 120$, $R^2 = 0.82$, coefficient $= 0.81$, intercept $= 1.03$), and (**c**) with watershedding to separate foci ($n = 120$, $R^2 = 0.92$, coefficient $= 0.89$, intercept $= 0.33$)

recapitulated by the automated count. However, it was apparent that counts for many images were underestimated. The RPA foci counted are highly abundant, typically 100–150 per nucleus, and as such, separation by thresholding alone is often not sufficient, resulting in foci fusing together in masks. Applying the watershedding step greatly improves the performance of the macro, producing an R^2 of 0.92 (Fig. 2c), demonstrating a high performance even when counting abundant foci.

3.3 Focus Position on Axis

Analysis of focus position along the chromosome axis is a measurement often generated for the crossover events that form well-established spatial patterns along the 1-dimensional axis. Similar patterns have also been reported for more abundant early recombination markers [9], but such measurements have rarely been performed [23], and analyses have been restricted to a selection of chromosomes. This positional measurement can be a laborious and manually involved process, poorly suited to these more abundant axial foci. Here we present a method to rapidly automate the calculation of this positional data (Fig. 3).

First, the image should be segmented using the Image_Segmentation macro; this will produce a labelmap image for the segmented foci (default name: "Watershed_Foci.tif"). To calculate the position of each segmented focus along the chromosome axis, the axes must also be traced using SNT, an Fiji tool that forms part of the Neuroanatomy plugin [17]. A guide to the installation of SNT and its use is included in the Materials section. The raw image being analyzed should be used as the template to generate these traces, with the axes and DAPI channels visible and the "Tracing channel" in SNT set to the channel with axial signal (*see* **Notes 10** and **11**).

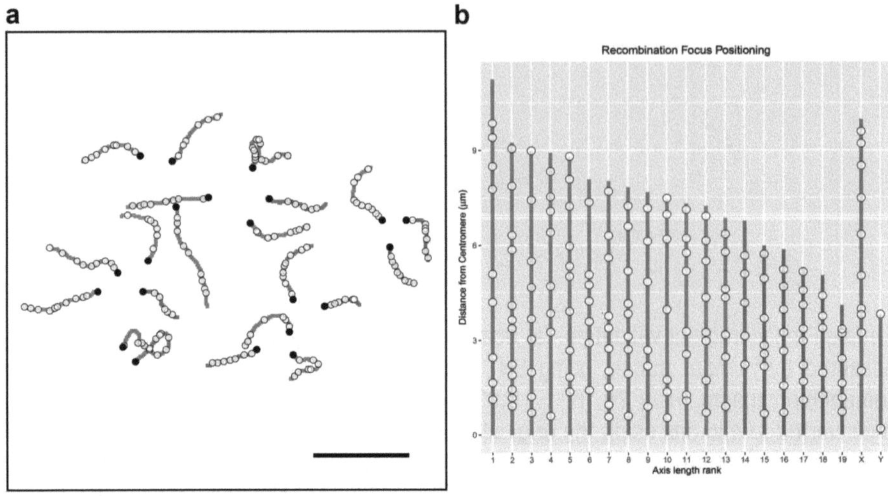

Fig. 3 Focus positioning on chromosome axes. (**a**) Plot displaying the 2D organization of chromosome traces as in the original image, oriented with centromeres marked in black circles. Opaque circles mark the centroids of segmented RPA foci selected for overlap with the segmented axis mask. 10 μm scale bar. (**b**) 1D plotting of chromosome traces and associated positioning of the segmented foci. Chromosome number is estimated by axis length ranking. Distances are converted from pixel values to μm using image capture metadata

Studying mice, which possess acrocentic chromosomes, we begin each trace at the DAPI-bright centromeric end of the axis to orientate the data. An alternative approach that would be applicable to non-acrocentric chromosomes would be to include a centromeric marker, which can be segmented as an axial focus in the Image Segmentation macro, and its position along traced axes subsequently calculated. This position can be used as a reference point for users to manually recalculate measurements. A further alternative is to trace from each centromere and rename the traces differently in SNT for each arm (e.g., 1p and 1q), these names will be preserved in the processing. Similarly, naming of sex chromosomes in these data can also be beneficial. Save the traces in the image's metadata folder as a ".traces" format to ensure it is identified in subsequent scripts in this toolkit.

An example batch analysis can be found in the "Focus_positioning_on_traces.py" file within "Example_Python_Scripts." Open this file in a text editor or using your preferred integrated development environment (IDE), and modify the paths to locate the necessary modules and images. The defaults are set to run example data and assume everything has been extracted to the desktop (*see* **Note 1**). The position of each focus along its closest traced axis will be added to the existing data relating to each focus, producing a file ending "_and_positioning.csv" in the metadata folder.

3.4 Comparing Observed Focus Positioning to Random Distributions

To explore whether observed localization patterns of segmented foci relative to axial or other focal segmented staining patterns reflect genuine associations, it can be useful to compare these measurements to those generated if the foci were randomly distributed [11, 19, 20]. The "Nuclear_focus_shuffling.py" script, in the "Example_Python_Scripts" folder, selects the foci in an image, filters out foci that are very small or very large, then assigns each focus in that image to a random location within the nucleus (defined by the nucleus mask) and determines the relative position of each shuffled focus and the axes. The example file for this script is stored in "Widefield-sample_data." The shape of each focus remains intact during this process, its centroid is repositioned, and new focus positions do not overlap other foci or extend beyond the nuclear boundary (Fig. 4). The script performs multiple rounds of shuffling to generate a set of shuffled distributions of foci that can be compared with the observed dataset. Before running the script, paths to modules included in this package, paths to the image metadata, the file names of segmented focal and axial masks to be analyzed, thresholds for foci filtering, and the number of shuffles need to be set according to the instructions in the script itself. The resulting random distributions can be used to determine whether foci are more frequently associated with axis, or closer to the axes, in each image than might be expected by chance [19, 20]. In addition to generating these measurements, the example script also saves images of the size-filtered segmented foci selected for processing ("Size-selected_foci_labelmap.tif"), and one representative shuffled-foci image ("Shuffled_foci_labelmap.tif").

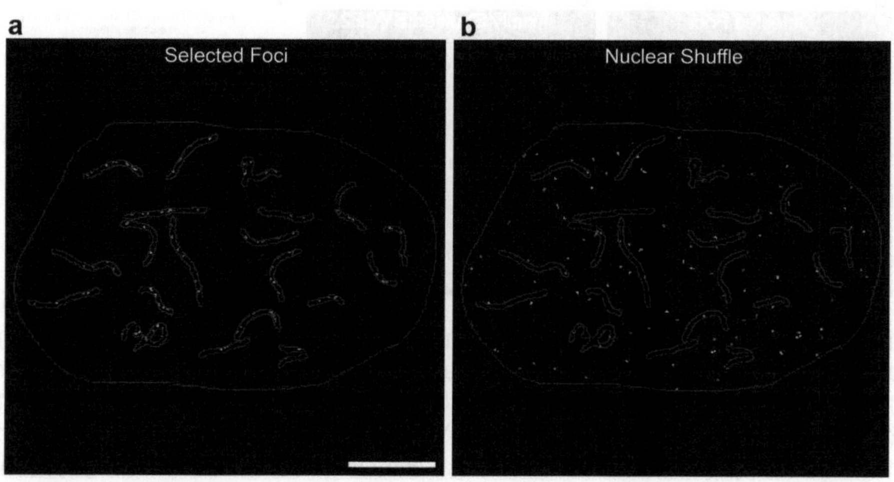

Fig. 4 Nuclear focus shuffling. (**a**) Segmented foci in their original positions. Separate foci colored distinctly. Nuclear and axial mask outlines in grey. 10 μm scale bar. (**b**) An example image showing foci from (**a**) after shuffling throughout the nucleus

When studying the proximity of two axis-associated proteins, it can also be useful to compare their observed proximity relative to that expected from each of these proteins being distributed along the chromosome axes at random [11]. A slightly different algorithm to the nuclear shuffling is required to achieve random shuffling along the chromosome axes, as the edges of foci often extend beyond the edge of an axial mask. The "Axial_focus_shuffling.py" example script performs a similar process to that described for nuclear shuffling, except that one set of foci centroids are randomly shuffled within that image's axes mask rather than its nuclear mask. In addition, the foci are permitted to extend outside the boundary of the axial mask to better reproduce their natural distribution (Fig. 5). To measure distances between two different foci labeled with different antibodies (A and B), distances from each shuffled B focus to the closest reference A focus are then measured (Fig. 5f, **Note** 7). As well as generating these measurements, the example script also saves images of the size-filtered segmented foci selected for processing, and one representative axis shuffled-foci image.

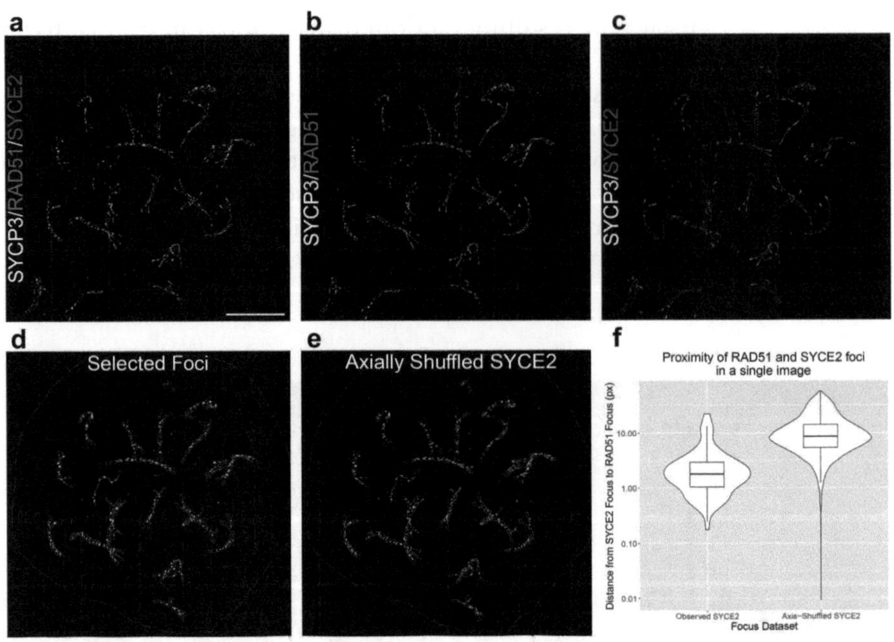

Fig. 5 Axial focus shuffling. Structured illumination microscopy (SIM) images of *Sycp1*$^{-/-}$ spermatocytes stained for the axial element protein SYCP3 (grey), and focal patterns seen by RAD51 (cyan) and SYCE2 (magenta) in this mutant. (**a**) Merged original image, 10 μm scale bar, (**b**) SYCP3 and RAD51 only, (**c**) SYCP3 and SYCE2 only. (**d**) Segmented RAD51 (cyan), and SYCE2 (magenta) foci in their original locations and (**e**) shuffled around randomly the chromosome axis. Axial and nuclear masks outlined in grey. (**f**) Violin plot demonstrating the focus distance measurements generated from this individual image

3.5 Paired Axis Proximity

Super resolution imaging has made it apparent that homologous chromosome axis proximity varies along the length of the synaptonemal complex (Fig. 6a, b). Quantification of these changes in proximity may reveal differences in synaptonemal complex structure and/or chromosome interactions. Pairing of chromosome axes is also seen in some synapsis-defective cells, such as synaptonemal complex mutants, where quantification of pairing could help to better understand and stratify different asynapsis phenotypes [19, 20].

We provide a tool to calculate axis proximity using chromosome trace input generated by the SNT plugin. Users must first use SNT to trace the individual chromosome axes from their original fluorescence image (*see* **Note 13**), renaming each member of a pair of axes with a shared number and a different letter, e.g., "1a" and "1b." These traces should be saved as a .traces file in the image metadata folder. The script recognizes the names of the chromosome pairs, and these are oriented, assuming the start position of the traces is the DAPI-bright centromere. For each pixel in each axis, the closest corresponding pixel in its partner axis will be identified and the distance to it calculated (Fig. 6a) (*see* **Note 14**). This nearest neighbor calculation involves organizing the coordinates of one axis into a k-dimensional tree to allow efficient querying, then using this to identify the pixel closest to each pixel in the partner axis. This process is repeated for each member of each axis pair.

An example script ("Paired_axis_proximity.py") for performing this task in a batch of images can be found in "Example_Python_-Scripts". This script will process the example image data in "SIM_-sample_data". To use this script, open the file and modify the paths to locate the necessary modules and images (as before, the image path should be a folder containing the metadata subfolders for each image, containing a .traces file of axis traces). The defaults are set to run example data and assume everything has been extracted to the desktop. The script will generate a spreadsheet of data titled "Axis_-Trace_Measurements.csv," which will be saved into each metadata folder.

We share an example analysis in Fig. 6 investigating the proximity of paired chromosome axes in image datasets of synapsed control pachytene spermatocytes and mouse mutants unable to form a synaptonemal complex due to deletion of the gene encoding the transverse filament protein SYCP1, but capable of distinctive chromosome pairing in a pachytene-like stage [21]. Calculating paired axis proximity in the synapsed *Sycp1*$^{+/-}$ controls reveals the axis separation varies, with density plots for axis separation peaking at 191 nm likely reflecting separation of the traced centers of SYCP3 axes at sites of complete synaptonemal complex assembly, and also peaking at 0 nm, indicating that the synaptonemal complex at many sites measured is twisted or collapsed (Fig. 6b).

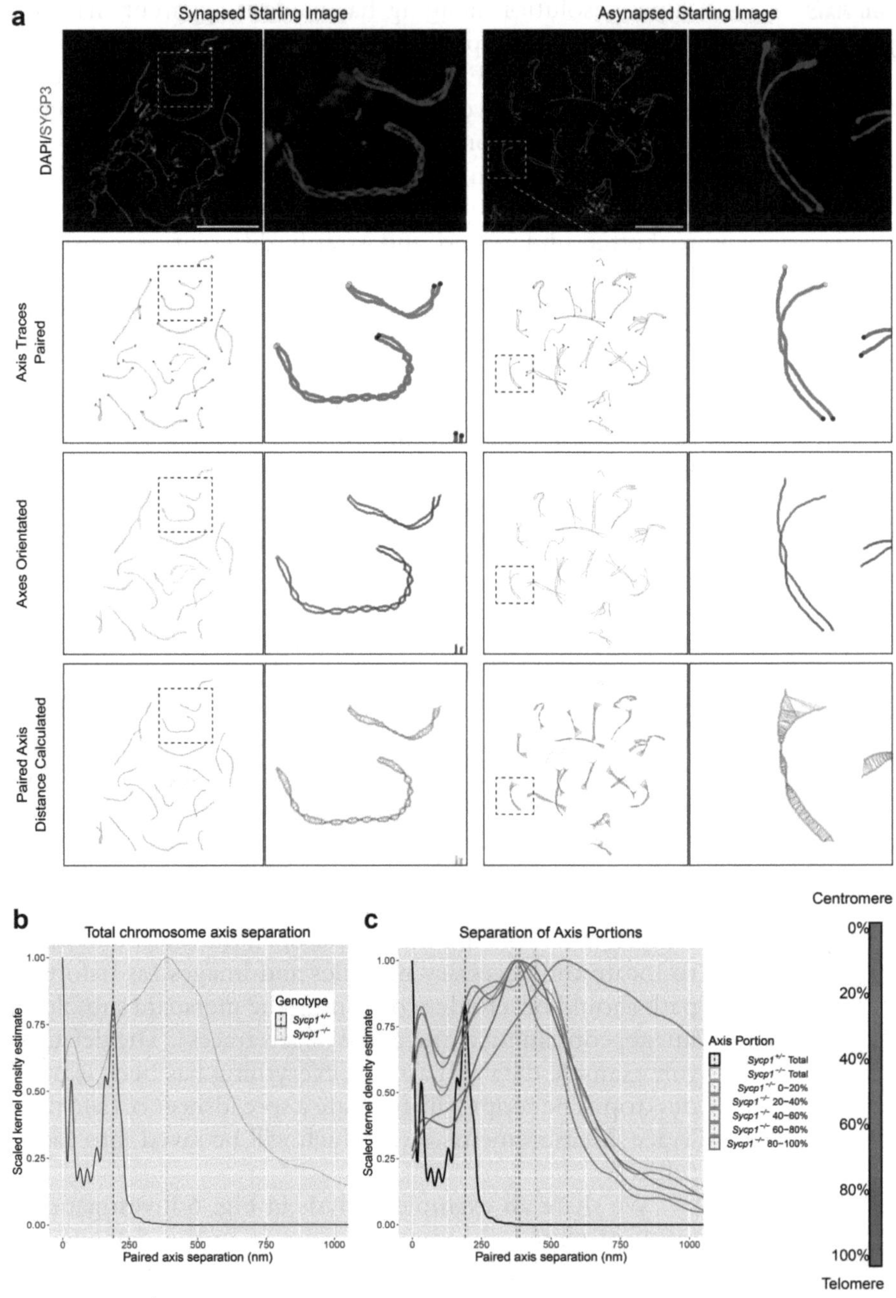

Fig. 6 Homologous axis proximity. (**a**) Synapsed (*Sycp1*$^{+/-}$) and asynapsed (*Sycp1*$^{-/-}$) pairs of homologous chromosomes in spermatocytes stained with DAPI (grey) and antibodies to SYCP3 (magenta) and imaged using SIM. 10 μm scale bar. Pairs of traced axes are recognized (colors paired) and orientated (centromere—red, to telomere —blue), and the minimum distance between each axis trace pixel and its partner axis is calculated (connections drawn and colored dark blue (close) to yellow (distant)). Distance coloring scales are normalised separately to each image. (**b**) Kernel density plot for paired axis separation for total traced axes from synapsed *Sycp1*$^{+/-}$ spermatocytes (9 images of synapsed pachytene spermatocytes from 2 mice, including 171 pairs of traced axes) and asynapsed *Sycp1*$^{-/-}$spermatocytes (2 mice, 18 cells with a minimum of 18 pairs of traced

Separation of *Sycp1*$^{-/-}$ axes is higher than controls, with density plots of SYCP3 axis separation peaking at 386 nm (Fig. 6b). However, it is clear from the images (Fig. 6a) that axis proximity varies greatly across the length of the chromosomes. To investigate whether axis proximity differs in any particular region, we split axes into five equal portions and asked how close the axes are in each of these (Fig. 6c). The axis proximity of most segments peaks in density plots within 60 nm of each other (384–444 nm), however, far greater separation is seen in the most centromeric fifth of the axes, which shows a peak at 560 nm, 116 nm greater than the next highest section. Therefore, while pairing of chromosomes persists in the absence of synaptonemal complex formation in this mutant, the degree of pairing that is achieved is not uniform, with the centromeric ends of the chromosome displaying dramatically lower association with one another.

Homologous pairing of chromosomes in mice is promoted by recombination [24], therefore differences in pairing of chromosome regions could be related to regional differences in recombination frequency. *Sycp1*$^{-/-}$ spermatocytes assessed for paired chromosome proximity were also co-stained for RAD51 (Fig. 7a), and the regional density of recombination foci can be investigated using the code described in Subheading 3.6 (Focus Position on Axis). The position of RAD51 foci along all autosomes included in the analysis of axis proximity was calculated, pooled, and plotted, with axes segmented into 10% bins (Supplementary Fig. 1). This reveals a consistent frequency of RAD51 foci along the center of the axes (mean 11.3% of RAD51 foci in each of the 8 × 10% segments from 10% to 90%). However, strikingly fewer foci are positioned in the terminal 10% of the axes, with 7.9% of foci in the most telomeric 10% of the axis, and only 2.0% of RAD51 foci in the most centromeric 10% of the axes. This centromeric depletion of RAD51-marked recombination sites may be responsible for the lower axis proximity seen for the most centromeric 20% of paired axes (Fig. 6c). This analysis also further highlights the value of the tools we are sharing and the data they generate.

Fig. 6 (continued) axes each). Distribution of these distances (above zero) peak at 191 nm (*Scyp1*$^{+/-}$) and 386 nm *Scyp1*$^{-/-}$) respectively, indicated by dasher vertical lines. (**c**) Kernel density distribution of paired axes separation in five equal portions of each *Sycp1*$^{-/-}$ axis length. Dashed vertical lines indicate the peak separation for each group (NB. Distances have been scaled from the original pixel data into nm using metadata from image capture). Peak separation of total *Sycp1*$^{+/-}$ axes = 191 nm, total *Sycp1*$^{-/-}$ axes = 386 nm, *Sycp1*$^{-/-}$ 0–20% = 560 nm, *Sycp1*$^{-/-}$ 20–40% = 444 nm, *Sycp1*$^{-/-}$ 40–60% = 373 nm, *Sycp1*$^{-/-}$ 60–80% = 384 nm, *Sycp1*$^{-/-}$ 80–100% = 388 nm. Schematic on the right indicates the separation of chromosome axes into five color-coded sections

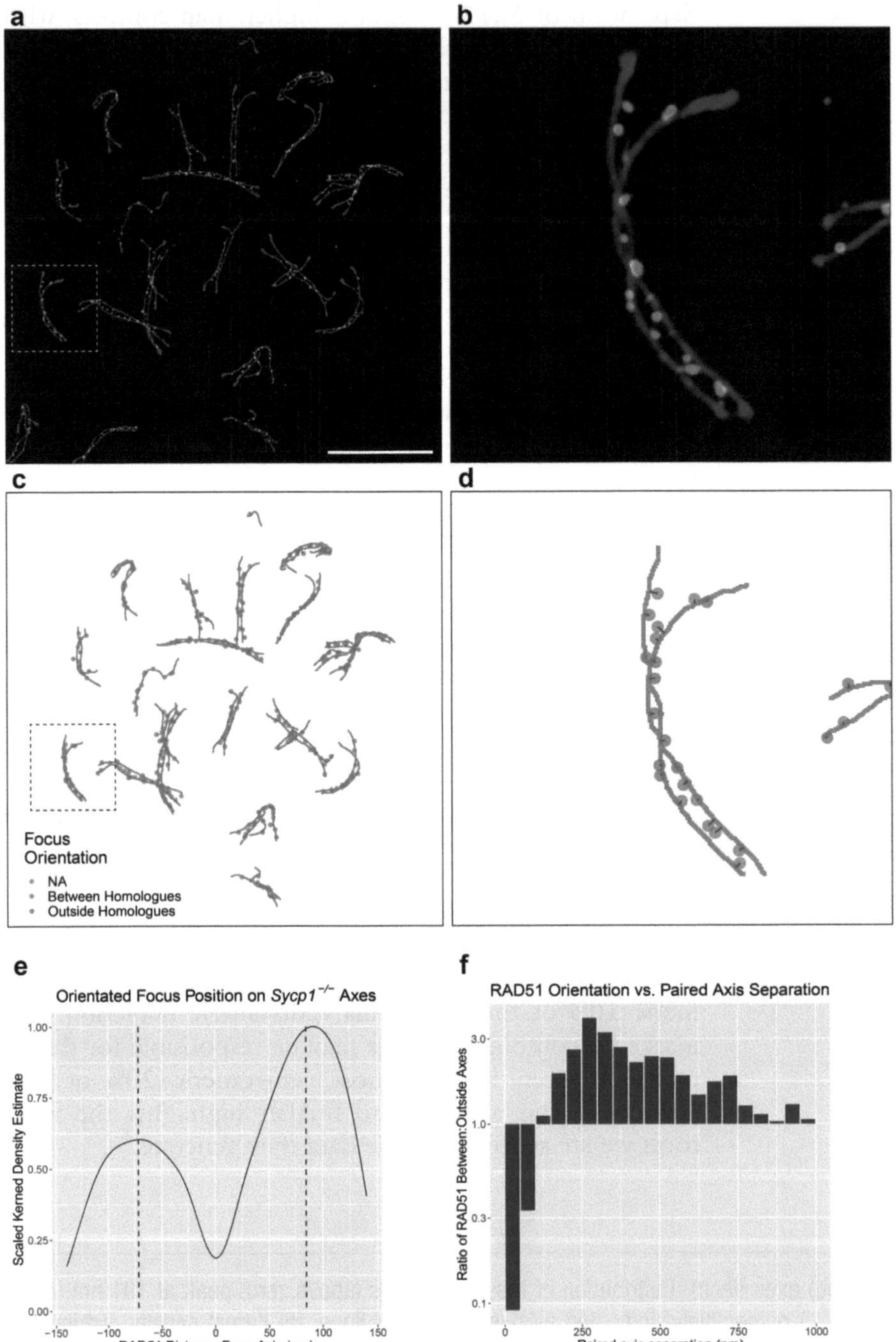

Fig. 7 Focus positioning relative to individual homologous chromosome axes. (**a**) Structured illumination microscopy (SIM) images of *Sycp1*$^{-/-}$ spermatocytes stained for the axial element protein SYCP3 (magenta), and focal patterns seen by RAD51 staining (green). 10 μm scale bar. (**b**) Zoomed example of two paired homologous axes. (**c**) Plotted coordinates of traced axes (grey) and the weighted centroids of segmented RAD51 foci colored based on their orientation relative to their closest axis par (red = NA i.e. closest axis is not paired with its homologue; green = focus between homologous axes; blue = focus outside homologous axes).

3.6 Focus Positioning Relative to Homologous Axes

New levels of insight into the structural intricacy of events taking place during meiosis are being provided by super-resolution microscopy [11, 12, 14, 25]. It is important to be able to measure these spatial relationships to best utilize these advances in imaging technology. As well as the existence of variation in proximity between homologous axes, the positioning of recombination foci within the paired chromosome axes also varies [25]. We share a method using a combination of axis trace data and segmented foci data to calculate the position of foci along the axes, their distance away from their closest axis, and whether the orientation of this position is between paired axes or outside of these.

To use this approach, focal staining patterns must first be segmented to generate a labelmap image. This can be done using the "Image Segmentation.ijm" macro described earlier. Paired axes must also be traced and named in the manner described for the calculation of paired axis proximity, with each member of a pair of axes having a shared name but a different number e.g., "1a" and "1b." These should be saved as a .traces file in the same image metadata folder, which contains the segmented focus labelmap image among other data generated during segmentation. The proximity of axes must first be calculated using the "Paired_axis_proximity.py" script to generate an "Axis_Trace_Measurements.csv" file, which will be read. The example Python script, "Focus_position_relative_to_individually_traced_axis_pairs.py" will use this data and calculate the closest axis trace position for each segmented focus centroid, efficiently searching their coordinates as k-dimensional trees. The distance to this axis and to its partner axis is calculated. If the distance between the focus-associated axis and its neighbor axis at this location is greater than the distance from the focus centroid to the same location in the neighbor axis, then the focus is judged to be between the axes. If this is not the case, then it is outside the axes (Fig. 7a–d). As with other scripts, the "Focus_position_relative_to_individually_traced_axis_pairs.py" script for batch analysis is set up to run on example data, assuming this data and necessary modules are located on the desktop. These paths should be changed if necessary. The name of the labelmap image to be analyzed should be defined, as well as the positioning of the focus staining channel in the original

Fig. 7 (continued) Red lines indicate the distance measured from the focus centroid to the closest pixel in the traced axis. (**d**) Zoomed example axes. (**e**) Kernel density estimate plot of RAD51 focus position relative to the centre of the closest axis trace. Focus distances orientated between paired axes are positive, distances away from a partner axis are negative. Only cells with a minimum of 18 pairs of traced axes are included (18 cells from 2 mice). Vertical dashed lines indicate the median positive and negative distances (Positive: 78.35 μm, Negative: total = −72.92 μm). (**f**) Barplot showing the ratio of RAD51 focus orientation being between paired axes, to outside paired axes, against the separation of paired asynapsed chromosome axes in *Sycp1*$^{-/-}$ spermatocytes, divided into 50 nm bins

fluorescence image, and the "Foci_name" you wish to use to identify the output data. Foci can also be filtered by a maximum- and minimum-area filter.

We have previously used this type of analysis to investigate focal localizations of proteins that could be involved in driving and/or maintaining interactions between axes in mutants unable to form a synaptonemal complex [19, 20]. As a further demonstration of the data generated using this method, we ran the analysis on RAD51 foci segmented from the *Sycp1*$^{-/-}$ image dataset previously analyzed for axis proximity. Plotting the distribution of foci away from the axes shows that RAD51 focus positioning is bimodal, typically falling onto one side of the axis or the other, but with a large bias towards foci being positioned between the traced chromosomes (Fig. 7e). Foci are positioned a similar distance from the axis with both orientations (between axes: median 78 nm, outside of axes: median 73 nm).

It has been reported from qualitative manual analysis of SIM images in wild-type mice that recombination foci formed by RPA move from being located along the center of chromosome axes in early meiosis to moving away from the axis into the space between paired axes after chromosome synapsis [25], possibly reflecting a structural change in the recombination intermediate as recombination proceeds. RPA binds to resected ssDNA at sites of double stranded DNA breaks during early meiosis, which are also bound by RAD51. RPA additionally appears to bind D-loop DNA after strand invasion [26], an event potentially located between chromosomes and possibly reflected in the latter positioning of RPA foci between homologous axes [25]. Therefore, it might be expected that the earlier localization of RPA tightly along chromosome axes would be shared by RAD51; however, this is not a pattern seen in our data, where rather than being positioned along the axis center, RAD51 foci in asynapsed *Sycp1*$^{-/-}$ spermatocytes are typically adjacent to the axes, most commonly being positioned between the paired axes (Fig. 7e). This difference in localization may reflect a distinction between RPA and RAD51 recombination foci or the cells analyzed in the two analyses (*Sycp1*$^{-/-}$ and WT), but the ability to objectively address this question highlights the value of our quantitative approach.

The preferential localization of RAD51 between paired chromosome axes in *Sycp1*$^{-/-}$ spermatocytes potentially reflects engagement of RAD51-containing single-stranded DNA ends in interhomolog repair events. Therefore, the frequency of these interhomolog repair events could potentially be associated with the distance between paired axes. Indeed, the data generated from our tool enables us to plot axis proximity at locations where RAD51 is present, either between or outside axis pairs. By plotting the ratio of RAD51 foci orientation between and outside axes for axes with different levels of separation, we see an enrichment of foci between axes that begins when axes are 800-1000 nm apart. This enrichment steadily grows until the axes are 250 nm apart (Fig. 7f). Beyond this proximity, the enrichment of RAD51 foci between paired axes steadily declines until below 100 nm axis separation, when foci are more likely to be positioned outside the axis pairs than between them (Fig. 7f). The 250 nm axis proximity where the orientation of RAD51 foci between axes peaks is strikingly similar to the 193 nm separation of traced axial element separation we measured in synapsed wild-type spermatocytes (Fig. 6b). This may therefore reflect an optimal distance for homologue engagement and the procession of recombination, a distance that is maintained by the synaptonemal complex in wild-type meiosis to facilitate this process. Furthermore, if inside:outside bias in focus positioning does reflect inter-homologue intermediates, then these intermediates seem to be able to link axes separated by as much as 800-1000 nm and could potentially initiate alignment and subsequent synapsis of homologous chromosomes. Together, the computational tools we share here make it possible to reveal and investigate diverse complex patterns such as these, facilitating the accumulation of a rich and detailed understanding of meiotic processes.

4 Notes

1. Note that when using OneDrive, a OneDrive Desktop is the system default. This location is not the default for our program. Rather, it is the user Desktop.
2. We recommend saving output metadata folders in the same folder as the images being analyzed (as in the Widefield_sample_data example). The macro will find existing metadata using this structure. However, because this existing data is used to populate some parameters used by the macro, to test full functionality of the script on the test image, first copy the image to a different directory and run there.
3. If additional channels are to be analyzed, remember to assign the output with a different name to existing focal/axial output so files are not overwritten.
4. Although we routinely use this option to distinguish autosomes and sex chromosomes during analysis of mammalian meiotic chromosome spreads, this option can also be used to separate images into two compartments based on any morphological or immunostaining features in the image.
5. An axes mask and focus mask must be set for the macro to work. It does not work with each input independently.
6. It is possible to build up masks and measurements for multiple axial and focal staining patterns within a single image by re-running the macro and changing the axes analysis title

and/or Foci analysis title as well as the associated channel number. This data will be added to any existing output metadata.

7. The Gaussian-filtered image can be inspected at this point by adjusting the threshold so the mask is removed.
8. If autosome analysis has been previously selected, then choose the ROI for the full axial mask, including sex chromosomes. The macro will recognize the autosome analysis selection in the data and run measurements to this selected ROI mask and the next ROI in the ROI manager, assuming this is the autosomal version of this mask.
9. If "autosome analysis" was selected earlier, then the removed sex chromosomes will reappear here. This provides an opportunity to also optimize the sex chromosome skeletons, so measurements relating to them can be calculated by subtracting the autosomal from the total measurements.
10. If both members of a pair of axes are being traced, name these traces with the same number and differing letters, e.g. 1a and 1b. This nomenclature is recognized in other tools herein for quantifying inter-axis distances and oriented focus positioning relative to axis pairs.
11. For tracing resolved super-resolution axes, it can be beneficial to reduce the "Enable Snapping within XY" value in SNT to allow separation of proximal axes. Alternatively, to generate a single trace for a pair of axes, the image can be processed using a Gaussian filter to blur the axes together.
12. During this analysis the inter-focus measurements generated are two-dimensional; these are not one-dimensional measurements of proximity along the chromosome axis.
13. It may be necessary to adjust the parameters of "Enable Snapping within XY" within SNT. Applying the "Sharpen" function to the axial image can also benefit tracing.
14. The closest pixel on the partner axis may not always correspond to the same position along the chromosome.

Acknowledgements

We wish to thank all the members of the Institute of Genetics and Cancer Advanced Imaging Resource team, for all their microscopy-related assistance, in particular Laura Murphy for sharing image-analysis advice. We are also grateful to Willy Baarends (Erasmus MC, Rotterdam) for sharing *Sycp1*$^{-/-}$ spermatocyte spreads, and C. James Ingles (University of Toronto, Canada) for sharing RPA antibody.

References

1. Eijpe M, Offenberg H, Jessberger R et al (2003) Meiotic cohesin REC8 marks the axial elements of rat synaptonemal complexes before cohesins SMC1beta and SMC3. J Cell Biol 160:657–670. https://doi.org/10.1083/jcb.200212080
2. Wojtasz L, Daniel K, Roig I et al (2009) Mouse HORMAD1 and HORMAD2, two conserved meiotic chromosomal proteins, are depleted from synapsed chromosome axes with the help of TRIP13 AAA-ATPase. PLoS Genet 5: e1000702. https://doi.org/10.1371/journal.pgen.1000702
3. Schalk JAC, Dietrich AJJ, Vink ACG et al (1998) Localization of SCP2 and SCP3 protein molecules within synaptonemal complexes of the rat. Chromosoma 107:540–548. https://doi.org/10.1007/s004120050340
4. Offenberg HH, Schalk JA, Meuwissen RL et al (1998) SCP2: a major protein component of the axial elements of synaptonemal complexes of the rat. Nucleic Acids Res 26:2572–2579
5. Zickler D, Kleckner N (2015) Recombination, pairing, and synapsis of homologs during meiosis. Cold Spring Harb Perspect Biol 7. https://doi.org/10.1101/cshperspect.a016626
6. Baudat F, Imai Y, de Massy B (2013) Meiotic recombination in mammals: localization and regulation. Nat Rev Genet 14:794–806. https://doi.org/10.1038/nrg3573
7. Novak I, Wang H, Revenkova E et al (2008) Cohesin Smc1beta determines meiotic chromatin axis loop organization. J Cell Biol 180: 83–90. https://doi.org/10.1083/jcb.200706136
8. Barchi M, Roig I, Di Giacomo M et al (2008) ATM promotes the obligate XY crossover and both crossover control and chromosome axis integrity on autosomes. PLoS Genet 4: e1000076. https://doi.org/10.1371/journal.pgen.1000076
9. de Boer E, Stam P, Dietrich AJJ et al (2006) Two levels of interference in mouse meiotic recombination. Proc Natl Acad Sci USA 103: 9607–9612. https://doi.org/10.1073/pnas.0600418103
10. Soriano J, Belmonte-Tebar A, de la Casa-Esperon E (2023) Synaptonemal & CO analyzer: a tool for synaptonemal complex and crossover analysis in immunofluorescence images. Front Cell Dev Biol 11:1005145. https://doi.org/10.3389/fcell.2023.1005145
11. Slotman JA, Paul MW, Carofiglio F et al (2020) Super-resolution imaging of RAD51 and DMC1 in DNA repair foci reveals dynamic distribution patterns in meiotic prophase. PLoS Genet 16:e1008595. https://doi.org/10.1371/journal.pgen.1008595
12. Koornneef L, Slotman JA, Sleddens-Linkels E et al (2022) Multi-color dSTORM microscopy in Hormad1-/- spermatocytes reveals alterations in meiotic recombination intermediates and synaptonemal complex structure. PLoS Genet 18:e1010046. https://doi.org/10.1371/journal.pgen.1010046
13. Schücker K, Holm T, Franke C et al (2015) Elucidation of synaptonemal complex organization by super-resolution imaging with isotropic resolution. Proc Natl Acad Sci USA 112: 2029–2033. https://doi.org/10.1073/pnas.1414814112
14. Morgan C, Fozard JA, Hartley M et al (2021) Diffusion-mediated HEI10 coarsening can explain meiotic crossover positioning in Arabidopsis. Nat Commun 12:4674. https://doi.org/10.1038/s41467-021-24827-w
15. Schindelin J, Arganda-Carreras I, Frise E et al (2012) Fiji: an open-source platform for biological-image analysis. Nat Methods 9: 676–682. https://doi.org/10.1038/nmeth.2019
16. Legland D, Arganda-Carreras I, Andrey P (2016) MorphoLibJ: integrated library and plugins for mathematical morphology with ImageJ. Bioinformatics 32:3532–3534. https://doi.org/10.1093/bioinformatics/btw413
17. Arshadi C, Günther U, Eddison M et al (2021) SNT: a unifying toolbox for quantification of neuronal anatomy. Nat Methods 18: 374–377. https://doi.org/10.1038/s41592-021-01105-7
18. Crichton JH, Read D, Adams IR (2018) Defects in meiotic recombination delay progression through pachytene in Tex19.1-/- mouse spermatocytes. Chromosoma 127: 437–459. https://doi.org/10.1007/s00412-018-0674-9
19. Crichton JH, Dunce JM, Dunne OM et al (2023) Structural maturation of SYCP1-mediated meiotic chromosome synapsis by SYCE3. Nat Struct Mol Biol 30:188–199. https://doi.org/10.1038/s41594-022-00909-1
20. Crichton JH, Dunce JM, Baarends WM, et al (2022) Parallel recruitment pathways

contribute to synaptonemal complex assembly during mammalian meiosis. https://doi.org/10.1101/2022.04.14.488335

21. de Vries FA, de Boer E, van den Bosch M et al (2005) Mouse Sycp1 functions in synaptonemal complex assembly, meiotic recombination, and XY body formation. Genes Dev 19:1376–1389
22. Kauppi L, Barchi M, Baudat F et al (2011) Distinct properties of the XY pseudoautosomal region crucial for male meiosis. Science 331: 916–920. https://doi.org/10.1126/science.1195774
23. Woglar A, Yamaya K, Roelens B et al (2020) Quantitative cytogenetics reveals molecular stoichiometry and longitudinal organization of meiotic chromosome axes and loops. PLoS Biol 18:e3000817. https://doi.org/10.1371/journal.pbio.3000817
24. Cahoon CK, Hawley RS (2016) Regulating the construction and demolition of the synaptonemal complex. Nat Struct Mol Biol 23:369–377. https://doi.org/10.1038/nsmb.3208
25. Yoon S, Choi E-H, Kim J-W, Kim KP (2018) Structured illumination microscopy imaging reveals localization of replication protein A between chromosome lateral elements during mammalian meiosis. Exp Mol Med 50:112–112. https://doi.org/10.1038/s12276-018-0139-5
26. Hinch AG, Becker PW, Li T et al (2020) The configuration of RPA, RAD51, and DMC1 binding in meiosis reveals the nature of critical recombination intermediates. Mol Cell 79: 689–701.e10. https://doi.org/10.1016/j.molcel.2020.06.015

Chapter 18

Identifying Chromosome Movement Patterns During Meiosis Using *ChroMo*

Jesús Pinto-Cruz, María Correas, Rodrigo Mendoza-Madrigal, Daniel León-Periñán, and Alfonso Fernández-Álvarez

Abstract

During meiosis, transient associations between the nuclear envelope and telomeres transmit nuclear movements to chromosomes, enabling their pairing and recombination. Recent advances in the field of quantitative cell biology allow a large volume of information about the kinetics of these chromosome movements to be extracted and analyzed with the aim of identifying biologically relevant movement patterns. To this end, we have developed *ChroMo*, a freely available application for the unsupervised study of chromosome movements in fission yeast meiosis. *ChroMo* contains a set of time series algorithms to identify chromosome movement motifs that are not easily observable by direct human visualization and to establish causal relationships between phenotypes. In this chapter, we present a detailed protocol for the processing of raw live imaging data from fission yeast and its subsequent analysis in *ChroMo*.

Key words Meiosis, *ChroMo*, Telomeres

1 Introduction

Of all the cellular events that safeguard genetic diversity, probably the most crucial is DNA recombination between homologous chromosomes to exchange genomic information. This process takes place during meiosis, the specialized cell division in which a single round of DNA replication is followed by two consecutive rounds of nuclear division that each ensure the correct distribution of chromosomes from one progenitor cell to four gamete cells (1–3). Meiotic DNA recombination is induced by controlled DNA damage that triggers DNA repair mechanisms (4), which require homologous chromosomes to pair in the nucleoplasm (5). The search for homologous chromosomes in the nucleus is facilitated by intense nuclear movements generated by cytoskeletal proteins, mainly actin and dynein (6–10). These cytoskeletal forces are transmitted to the chromosomes by the formation of the telomere

Jesús A. Carballo (ed.), *Meiosis: Methods and Protocols*, Methods in Molecular Biology, vol. 2818,
https://doi.org/10.1007/978-1-0716-3906-1_18,

bouquet and the grouping of telomeres in specific regions of the nuclear envelope, creating a meiosis-specific chromosome polarization (11–16). Studies to decipher whether the trajectories of chromosomes during telomere bouquet formation are stochastic or follow predictable behavior have been difficult to perform for a number of reasons, including the difficulty in manipulating meiosis in most eukaryotes, the limited resolution of imaging of the bouquet, and the computational cost of analyzing a large number of trajectories. However, recent advances in the field of quantitative cell biology allow us to improve our understanding of the kinetics of chromosome movement during the bouquet stage (17–19).

The behavior and extent of meiotic chromosome movements have been explored most intensively in the fission yeast *Schizosaccharomyces pombe*, as it exhibits dramatic chromosome movements in which the nucleus is pushed back and forth by microtubules that generate an oscillatory force, causing it to take the shape of a horsetail (10, 11, 20–23). These nuclear movements take about 2–3 h and are followed by a stage in which the nucleus slows down in the middle of the cell for about 30 min before the first cell division; the telomere bouquet is formed during both phases. Hence, the meiosis prophase in fission yeast can be divided into two stages: the horsetail movement stage, when the nucleus oscillates more intensely towards the cell poles; and the post-horsetail stage, when the nucleus no longer moves (24)(Fig. 1).

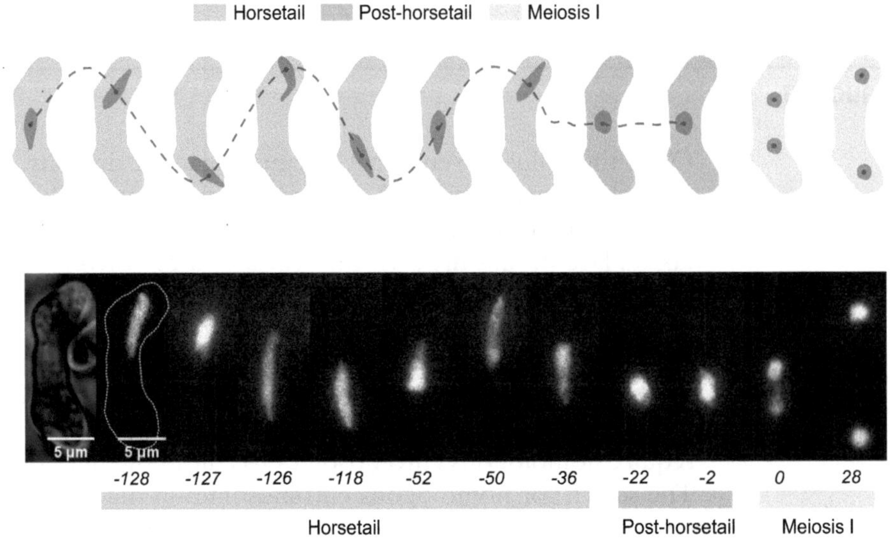

Fig. 1 A schematic of meiotic progression in fission yeast based on frames from films of *h90 S. pombe* cells carrying Hht1-CFP (histone H3 tagged at one of the two endogenous *hht1*$^{+}$ loci). This protocol is focused on the analysis of the horsetail movement and post-horsetail stages

Thanks to the special characteristics of the horsetail movement itself (a dramatic movement observable by live meiosis), we can quantify the kinetics of telomere trajectories and the morphological descriptors of chromosomes during meiosis in fission yeast and then use time-series analysis to search for patterns in chromosome motility that might be biologically relevant to meiotic progression (19, 25). To facilitate a more detailed analysis of chromosome movements during meiosis in fission yeast, we devised *ChroMo* (Interactive and Unsupervised Analysis of Chromosome Movements in Meiosis). *ChroMo* is an open access platform that is easy to use for non-computational biologists (https://chromo.cloud). *ChroMo* contains a set of algorithms to identify patterns in the kinetics of chromosome movements and to establish causal relationships between phenotypes (18). Here, we describe a step-by-step protocol that includes the Fiji, R, and Python scripts required to process raw microscopy data for analysis by *ChroMo*.

2 Materials

2.1 Strains and Media

The media employed were described in (26). *S. pombe* homothallic (*h90*) cells were initially grown on yeast extract-supplemented (YES) agar plates and then incubated for 24 h at 32 °C. Subsequently, they were transferred to a fresh plate and incubated for another 24 h at 32 °C. To induce meiosis, the cells were repositioned on sporulation agar (SPA) medium and incubated for approximately 6–6.5 h at 28 °C, which is the optimal temperature for mating and meiosis. Following induction, the meiotic cells were immobilized using lectin (0.2 μg/mL, Sigma-Aldrich, L1395) at the bottom of a μ-Dish glass-bottom 35 mm uncoated dish (81151, ibidi Gmbh), washed, and then covered with a total of 3 mL of EMM2 minimal medium lacking nitrogen to ensure the continuation of meiosis.

2.2 Time-Lapse Image Data

Raw images of fission yeast meiosis can be obtained using a microscope equipped for live fluorescence imaging (for example, the DeltaVision and Spinning Disk systems have been used successfully to test this methodology). To visualize chromosome movement in fission yeast, we used CFP-tagged histone H3 at one of two endogenous *hht1*$^+$ loci, as described previously (27, 28). To obtain sufficient resolution for subsequent time-series analysis, images should be taken every 60 s for at least 3 h. The format of the image files must be digital video (DV) or tag image file (TIF). The microscope setup must be adjusted to acquire the images without disturbing the faithful progression of meiosis.

2.3 Hardware

A high-speed computer is not essential, as the Fiji, R, and Python scripts use Z-projected images, and the *ChroMo* analysis can be performed online.

2.4 Software

Raw image processing for *ChroMo* is performed with the image analysis program Fiji version 1.53q (https://imagej.net/software/fiji/downloads). The MorphoLibJ_-1.4.3 and StackReg plugins (29, 30) (https://github.com/jpincru/protocol/) should be installed in the ImageJ plugins folder. The comma-separated values (CSV) files generated by Fiji are processed using either R or Python. The first approach requires the installation of R-4.2.0 (https://www.r-project.org/) and an integrated development environment such as Rstudio 2022.02.1 Build 461 (https://posit.co/download/rstudio-desktop/). On the other hand, the Python user interface (UI) does not require any external dependencies to be installed beforehand. However, if desired, the Python script is functional and included (https://github.com/jpincru/protocol/), which requires the installation of Python 3.11.4 and an integrated environment such as Jupyter Notebook. These are included in the Anaconda distribution, which we recommend downloading (https://www.anaconda.com/download).

3 Methods

The following procedure describes how to generate a readable CSV file for *ChroMo* from time-lapse images of fission yeast meiosis. First, the captured images are pre-processed. Next, a Fiji script performs image segmentation and data extraction for several morphological and kinetic chromosome descriptors. Finally, either two R scripts or a Python UI process these data and generate a final CSV file for *ChroMo* analysis. We include a guide to the most common configuration settings for *ChroMo* analysis (Fig. 2).

3.1 Time-Lapse Image Preprocessing

To ensure that the time-lapse images can be processed correctly, the following conditions must be satisfied in the original raw images:

1. Ensure that the chromosomes of interest are isolated as much as possible and do not touch other particles without leaving the region of interest (ROI) during the entire analysis. In cases where several particles are in close proximity, manually crop the image to isolate the chromosome of interest (Fig. 3a).
2. Keep the signal intensity of the analyzed chromosomes constant throughout the time series; intensity fluctuations can cause the mask to be generated incorrectly. An acceptable signal/background (S/B) ratio value is >1.45 (Fig. 3b).

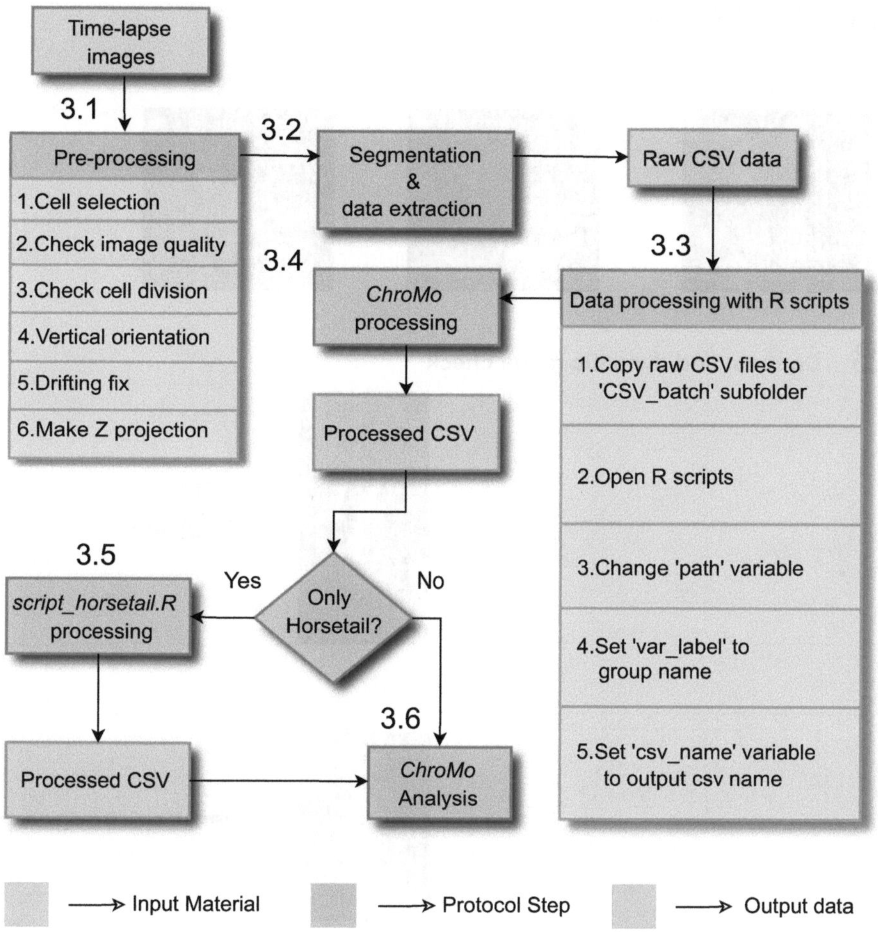

Fig. 2 Workflow for time-series analysis of chromosome movements using *ChroMo*. The numbers indicate the sections of this chapter in which the relevant steps of the protocol are described

3. Confirm that the cell division event in meiosis I is clearly visible, showing two chromosomal masses, since this will be established as time 0 in the *ChroMo* analysis.

Once all these requirements are met, the resulting images must be pre-processed following these recommendations:

1. To ensure a consistent comparison of the movement of the chromosomes, it is necessary to orient the images vertically so that the main axis of oscillation is aligned in all cells that will be analyzed. In other words, our goal is to compare oscillations in a uniform direction by orienting all cells accordingly (Fig. 3c).
2. Any cellular drift (variation in the position of chromosomes during image acquisition) should be corrected. Failure to correct for cellular drift could introduce false movement components into the analysis of chromosome movement. One

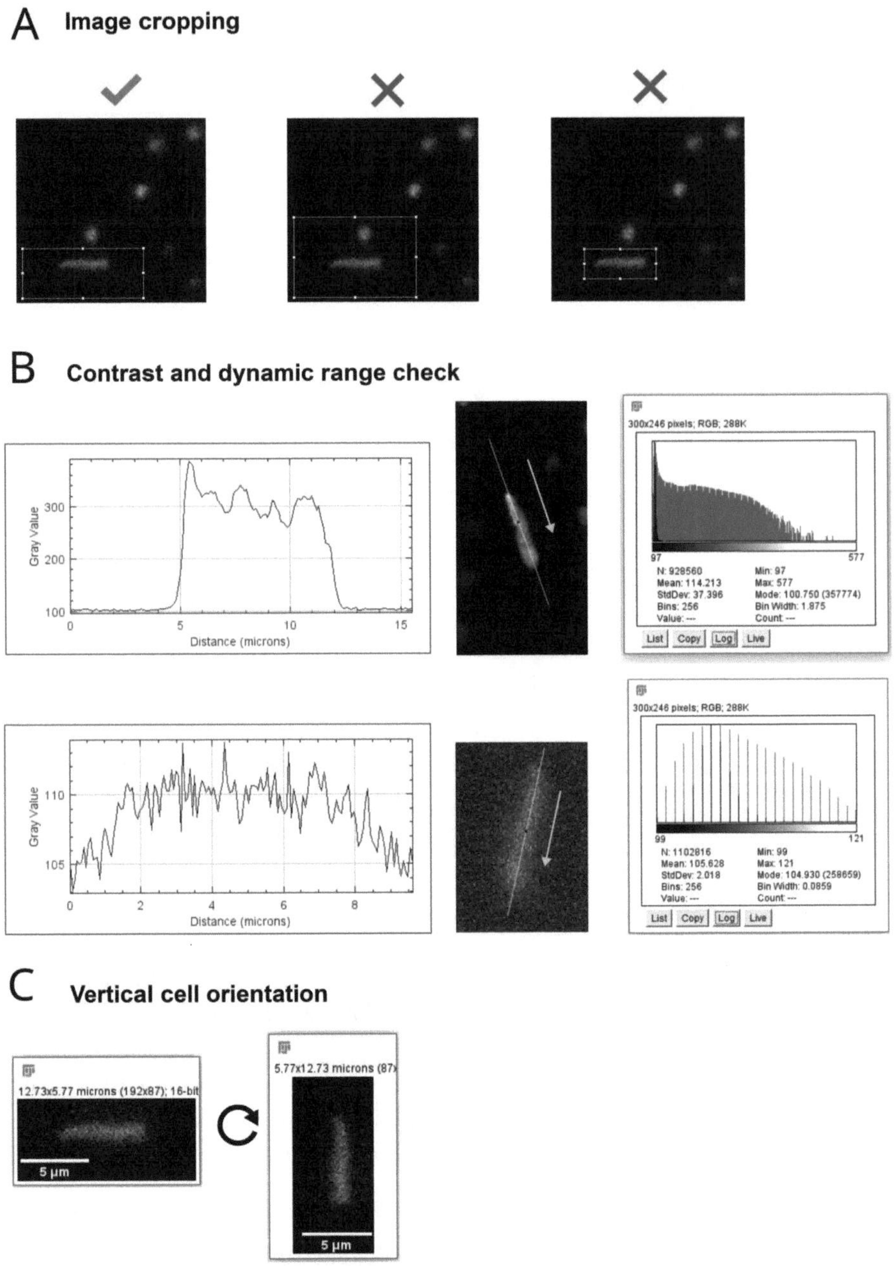

Fig. 3 Important aspects of image preprocessing. (**a**) Images should be cropped to isolate the particle of interest. (**b**) Examples of suitable (S/N ratio > 1.45) (top) and poor (S/N ratio < 1.45) images for dynamic range and contrast (bottom). (**c**) An example of rotating the selected cell on the y-axis

effective method of correction is to use the ImageJ StackReg plugin (30).

The image to be analyzed in the next steps must be a Z projection of the raw image.

3.2 Image Segmentation and Data Extraction

Image segmentation and chromosome descriptor measurements are performed using the Fiji macro script (*fiji_segmentation_macro.ijm*), which is included in the "preprocessing_for_ChroMo" folder (https://github.com/jpincru/protocol/). This main folder has five subfolders: CSV, Mask, ROIMasks, ROIs, and Z_projections. To avoid path errors, place the image files in the correct folders, as instructed.

Processing steps:

1. Place the images obtained from the microscope (TIF or DV format) after the image pre-processing (epigraph 3.1.) in the "batchmacro" folder inside the "Z_projections" folder. The images in the "batchmacro"' folder will be the files processed by the Fiji macro.
2. Run the *fiji_segmentation_macro* script from the "preprocessing_for_ChroMo" folder (*see* **Note 1**).

A correct segmentation consists of only the desired particle (chromosomal mass); if any of the time-lapse frames before meiosis divisions have more than one particle, that time-lapse is invalid. Note that *fiji_segmentation_macro* automatically excludes small particles that affect the measurement only above a certain size in microns.

Running the macro produces:

- One CSV file per analyzed cell containing information on 14 kinetic and morphological descriptors of chromosome movement (area, X, Y, XM, YM, Perim., Major, Minor, Angle, Circ., Slice, AR, Round, Solidity). These files, which are saved in the "CSV" folder with the filename format "Results_" + original filename, are the ones needed for the next step in the data processing.
- ROI masks for each time-series frame, saved in the "ROIMasks" folder. We can then manually check if there are any issues with the segmentation.
- ROI data tables, saved in the "ROIs" folder.

3.3 Data Processing for ChroMo

Once the CSV files are generated by the Fiji macro, they need to be processed either with the two R scripts called *CSVprocess.R* and *CSVmerge.R* or a Python UI called *UI4ChroMo.exe* (included in the main folder). First, this processing with the *CSVprocess.R* script subdivides only the data corresponding to the fission yeast meiosis prophase, removing the information after the first cell division (meiosis I). The R script automatically identifies when each cell is split by detecting when a second particle appears, using the particle column data provided by Fiji. This step is very sensitive; if the image is not clear or if other particles appear in the image over time, the script may falsely detect a split. This can be identified by some output messages indicating the frames where a split is detected by the script and manually checking if the split corresponds to the cell division in the movie files.

The *CSVprocess.R* script then changes the column names from the Fiji names to the default names recognized by *ChroMo*: frame, particle, area, x, y, major, minor, circularity, and convexity.

ChroMo analysis necessitates the creation of a unified CSV file that encompasses all the required data, which includes the number of particles for two or more strains or experimental condition. This file serves as the input for the *ChroMo* analysis algorithm, which analyzes the chromosome movement in each experimental conditions. By having all the data in one file, the algorithm can perform comparisons and identify differences between the conditions more efficiently. To achieve this, we combine all the processed CSV files situated in the "batchmacro" directory into a sole CSV file that comprises all the data utilizing the *CSVmerge.R* script. The script merges all files that share a common string in their names, as defined by the variable "pattern" in the script. By default, the value of "pattern" is set to "Results", which is the default output file name obtained from Fiji for each particle. When merging conditions, ensure that you have assigned a shared name for all the CSV files belonging to a specific condition, such as "Mutant" or "Condition". This allows the script to identify and merge the correct files and produce a single CSV file with the desired data for each experimental condition.

Processing steps:

1. Copy the generated CSV files in the "CSV" folder to the "CSV_batch" subfolder (*see* **Note 2**).
2. Open the script *CSVprocess.R* with RStudio to set some variable values in the next steps.
3. Establish within the script the folder path where the CSV files generated by the Fiji macro are located, e.g.
   path <- ("C:\\User\\preprocessing_for_ChroMo\\CSV\\CSV_batch\\").
4. Set the value of "var_label" to the name you want to associate with the cell group. By default, this is the name of the strain (e.g. "var_label" <- "Mutant A").
5. Run the *CSVprocess.R* script (*see* **Note 3**). This process involves modifying the CSV files by subsetting the frames before cell division and adjusting the column names. Specifically, we extract data for the frames corresponding to the pre-division stage of cells, as this is the relevant stage for our analysis. Additionally, we update the column names to reflect the data for pre-division cells and ensure consistency across all the CSV files.

6. Open the script *CSVmerge.R* with RStudio to set some variable values in the next steps.
7. Establish within the script the folder path where the CSV files generated by the Fiji macro are located, e.g.:
   path <- ("C:\\User\\preprocessing_for_ChroMo\\CSV\\CSV_batch\\").
8. Set the variable "csv_name" to the desired final name of the CSV file with all cells in the group merged (e.g. "csv_name" <- "Mutant_A.csv").
9. Run the *CSVmerge.R* script. The result is a CSV file with the previous individual CSV files merged, this can be used in the *ChroMo* application.

To analyze multiple strains or conditions, follow **steps 1–9** repeatedly, save the resulting CSV files in the pre-existing "CSVbatchmacro" folder, and run the *CSVmerge.R* script again. This time, set the "pattern" variable to the common name of the files. For instance, if comparing the strains "Mutant_A" and "Mutant_B", set the pattern variable as "Mutant" (with case sensitivity). The output will be a single CSV file with both strains that can be used in *ChroMo*.

Alternatively, a Python UI is provided to achieve the same results without requiring any external dependencies, along with its respective script. The processing and merging steps in this UI are very similar to those of the R scripts.

First, let's consider the *Process Files* function, which requires four entries (Fig. 4a):

1. "Path": This is the location where the files to be processed are stored. For R, this would be the "CSV_batch" subfolder. However, for the UI, you can choose any desired folder (e.g., "C:/User/preprocessing_for_ChroMo/CSV/CSV_batch").
2. "Save Path": This entry specifies the path where the processed files will be saved (e.g., "C:/User/preprocessing_for_ChroMo/CSV/Processed_mutants").
3. "Variable Label": Enter the label you wish to associate with the cell group (e.g., "Mutant_A").
4. "Pattern": Specify the common pattern used in the filenames of the files to be processed (e.g., "Results").

Additionally, the *Process Files* function adds an option for a column with the calculation of vectorial velocity and an option for time normalization if selected. These functions can also be performed by *ChroMo*. The frames where the script detects a split in each cell are displayed in the window below the entry fields, along with the newly generated files. To execute the code, simply click the "Process Files" button.

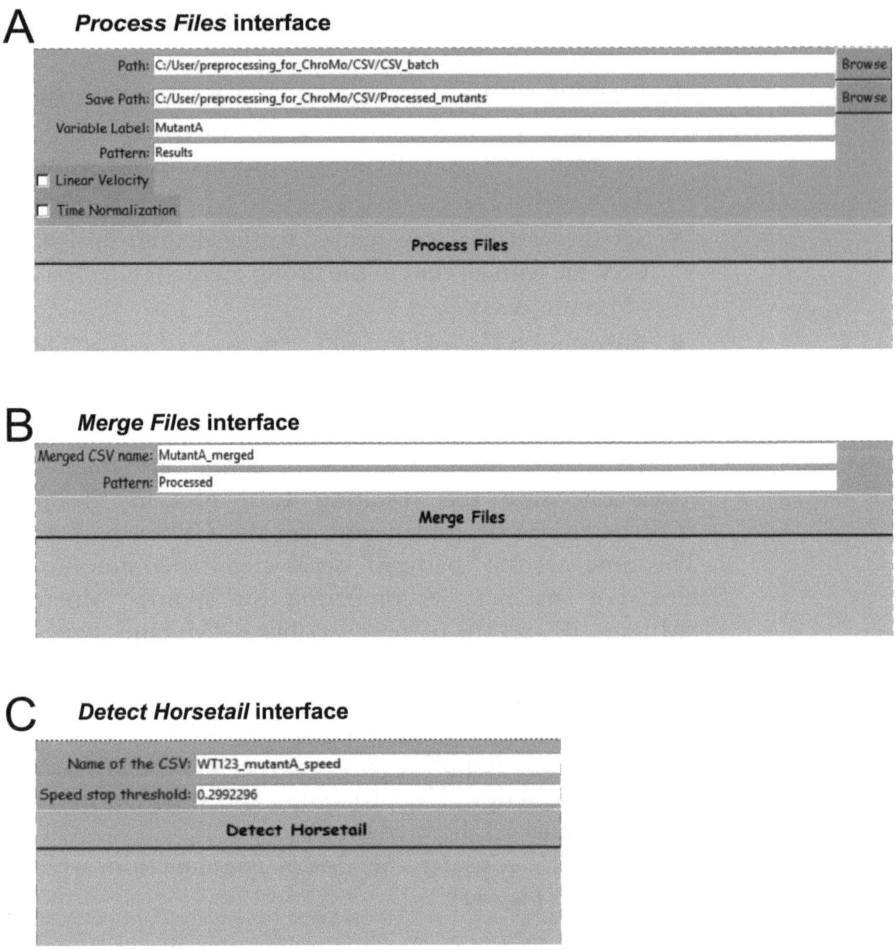

Fig. 4 Screenshots of the Python UI. (**a**) Display of the *Process Files* interface featuring the four required entry fields with example settings and the checkbox option for time normalization and linear velocity calculation. The results are presented in the window below. (**b**) Display of the *Merge Files* interface, along with the required entry fields and example settings; the window below indicates the location of the merged files. (**c**) Display of the *Detect Horsetail* interface with the example and default settings. The display window shows the location of the files

On the other hand, the *Merge Files* function requires only two entries (Fig. 4b):

1. "Merged CSV name": Specify the desired name for the merged CSV file (e.g., "MutantA_merged").
2. "Pattern": Enter the common pattern used in the filenames of the files to be merged (e.g., "Processed").

To execute the code, simply click the "Merge Files" button. It is important to note that the files to be merged must be in the same directory as the "Save Path" or manually select the correct path. Like the R scripts, this process can be repeated to analyze multiple strains or conditions. For the Python script provided in a .ipynb extension, it should be opened and executed using Jupyter Notebook or a similar environment.

3.4 ChroMo Processing and Analysis

ChroMo compares the behavior between two or more strains, so the input should be two different sets of processed CSV files (e.g. wild type versus mutant). Here, we present the step-by-step protocol to optimize analysis in *ChroMo* (Figs. 5 and 6):

1. Choose the Large(r)-tier instance, *ChroMo* L (v-1.3), of the latest version of *ChroMo* at https://chromo.cloud/.
2. Using the "BROWSE" button, select the processed and merged CSV file obtained from the R script in Subheading 3.3.
3. Check the boxes "Add vectorial and angular velocity" and "Add cumulative distance". In the "Time normalization" section, select "Last is common and zero" (Fig. 5b).

   The "Add vectorial and angular velocity" checkbox adds two new columns with data variables that are essential for analysis: linear velocity, "cal.speed"; and angular velocity, "ang.speed".

   As the duration of each time series until the cell is split (the number of frames) is variable, time normalization is used to set the split event as zero time so that different time series can be compared and processed.
4. Download the output data with the "DOWNLOAD DATA" button at the top of the page.

   This CSV file contains the data for the entire time series up to the first cell division. To analyse only the movement of the horsetail movement stage (every frame until the cell stops moving), the time series can be split by running the *script_horsetail* script using either the *ChroMo* output file or the Python output file with the "cal.speed" column.

3.5 Extracting the Horsetail Movement Stage from the Rest of the Meiosis Prophase

The horsetail movement and post-horsetail stages in fission yeast have different features that must be analyzed separately (Fig. 1). To differentiate between the two phases of meiosis prophase, it is necessary to define some condition to establish the end of the horsetail movement stage and the beginning of the post-horsetail stage. The linear velocity of the oscillations is high in the horsetail movement stage, while the movement of the chromosomes is minimal until their division during the post-horsetail stage. We measure the change in linear velocity to establish a threshold value that can be used to separate the time series into the horsetail

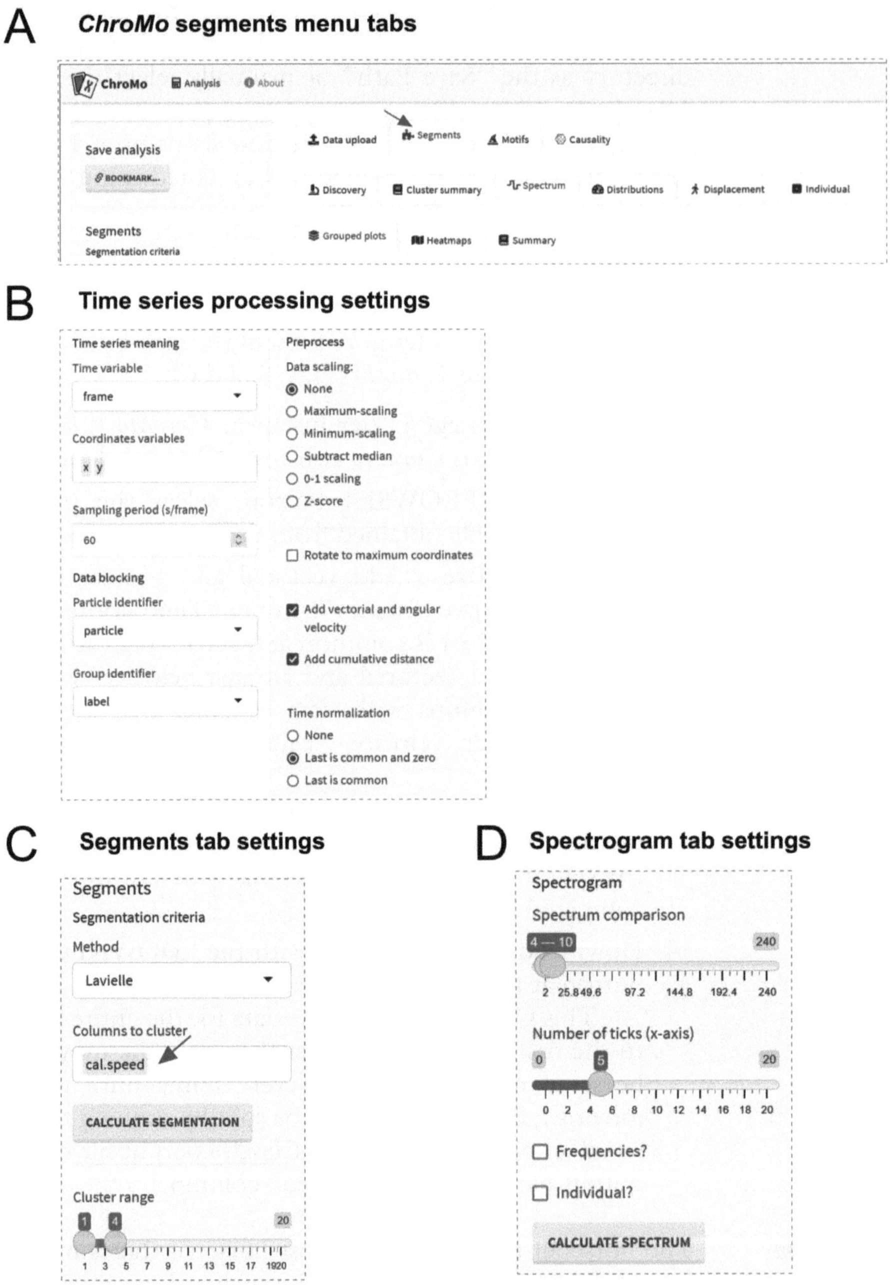

Fig. 5 Screenshots of the *ChroMo* interface I. (**a**) Display of tabs when the Segments tab is selected. (**b**) Selection of settings for data processing. The default settings are displayed. Selecting the "Add vectorial and angular velocity" and "Add cumulative distance" checkboxes generates columns of velocity data, and "Time normalization" is set to "Last is common and zero". (**c**) The Segment tab interface with default settings. The default values for "Columns to cluster" are 1–4 for whole prophase analysis and 1–2 for horsetail-only analysis. (**d**) The Spectrogram tab interface with default settings

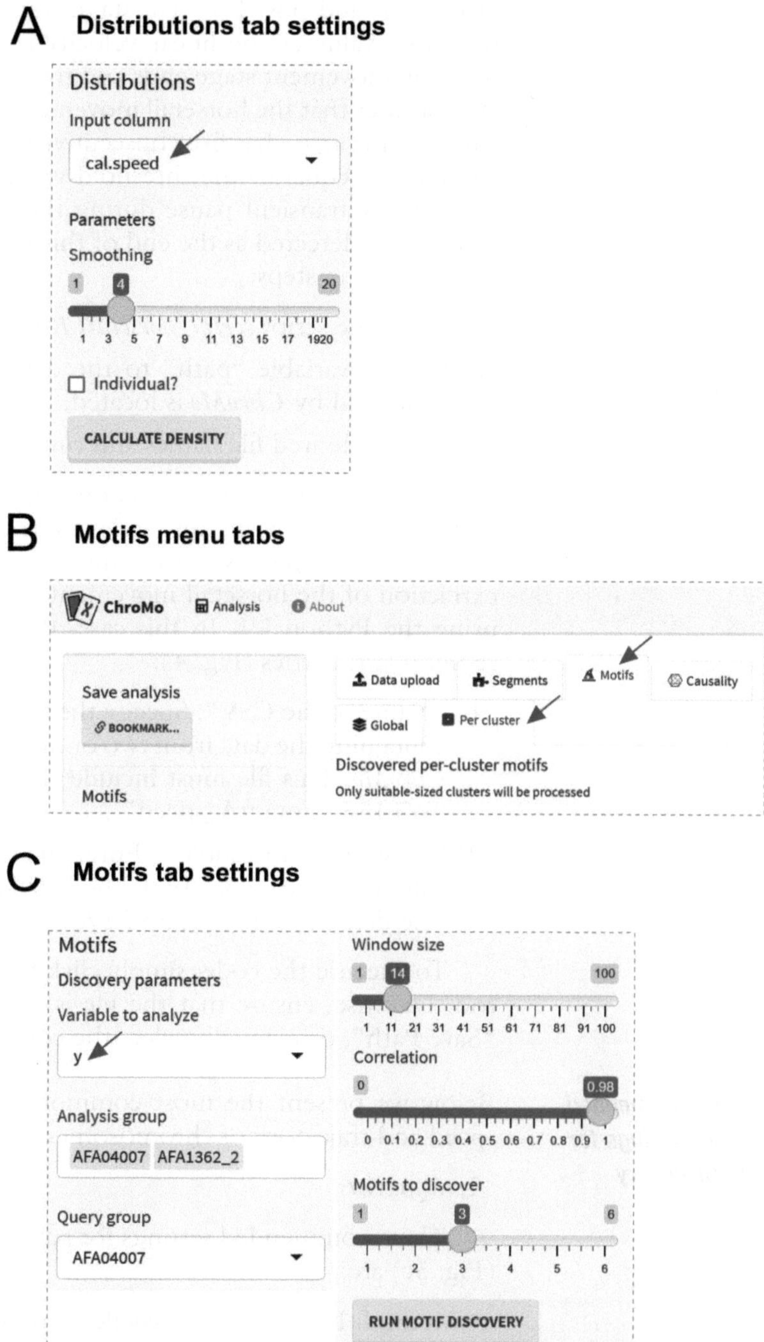

Fig. 6 Screenshots of the *ChroMo* interface II. (**a**) The Distributions tab interface with default settings. (**b**) Display of tabs when the Motifs tab is selected. (**c**) The Motifs tab interface with default settings

movement and post-horsetail stages (*see* **Note 4**). We define this threshold value as the linear velocity of chromosomes when the horsetail movement stage ends and the post-horsetail stage begins. We consider that the horsetail movement stage has ended when the studied time series has five consecutive frames with a linear velocity less than or equal to the threshold value. This condition is set to prevent any transient pause during the horsetail movement stage from being detected as the end of this stage.

Processing steps:

1. Open the script *script_horsetail.R* in RStudio.
2. Set the variable "path" to the directory where the CSV file generated by *ChroMo* is located.
3. Set the desired file names and run the script.

The generated CSV file will only include data for the horsetail movement stage of the time series to be analyzed by *ChroMo*.

Just like the processing and merging steps discussed earlier, the extraction of the horsetail movement stage can also be performed using the Python UI. In this case, the *Detect Horsetail* function requires two entries (Fig. 4):

1. "Name of the CSV": Specify the name of the merged CSV file containing the data from two or more conditions to compare in *ChroMo*. This file must include the "cal.speed" column (e.g., "WT123_mutantA_speed").
2. "Speed stop threshold": Enter the desired velocity value that will be used to separate the horsetail and post-horsetail stages. The default value is 0.2992296.

To execute the code, simply click the "Detect Horsetail" button. Likewise, ensure that the file is in the same directory as the "Save Path", or manually select the correct path.

3.6 Recommended ChroMo Settings for Motif Discovery

Below we present the most common settings for evaluating the speed and trajectory of chromosomes.

1. *Segments*

The recommended settings for parameters in the Segments tab (Fig. 5c) are:

- Method: Lavielle. The Lavielle method is best for analyzing the speed and trajectory of the main axis of chromosome movement (31). To analyze more than two variables, choose the multivariate option (32).
- Columns to cluster: cal.speed.
- Cluster range: 1–4.
- Significance test: Likelihood Ratio.
- P-value adjustment: None.

2. *Spectrogram*

The segmentation from the previous step is necessary to generate the spectrum diagrams.

First, click the "Calculate spectrum" button with any value on the slider, then change the slider value to the desired period interval to obtain the "difference significance data" for that interval (Fig. 5). The values of the "Spectrum comparison" slider correspond to the period interval of study (Fig. 5d). The recommended setting is 4–10. The "Number of ticks (x-axis)" slider changes the display of period values on the x-axis and is a visual function only.

3. *Distributions*

The "Distributions" tab enables users to compare selected variables between groups with a Wilcoxon rank sum test and to determine the "difference significance globally" and in each of the segments generated previously (Fig. 6a). The recommended settings are:

- Input column: cal.speed.
- Smoothing: 4. Smoothing calculates a moving average of the selected number of time points.

4. *Motifs*

ChroMo performs this analysis using algorithms for matrix profile generation and motif discovery (33, 34). To run the motif discovery, select the "Per cluster" tab and press the "Run motif discovery" button in the bottom left corner (Fig. 6b).

Recommended motifs settings are:

- Variable to analyze: y.
- Analysis group: strains or conditions.
- Query group: one of the strains or groups.
- Window size: 14.
- Samples to compute: 1.
- Correlation: 0.98.
- Motifs to discover: 3.
- Motifs checkbox: checked.

The variable analyzed is the position variable, which in the case of studying chromosome movement in fission yeast corresponds to the "y" variable. Analysis groups are determined by the "label" variable. The window size determines the length in frames of the detected motifs; the default value is 14, but any value between 12 and 16 can discover useful motifs for fission yeast.

4 Notes

1. Initial Verifications:
   - Make sure there are no opened images or windows in Fiji before running the *fiji_segmentation_macro*.
   - Confirm that *MorphoLibJ_-1.4.3.jar* and *Stackreg.jar* are placed in the plugins folder in Fiji.
   - Images must be calibrated in microns to accommodate variations in image capture between microscopes with different specifications.
   - The script can be run by directly dragging the file onto Fiji, which will open the window where it is possible to execute it.
2. If any CSV files are generated incorrectly (i.e., CSV with few (<30 rows per cell) or no input), *ChroMo* will fail when generating new columns or calculating segmentation. Check the file size of the generated CSV for unusually low file size to identify incorrectly generated CSV.
3. To run the script, you can either use the keyboard shortcut Ctrl + Alt + R or select the entire document and click on the "Run" button located in the top right corner (the keyboard shortcut for this is Ctrl + Enter). It is worth noting that this action runs the current line or selection, which is why it would be necessary to select the entire document to execute all lines together.
4. To calculate the threshold velocity, 56 time-lapse recordings were collected for 14 cells from each of four strains. In each of these 56 recordings, the last frame of the horsetail movement stage was manually identified. Using the time series of the recordings for each of the 56 cells, the mean of the linear velocities corresponding to that frame and the next five frames was calculated (the next five frames are included to avoid, as far as possible, any subjectivity in the choice of stop frame). These values correspond to the velocity recorded at the beginning of the post-horsetail stage. Finally, to calculate the threshold speed, the average of these 56 mean linear velocities was taken.

$$\text{Threshold speed} = \frac{\sum_{n=1}^{56} \text{speed}_n}{56}$$

Acknowledgments

We thank all lab members for their critical comments on the manuscript. This work was supported by PID2021-127232NB-I00 funded by the Ministerio de Ciencia e Innovación, Agencia Estatal de Investigación and by the "FEDER, Una manera de hacer

Europa," awarded to AFA. MC and RMM were supported by Programa JAE intro 2021 and 2022, respectively. The IBFG is an institution funded by Consejo Superior de Investigaciones Científicas (CSIC), Regional Government of Castilla y León and University of Salamanca.

References

1. Hunter N (2015) Meiotic recombination: the essence of heredity. Cold Spring Harb Perspect Biol 7(12):10.1101/cshperspect.a016618
2. Klutstein M, Cooper JP (2014) The chromosomal courtship dance-homolog pairing in early meiosis. Curr Opin Cell Biol 26:123–131. https://doi.org/10.1016/j.ceb.2013.12.004
3. Petronczki M, Siomos MF, Nasmyth K (2003) Un menage a quatre: the molecular biology of chromosome segregation in meiosis. Cell 112(4):423–440. https://doi.org/10.1016/s0092-8674(03)00083-7
4. Keeney S, Giroux CN, Kleckner N (1997) Meiosis-specific DNA double-strand breaks are catalyzed by Spo11, a member of a widely conserved protein family. Cell 88(3):375–384. https://doi.org/10.1016/s0092-8674(00)81876-0
5. Lam I, Keeney S (2014) Mechanism and regulation of meiotic recombination initiation. Cold Spring Harb Perspect Biol 7(1):a016634. https://doi.org/10.1101/cshperspect.a016634
6. Koszul R et al (2008) Meiotic chromosomes move by linkage to dynamic actin cables with transduction of force through the nuclear envelope. Cell 133(7):1188–1201. https://doi.org/10.1016/j.cell.2008.04.050
7. Lee CY et al (2020) Telomere-led meiotic chromosome movements: recent update in structure and function. Nucleus 11(9):111–116. https://doi.org/10.1080/19491034.2020.1769456
8. Trelles-Sticken E, Adelfalk C, Loidl J, Scherthan H (2005) Meiotic telomere clustering requires actin for its formation and cohesin for its resolution. J Cell Biol 170(2):213–223. https://doi.org/10.1083/jcb.200501042
9. Wynne DJ, Rog O, Carlton PM, Dernburg AF (2012) Dynein-dependent processive chromosome motions promote homologous pairing in C. elegans meiosis. J Cell Biol 196(1):47–64. https://doi.org/10.1083/jcb.201106022
10. Miki F et al (2002) The 14-kDa dynein light chain-family protein Dlc1 is required for regular oscillatory nuclear movement and efficient recombination during meiotic prophase in fission yeast. Mol Biol Cell 13(3):930–946. https://doi.org/10.1091/mbc.01-11-0543
11. Chikashige Y et al (1994) Telomere-led premeiotic chromosome movement in fission yeast. Science 264(5156):270–273. https://doi.org/10.1126/science.8146661
12. Chikashige Y et al (2006) Meiotic Proteins Bqt1 and Bqt2 Tether Telomeres to Form the Bouquet Arrangement of Chromosomes. Cell 125(1):59–69. https://doi.org/10.1016/j.cell.2006.01.048
13. Conrad MN et al (2008) Rapid telomere movement in meiotic prophase is promoted by NDJ1, MPS3, and CSM4 and is modulated by recombination. Cell 133(7):1175–1187. https://doi.org/10.1016/j.cell.2008.04.047
14. Scherthan H (2001) A bouquet makes ends meet. Nat Rev Mol Cell Biol 2(8):621–627. https://doi.org/10.1038/35085086
15. Rubin T, Macaisne N, Huynh JR (2020) Mixing and matching chromosomes during female meiosis. Cells 9(3). https://doi.org/10.3390/cells9030696
16. Fernandez-Alvarez A, Cooper JP (2017) Chromosomes orchestrate their own liberation: nuclear envelope disassembly. Trends Cell Biol 27(4):255–265. https://doi.org/10.1016/j.tcb.2016.11.005
17. Nozaki T, Chang F, Weiner B, Kleckner N (2021) High temporal resolution 3D live-cell imaging of budding yeast meiosis Defines discontinuous actin/telomere-mediated chromosome motion, correlated nuclear envelope deformation and actin filament dynamics. Front Cell Dev Biol 9:687132. https://doi.org/10.3389/fcell.2021.687132
18. Leon-Perinan D, Fernandez-Alvarez A (2021) ChroMo, an application for unsupervised analysis of chromosome movements in meiosis. Cells 10(8):10.3390/cells10082013
19. Fujita I, Kimura A, Yamashita A (2023) A force balance model for a cell size-dependent meiotic nuclear oscillation in fission yeast. EMBO Rep: e55770. https://doi.org/10.15252/embr.202255770

20. Tanaka K et al (2005) Hrs1p/Mcp6p on the meiotic SPB organizes astral microtubule arrays for oscillatory nuclear movement. Curr Biol 15(16):1479–1486. https://doi.org/10.1016/j.cub.2005.07.058
21. Saito TT et al (2005) Mcp6, a meiosis-specific coiled-coil protein of Schizosaccharomyces pombe, localizes to the spindle pole body and is required for horsetail movement and recombination. J Cell Sci 118(Pt 2):447–459. https://doi.org/10.1242/jcs.01629
22. Chikashige Y et al (1997) Meiotic nuclear reorganization: switching the position of centromeres and telomeres in the fission yeast Schizosaccharomyces pombe. EMBO J 16(1): 193–202. https://doi.org/10.1093/emboj/16.1.193
23. Niwa O, Shimanuki M, Miki F (2000) Telomere-led bouquet formation facilitates homologous chromosome pairing and restricts ectopic interaction in fission yeast meiosis. EMBO J 19(14):3831–3840. https://doi.org/10.1093/emboj/19.14.3831
24. Moiseeva V et al (2017) The telomere bouquet facilitates meiotic prophase progression and exit in fission yeast. Cell Discov 3:17041. https://doi.org/10.1038/celldisc.2017.41
25. Chacon MR, Delivani P, Tolic IM (2016) Meiotic nuclear oscillations are necessary to avoid excessive chromosome associations. Cell Rep 17(6):1632–1645. https://doi.org/10.1016/j.celrep.2016.10.014
26. Moreno S, Klar A, Nurse P (1991) Molecular genetic analysis of fission yeast Schizosaccharomyces pombe. Methods Enzymol 194:795–823. https://doi.org/10.1016/0076-6879(91)94059-l
27. Fennell A, Fernández-álvarez A, Tomita K, Cooper JP (2015) Telomeres and centromeres have interchangeable roles in promoting meiotic spindle formation. J Cell Biol 208(4): 415–428. https://doi.org/10.1083/jcb.201409058
28. Fernández-Álvarez A et al (2016) Mitotic nuclear envelope breakdown and spindle nucleation are controlled by interphase contacts between centromeres and the nuclear envelope. Dev Cell 39(5):544–559. https://doi.org/10.1016/j.devcel.2016.10.021
29. Legland D, Arganda-Carreras I, Andrey P (2016) MorphoLibJ: integrated library and plugins for mathematical morphology with ImageJ. Bioinformatics 32(22):3532–3534. https://doi.org/10.1093/bioinformatics/btw413
30. Thevenaz P, Ruttimann UE, Unser M (1998) A pyramid approach to subpixel registration based on intensity, vol 7. IEEE Trans Image Process, In, pp 27–41
31. Lavielle M (1999) Detection of multiple changes in a sequence of dependent variables. Stochastic Process Appl 83:79–102
32. Patin R et al (2020) Identifying stationary phases in multivariate time series for highlighting behavioural modes and home range settlements. In: Street G (ed). J Anim Ecol. https://doi.org/10.1111/1365-2656.13105
33. Yeh C-CM et al (2016) Matrix profile I: all pairs similarity joins for time series: a unifying view that includes motifs, discords and shapelets. In: 2016 IEEE 16th International Conference on Data Mining (ICDM). IEEE
34. Zhu Y et al (2018) Matrix profile XI: SCRIMP ++: time series motif discovery at interactive speeds. In: 2018 IEEE International Conference on Data Mining (ICDM). IEEE

INDEX

Jesús A. Carballo (ed.), *Meiosis: Methods and Protocols*, Methods in Molecular Biology, vol. 2818, https://doi.org/10.1007/978-1-0716-3906-1,

M

N

O

P

Q

R

MIX
Papier aus verantwortungsvollen Quellen
Paper from responsible sources
FSC® C105338

If you have any concerns about our products,
you can contact us on
ProductSafety@springernature.com

In case Publisher is established outside the EU,
the EU authorized representative is:
Springer Nature Customer Service Center GmbH
Europaplatz 3, 69115 Heidelberg, Germany

Printed by Libri Plureos GmbH
in Hamburg, Germany